The Science of Possibility

Patterns of Connected Consciousness

Jon Freeman

With

Juliana Freeman

www.scienceofpossibility.net

Published by: SpiralWorld

3 Merrifield Road

Ford

Salisbury SP4 6DF

www.spiralworld.net

Printed on demand at the nearest print location

ISBN 978-0-956-0107-3-5

Contents

For the cherub with his flaming sword is hereby commanded to leave his guard at the tree of life, and when he does, the whole creation will be consumed and appear infinite and holy whereas it now appears finite and corrupt.

This will come to pass by an improvement of sensual enjoyment.

But first the notion that man has a body distinct from his soul is to be expunged; this I shall do, by printing in the infernal method, by corrosives, which in Hell are salutary and medicinal, melting apparent surfaces away, and displaying the infinite which was hid.

If the doors of perception were cleansed every thing would appear to man as it is, infinite. For man has closed himself up, till he sees all things thro' narrow chinks of his cavern.

William Blake: Marriage of Heaven and Hell

Introduction – The First Trumpet

*Behold, I tell you a mystery: We shall not all sleep, but we shall all be changed --
In a moment, in the twinkling of an eye, at the last trumpet. For the trumpet will
sound, and the dead will be raised incorruptible, and we shall all be changed.*

First Letter to the Corinthians, 15: 51-52

What if almost everything that you have ever learned, experienced or been told about our world has been skewed, misaligned and misunderstood? What if our ways of perceiving who we are, what the world is, and how we relate to it are somehow inside-out and back-to-front?

Are you willing to suspend your current viewpoint for long enough to engage with something as radically different as that? We are ready to take you on a journey through a world that you know very well and have you see it for the first time. We hope that you will come with us.

We could fairly call this "The Greatest Story Ever Told". It spans life from the smallest particle to the hugest expanse of the cosmos. It encompasses the innermost being of every human and the beingness of all creation. There is no part of our existence that it does not touch. Truly.

Although we are about to tell a story, it isn't fiction. It is filled with paradox, containing both fact and mystery. It is at once deeply scientific, and fundamentally challenging to our view of science. It is rooted in spirituality yet often unsupportive of religion. It is intensely rational and at the same time rooted in the heart; the rational mind has for too long sat on a throne, a king without a consort.

For all of human existence we have sought answers to the most basic of questions. Who are we? What are we? Why are we here? Now we are here, what do we do? Any book which attempts to address such questions is by its nature ambitious, risks great arrogance and dooms itself to incompleteness. Since the only full container for answers is the cosmos itself, we offer a candle towards the light of the sun. Nevertheless we will bring possibilities, seeds of answers which we see as profound and transformational, answers which leave the questions more open than ever because they address the heart of creation itself, and creation is incomplete. The Universe itself is work in progress. We are all a work in progress.

Humanity knows more now than it has ever known and each decade adds more to its library than had been produced in all of the preceding centuries. This book is possible now because of that breadth of knowledge. You know more than all earlier humans. Your visibility encompasses the globe and near space through cameras and via satellites. Within days (and often within hours) you could potentially shake hands with almost any one of seven billion humans. From your computer screen you can access more information than any of us can handle. Science has brought us much of that knowledge and the capability to access it. It has also become the authority for our times. Science tells us what is real and what is not.

Before science, there was religious knowledge and Divine authority telling us what is real. Sometimes it was based in spiritual experience and exploration and sometimes it was filled with miracle and revelation. Sometimes religions contained the written encoding of the best of human understanding at their time, often invested with the authority of a Divine figurehead and presented to the peasant or labourer by hierarchies of intermediate priesthoods.

Humanity needed codes to live by, and still does. At the same time, these two authorities are both failing to meet our needs, and they are in intense and deep conflict. We don't wish to enter that debate but it contains the history that inhabits the context of our conversation. Though we touch on it, our purpose is to take you far beyond it because it is old, sterile and futile.

Both views are creations of humanity's desire to understand the world, to get our minds around it. Philosophers sought to know the mind of God. Religions sought to know and live by God's will. Scientists have believed that if we knew enough about the mechanisms, we could predict what the mind of God would do next. For many generations now the human rational mind has been the vehicle for these explorations, and in the scientific age it has become the only one acceptable. The brain is the most important organ in the body – according to the brain!

From Reality to Possibility

So what if our world is unlike either of these viewpoints, or simultaneously like both and yet fundamentally different? What if we humans have trapped ourselves by asking the wrong kinds of questions for the last two thousand years and more? What if we have used our brilliant rational minds to think ourselves into a corner?

From the ancient Greek philosophers to our present-day scientists we have been seeking to know and understand what "reality" is. We have learned a lot about material existence and given ourselves great power to manipulate it physically and shape it to our ends. Now, that material "reality" is all that is usually seen. While we know even from the scientific perspective that what we look at as solid is 99.999% space, populated by tiny particles and humming with energy masquerading as matter, we still adhere to the notion that there is such a thing as "reality". What if we are not actually living in reality, but in possibility?

Who are you? Are you a body, carrying around a mind, having an experience that is called consciousness? That is what science will tell you. It sees consciousness as an "emergent property of matter". To science you are not really you at all, instead you are an accident of neurology and brain chemistry; your brain just imagines that you are you as some kind of artifact. Science just sees you as a body. Does that make you feel good?

Conversely most traditional religious perspectives perceive you as a body that is inhabited by a soul or spirit. You, whoever and whatever you are, exists inside your physical form like a digger operator looking out through the cab window and pulling the levers that make you do things. At the end of the shift it will clock off and go home. Religions would often like you better without your body.

Are you willing to engage with the possibility that both these views are the inverse of what is really going on? What if you are already home? What if your body is inside of who you are? What if your body is an emergent property of you – of who you truly be? What if you are the creator of this form?

These are big and fantastic notions to take on. But this is not a fantasy. Throughout the body of this book we will present the evidence, including the scientific evidence, that supports this alternative way of seeing and being. Although what we say is also based in our own experience we don't expect you to believe us. There are facts too – good, hard, objective scientific facts. The only reason that science is not yet telling you the story we will tell you, is that it finds these facts uncomfortable because they demand changes to scientific theories which could mess up careers, research grants and possibly whole industries.

We encourage you to see this as good news. We have built today's world with religion and science and it self-evidently isn't working well. Neither of them is complete or adequate to the challenges facing the planet. Our future requires that we find something more comprehensive, more coherent, more potent and more empowering.

There are things we must let go of if we are to engage with that completeness. As Mark Twain wrote *It ain't what you don't know that gets you into trouble. It's what you know for sure that just ain't so.* Some of the "ain't so's" that we will meet are deeply embedded in our culture, our education systems and the whole reality structure that we all share as common knowledge. We ask that you are ready to question all of your preconceptions, at least for long enough to complete the journey. How much of the past structures are you willing to let go of in order to make space for possibility? Because it is a big picture, it will take time to get all the pieces of the jigsaw in place. Again, we ask that you suspend your judgement for long enough to enter our world. At the end you will of course be free to accept or reject our viewpoint. After all, you can't lose anything that is true.

Some "ain't so's" may be precious to you. We don't wish to be confrontational and we apologise if at any time the strength of our views causes offense. The search for truth involves challenge. Often the questions are more powerful than the answers and more important. One of our themes is that the world is often paradoxical. Paradox is what we get when two truths are seen side by side, and appear incompatible. In a world where humans have believed that it is essential to systematise what is right and wrong, what is good or bad, paradox causes discomfort. Sometimes it is funny too. We will inevitably contradict ourselves. Sometimes we will do so on purpose.

The upside of paradox is liberation and creativity. It will be obvious as the story unfolds that we do not believe in certainties. Not only do we see the goal of predicting how the Universe will unfold as an unachievable and daft fantasy; we also see it as dreadful and undesirable. Paradox is juicy and uncertainty makes it possible for the world to shape itself in every moment. As indeed it already does. In the contradictions lie the entrances to a world that knows that either-or responses are restrictive, that the potency is found in the flexibility of both-and.

And that will be our story. All religious contexts embrace some form of awareness that something meaningful is present beyond the solidity of the world. Science has reached the conclusion that there is only energy which can also be matter but has no meaning. Protagonists and spokespersons continue to debate these two perspectives. In the absence of further progress they could continue to do so until hell freezes over or entropy runs everything down to stasis. We will show and present substantial evidence for the both-and, that neither is right, and neither is entirely wrong. What we will describe is the complete transformation that comes when we open our eyes and hearts to the full picture.

Our ultimate conclusion doesn't deny science. Rather, what we do is turn it inside out. Inverting the view that we are inside of our bodies is a small example[1]. We

---

[1] We are grateful to Gary Douglas, Founder of "Access Consciousness" for this neat formulation

also invert the whole of creation and tell you that Consciousness came first. The whole material Universe is an emergent property of consciousness and the scientific, material reality sits within consciousness. Matter is formed from energy by consciousness and shaped by it. We will give you the physics for that. We will do it painlessly; you won't need an anaesthetic. As for religion, when we understand what consciousness really is and does, spiritual experience too finds a new context, a container which justifies much of experience and perception. We apologise if our karma runs over anyone's dogma, but dogma and paradox are not good friends. Our perspective is meta-scientific and meta-religious. It transcends and includes both. It is more powerful than both together.

Even though their experiences are genuine and their conclusions may be accurate, this will not be one of those fluffy books that justifies the spiritual or religious point of view by describing experiences and asserting that this means science is wrong. While there will be some examples that are based on experience these are given as illustrations rather than as justifications. The science here will stand on its own because there is much more hard data and experimental evidence than you are usually presented with. That said, we now want to tell you a personal story, because it explains why this book ever came to be written.

A personal background

A book about spirituality has a personal core, however strong our central intention to address scientific questions may be. The path we are taking mirrors our personal journeys. It will support your understanding to know who is writing and why and this knowledge is particularly relevant to our starting point.

Jon's personal journey began with a background of atheistic upbringing, an intellectually challenging father and an interest in sciences which led via A-levels in maths and physics to the study of Human Sciences at Oxford. This subject included elements of genetics and evolution, ethology (a young Richard Dawkins being one of his lecturers), Human Biology plus various other disciplines with mankind as their central subject. Mysticism, religion and spirituality were not part of this package. Jon's extra-curricular focus was a combination of student politics and classical music amidst a life embedded in early seventies student hippydom with its long hair and flared trousers.

Clearly something happened to change this atheist, but there was no voice from the heavens, no blinding experience on the road to Damascus, no conversion by charismatic evangelists. The experience which brought this shift was low-key and undramatic while at the same time being ultimately as powerful and life-changing as any of those options. Jon describes it as follows.

"After leaving University, having skirted one path that led to research work in chimpanzee behaviour and another in the Philosophy of Science, I pursued a fascination with the world of computers, then in its infancy. An initial impulse towards the use of computers as an educational tool had been displaced by learning what they could do in the commercial world and I was enjoying the experience of transforming organizational life through well-designed application systems. I was also on the conventional life-path with a wife and two children.

Another part of this conventional path was stress, both in over-work and from the marriage which was not going well and this led me to take a course in relaxation techniques and creative visualisation. At least, that is what I thought it was. Over the four days it became apparent that there were much deeper intentions in the training which had to do with extending sensing abilities towards intuition in order to assist people to address their own and other people's health issues. By the end of the course, I was told, we would experience ourselves as having the ability to gain knowledge psychically.

I did not believe this possible, at least not through training. I might have allowed the possibility of extraordinary gifts but saw myself as the least likely person to be capable of such a thing. Even so, I went along with the training process in an open-minded and curious way. On the last afternoon we each went through exercises, using the techniques we had been taught. In these exercises we would be given the name, age and approximate location of an individual, known to one other of the participants who had written down some basic facts about that individual. This information was in the hands of a third person.

My experience during the exercise was of "tuning in" to the subject individual (David, a 27-year-old male living in Devon) using the techniques we had practiced. For a long time I was getting nothing until I tried a particular approach in which rather than trying to "see" the person or "hear" information about them in my mind, I visualised what it would be like to be that individual. The effect was instant. I had the strong feeling of there being pain in the rear left side of my head and spontaneously grimaced and twisted my head. I asked the observer if the person had pain from some condition like a brain tumour and was told that this was exactly what was written down. I used other techniques to attempt to send healing (this being the real purpose of the training) before bringing my attention back and participating in the experiences of other students.

Of the twenty or so people in the room, all but two had some kind of success in this kind of detection – enough that they knew with confidence that they had detected something that they could not have known by any "normal" means. I therefore had to recognise not just my own experience but that of several others, none of whom had given any indication of previous training or gifts in such areas.

For a while this experience was simply a curiosity and I did not know what to do with it.   Later it began to bother me. All my scientific training told me that there was no known mechanism by which such things can happen. There was no physical medium by which such information is known to pass. There was no way known to science in which the information itself is "published".   There was no way to understand in any scientific terms how I could have located the correct individual based on just a name, age and location. Feeling what someone else was experiencing was simply not in the frame, not catered for. Yet in true scientific spirit my questions were "evidence-based".

Basically, this meant that the science I had been taught had to be flawed in some way. Although I subsequently observed other courses in which larger numbers of people went through a similar experience, and even though I later trained as a teacher for the method and saw those whom I taught having the similar success, it was the personal experience that was transformational.   I have been busy ever since pulling the loose thread in the scientific knitting. The explorations which that led to are central to all that follows."

Over three decades have passed since that time. That single experience opened a doorway to explorations of many other aspects involving "alternative" views of our reality.   Juliana's pathway, which she will introduce later, was different, but for more than two decades we have explored these new worlds on differing but parallel paths. Each of us could easily fill a book with anecdotes if we chose to, things experienced or witnessed, piling layer upon layer of personal confirmation and perhaps delivering more entertainment. Our purpose here has to be different, so maybe some other time.

Scientists and atheists alike have attacked religions for their dogmatic and often damaging adherence to beliefs which may have been helpful in earlier times, but now appear ignorant. My friend Cindy Wigglesworth, a pioneer in the arena of Spiritual Intelligence coaching, has a great illustration of this. If we ask you whether you think that the Old Testament phrase "an eye for an eye, a tooth for a tooth" provides a good justice system, we anticipate that you will not think so. But consider this. Prior to that time, the way of thinking would have been "If you take out my eye, I will take out your eye, then your heart, and then those of your wife, your children and your goats!" Against this background, the Old Testament offers a step forward towards proportional justice. Human thinking evolves.

Old dogma, New tricks

Thus each age has its dogmas and the scientific age is no different. Arguably the New Testament contained some improvements on the old, and most of us think that we have improved further since then. But we also know that the centuries of technology have also brought challenges. The industrial age has delivered excesses of greed, exploitative industrial empires and unsustainable attitudes to the planet. These are the consequences when you believe that it is only material outcomes that matter.

Similarly, many people experience the impact of scientific materialism on human connection and bonding. It is a theme for our times that we shift the balance toward caring and sharing and this sometimes produces a wish to attack science and destroy capitalism. We cannot afford destructive reactivity and we need a way forward. Both capitalism and science need to evolve if we are to survive our current crisis. We will offer that expansion, presenting an integration that preserves the benefits of science and materialism while simultaneously stepping into choice, possibility and creative empowerment that is beyond anything we have conceived of.

There are many benefits and possibilities that are currently being denied to most of humanity, the benefits of greater health, greater ease, increased knowledge of how to bring about happiness, a barely imaginable expansion in possibility for the future we create together. There are planetary problems which cannot be solved using the old ways of thinking. It is with good reason that Einstein is quoted for his recognition that new levels of thinking are required because you cannot solve problems using the same thinking that created them. Buckminster Fuller similarly said "In order to change an existing paradigm you do not struggle to try and change the problematic model. You create a new model and make the old one obsolete."

Our new model is not a luxury but an essential. This is not because we have all the answers. Rather it is because in this new way of thinking we are able to ask the right questions and produce an open space into which new choices can emerge, creatively and moment by moment. Indeed, it is more about being than thinking, more about awareness than knowing. We are all living through the shift out of a linear and predictable world in which there was some hope that our choices would have reliable outcomes where "If I do this, that happens". You are probably noticing that there are levels of turbulence and unpredictability associated with a complex, global, fast-paced world and that these are confounding our rational attempts to command and control events. Instead we are like surfers on the wave, needing to respond moment by moment, combining knowledge and expertise with instinct that responds faster than the speed of rational thought. We need to equip ourselves for the world in which anything can happen, and generally does. In recent survey by IBM of global CEO's, almost three-quarters placed unpredictability as their number 1 challenge. The old model can't deal with this.

Above we hinted at the new aspect of our thinking that turns science inside out. This may not have sounded to you like a very big deal. Even if consciousness comes first – so what? As we unfold this story in full we intend to show that in fact it is an absolutely huge deal, one which can transform the way that we think about life, about ourselves, about our world and about what is possible. We say this mindfully because we are experiencing that transformation. From experience we know that some of the old ways of thinking are deeply embedded in our habits and ways of being. The rational, scientific, materialist perspective, the view that our analytical minds deliver the keys to our success can be quite hard to dislodge. But it is where we must start. Since science has taken on such huge authority we must understand its weaknesses. If we are to turn it inside out, we will need to show how it flexes. We come not to bury science, but to raise it.

This is emphatically not an anti-science book. We are passionate about the value of science. There is great value in requiring some common sense, intellectual rigour, and up-to date recognition of the way the physical universe operates. But there is a belief system of "scientism" which attaches to science just as religion attaches to spirituality. Fundamentalism is a deep problem in all religions and science has its own fundamentalists. There is so much in science that we can and will endorse and plenty of bathwater worth removing from religion, but some fundamentalist scientism is throwing out the baby. The baby is still breathing. Just. It is time for that baby to begin its journey towards adulthood.

The baby is the universe of spiritual consciousness. It's a big baby; indeed we are saying that they don't come any bigger, since science inhabits the baby's universe. They are together, intertwined and inseparable and we intend to show how, to reveal a whole network of missed information and rich connectivity.

We don't seek to prove all of science wrong; much of it is brilliant. We do wish to re-frame it though and to call for change in the areas where flaws are holding it back. There is only one universe in which scientific humans and spiritual humans share one world, not two. In that world, both scientism and those who recognise the existence of a spiritual consciousness have portions of the truth to argue over. It is time for these two viewpoints to blend and for both the world and for the human beings who inhabit it to be made whole. The text of this book is devoted to filling the gulf, to placing the collagen fibres over the wound that divides our world.

This would be an impossible task were it not for the many brilliant people who have already described these fibres. Like Newton, if we can see far it is because we stand on the shoulders of giants. Our story will be told largely through their many voices. Our aim is not to claim any authority for ourselves but to weave their strands into a coherent and healing whole.

Let the journey begin!

1. The Journey - So what else is possible?

The truth you believe and cling to makes you unavailable to anything new.

Pema Chodron

The man who never alters his opinion is like standing water, and breeds reptiles of the mind.

William Blake

Theme
This chapter provides an overview of the book's contents together with some maps and overview concepts that assist in navigating the conceptual framework of the new worldview that is being presented.

The territory that we are entering into is vast. Each chapter represents a whole area of study. As a result we can only pull out core concepts, illustrate a few crucial facts and in doing so hope to construct a road map through several of the larger countries. We can post some signposts, describe a little landscape. It's a "Rough Guide" to the conscious universe. But we do know where we are going and we will arrive home safely with broadened minds.

Chapter 2 describes our scientific start-point. It explores what science is, in terms of what it aspires to be and the way it currently functions. We explore the boundaries between science as a discipline and "scientism" as a fundamentalist belief system, pointing out some of the supposed "truths" which are deserving of closer examination. We see how these lead to conflict as science encounters other aspects of reality that lead to religious and "spiritual" perspectives.

Chapter 3 presents our review of the territory in which spiritual experience operates along with some exploration of phenomena which have labels such as intuition and psychic awareness. We will put some definition around these, some examination of what evidence is to be found and what this means for our use of words like knowledge, belief, faith, certainty and truth. We start here because the core to all threads in our tapestry lies in understanding what "consciousness" is. Consciousness and our own inner experience exist at the boundary between subjective and objective experience. There is an information flow which pervades biological and physical worlds. Consciousness lives in the space between what has previously been termed "physical" and "spiritual". This is both why it has been missed and why it is central to our text. This book sets out to illustrate, rather than attempt a precise definition of what we mean by this.

This will lead us in chapter 4 into discussions of the relationship humans have as biological organisms with their physical environment. We will redefine the understanding of the spiritual experience as an integral part of that relationship. Scientism tends to treat the body as a biological machine. In doing so it oversimplifies the data and narrows the scientific possibilities. There is a physical component to the way in which consciousness is experienced and the relationship between subjective experience and biological mechanisms is highly complex. Humans are capable of so much more than we have imagined or allowed.

One area of major conflict between science and the world of spiritual alternatives is that of complementary medicine. We will explore the mind-body-spirit relationship in this context and show how the attack on spirituality and the attack on holistic medicine arise from the same misconceptions and use the same flawed methods. In chapters 5 and 6 we will look at the shamanistic view of the world and what it tells us about the nature of "reality" as well as seeing what it means for the nature of health and healing.

The boundary between spiritual healing and alternative medical methodology is frequently blurred. There are good reasons for this but it does lead to confusion. The involvement of human consciousness and spirit in the healing process is powerful, complex and sometimes subtle but it is also erroneously dismissed with the phrase "placebo effect" – a phenomenon which neatly hides the keys to the real truth. There is so much more involved. In general, science has taken no trouble whatsoever to investigate the evidence or to understand the theoretical under-pinning of holistic medical theory. Frequently and at worst its misrepresentations amount to ignorant ridicule. At best it ignores huge quantities of systematic evidence-gathering. In Chapter 7 we will use the science of homeopathy as a specific illustration of this general trend, and as a template for understanding "energy" medicine. In Chapter 8 we will give some further illustrations of this territory and draw on personal experiences of it as well as authoritative work by others.

This will lead us naturally into the biological sciences, which are the subject matter of section 2. We will look first at the wondrous complexity of processes which build and maintain bodies like ours. We then follow the thread of inheritance between generations and change over millennial timescales with a discussion of genetics, inheritance and evolution. We are not about to argue for the biblical creation story but neither do we accept the purely mechanistic view of life's emergence which books like "The Selfish Gene" and "The Blind Watchmaker" try to persuade you is the whole truth. In chapters 9 and 10 we will challenge some serious over-simplification in the way that science presents the role of the gene and in chapters 11 and 12 provide a robust alternative to the view that undiluted randomness lies at the core of evolutionary change.

Consciousness enters into this world too, so these chapters also explore the relationship between organisms and their environment to know deeply how our experience of consciousness arises and how its relationship with our environment is maintained. Evolution has supported and selected for human spiritual experience not as the "side-effect" suggested by scientism but because it provides a genuine adaptive advantage which we have so far barely exploited. There is empowerment and creativity awaiting us when we know who we really are.

In section 3 our perspective switches from the human to the cosmic as we enter the world of the tiniest particles and the largest scales of time and space. This is the realm of other best-sellers such as Stephen Hawking's "Brief History of Time". In the face of the failure to unite these extremes through a physical "Theory of Everything" we will once again be placing consciousness in the middle. Albert Einstein famously (if metaphorically) said that "God does not play dice with the Universe". In these chapters we will show how an active and creative consciousness sits with quantum randomness in the realm of the small and how chaos theory unites the impact of the small with the unfolding of the large. We will engage with the historical choices and conceptual challenges that have taken physics down what appears to be a blind alley and reveal the real role of harmonic form in the shaping of creation.

In section 4 our final chapters return to our starting-point in order to draw together all these scientific strands with regard to the wide varieties of human experience. We will go deeper into our way of defining "reality" and into social and cultural aspects of our relationships with the world. Having laid down the foundations for how consciousness is threaded through biological and physical realities we will show how we fashion our personal and collective viewpoints and how our perceptions of consciousness and reality frame the ways that the world works. It will be a major achievement here if we can demonstrate how foolish has been the scientific dismissal of many varieties of human spiritual experience, but this is our secondary goal. More than that we wish to crack open the current belief system in order to weave all the strands of our thinking into a framework for future scientific understanding and investigation. Our primary goal is to enable the emergence of a tolerant, inclusive, broad, expanded, empowering and flexible basis for choices about how to live life.

The Evolution of Belief

Human religion has had its own evolutionary path. Historically, religions have evolved to meet a variety of needs. They created tribal bonds and community cohesion, and provided moral guidance to curb individual extremes. At their best these developed into governance and moral cohesion to unite countries and continents. At their worst they became the vehicles for many extremes in the exploitation of power and dominance or the fight for territory. Religion has retained too much of the primitiveness of its roots and quite often still exhibits an intolerant form of tribalism. It has not so far evolved at a large scale in a way which accommodates scientific knowledge or modern forms of thought, though valuable attempts at such an evolved viewpoint have been made, for example by William Bloom in his book "Soulution – The holistic manifesto".

The attacks from scientism and atheism are part of the process by which religion and spiritual thinking need to evolve. Thinking which is primitive and not equipped to survive deserves to die. Richard Dawkins, Christopher Hitchens and others have not held back from eloquently delivering the killing blow to maladaptive thoughts and values. At the same time the implied superiority of intellectual rigour and the Westernised view of Truth are themselves subjective values. They are points of view that we can choose, or not choose.

Scientific Truth is not Absolute Truth. The rational mind is not the whole of human intelligence and our excessive esteem for reasoning has deprived us of other forms of knowing. For individuals the effects are alienating limiting and disempowering. On the larger scale these hyper-rational values are not global and the worst of their unbalancing effects are seen in environmental exploitation, cultural extermination and economic imperialism. Scientism will be felt by many who are at the effect of these forces as another part of the attack on who they are and resistance will become a matter of survival for their very identity. While undoubtedly not the scientists' intention, it is no less dangerous for that.

Nearly all discussion of these subjects have until now been embedded in an "either-or" that lives in the gulf between science and spiritual experience, founded on a definition of spirituality that puts it in a non-physical realm, beyond the reach of science. As a result the debate can feel warlike – for sure it produces similar levels of passion in some individuals. We may feel personally that the battle is only between ideas. But beneath this whole cultures are at stake. If any of us expect others to give up their forms of what we regard as primitive thought, we must first abandon our own dogma, deal with the mote in our own eyes and the impairment of our own vision.

"The God Delusion" continued the missionary tradition with scientism as its new religion. Having barely survived the Christian missionaries, the world may not be keen for a new form of cultural brain-laundering. No doubt the West does have something to offer with its science and democracy but if others are expected to surrender their existing cultures and identities then we must allow them to develop their own replacement. It is unrealistic, arrogant and dangerous to require that they simply follow us into our limited version of the Promised Land. We would not wish anything we say in this book to be felt as prescriptive. We hope that the closing chapters of this book will be seen as a template within which all cultures may develop in ways which honour their traditions and evolve toward whatever species of culture they need to become.

There is one further thread in these last chapters and it is possibly the most challenging of all to the world of scientism. One of the fundamental activities among religious and spiritual humans is prayer. We referred above to a human engagement with the realm of consciousness as "creative". We will look at what this creative engagement really involves and at what it could mean for a genuinely inclusive, empowered and peaceable human race to live integrally through spiritual science and scientific spirituality. We do not claim that this book will change your life, but the opportunity is there for you to change it.

Because the territory we are attempting to cover is huge, inevitably we will skim. We will have to over-simplify if we are not to be bogged down. This is a book about scientific thinking. For narrative simplicity we will not adopt a text-book approach to references. Even so, everything we say has back-up and the details can be found in the major texts that we refer to.

Where we delve into scientific detail, we are aware that we risk "losing" lay readers. We believe that we have avoided this most of the time, but there are perhaps five chapters that may provide an intellectual challenge to non-scientists. We have attempted to provide some graphical support, but if it still feels too difficult we encourage readers to skim rather than give up. Sometimes the penny will drop later. But you can gain great value even if you don't grasp everything and we believe that our overall theme can be understood even if some of the detail is fuzzy. We also believe that it is of psychological value to know that the underpinning is there and could be revisited if necessary.

Oneness to Complexity Spiral

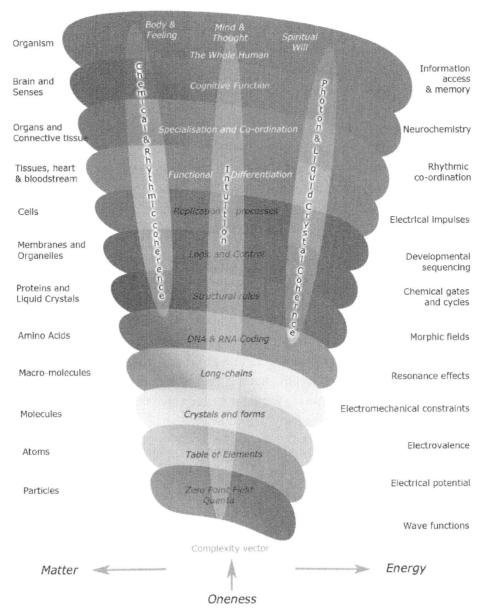

This spiral is not intended to delineate exact boundaries. It illustrates the balance between the energy and matter aspects of creation (right-left) and depicts layering of increased complexity (bottom-top). The central pillar of labels indictates the successive addition of information content layers to the story of creation and represents the front-to-back dimension as the spiral loops around. Correspondences between these layers and the material or energy aspects to either side are indicative approximations which inevitably overlap and are superceded and included within later layers.

Maps to guide us

Every journey is easier with a map and we will use a pair of images as a skeleton for our presentation. These two images, belong together and depict several elements of the world we are about to describe. They are in black and white here, but we encourage you to download and print larger versions from our website[2]. The texture will emerge as our narrative develops, and we present the outline here. You will find it helpful to refer to the diagrams as we explain.

The images are a spiral with multiple layers and a disc of concentric rings. We would like you, if you can, to imagine them as if the disc is horizontal and the spiral is vertical, placed at its centre, like a cup and saucer or an ice-cream sundae glass. It is helpful to have a sense of the three-dimensionality of this image and also of its flows. While it is represented on a flat printed page, it has movement side-to-side, front-to-back and bottom-to-top.

Human ways of thinking are often constructed in polarities. We tend to see and speak in opposites, black or white, visible or not visible, good or bad. Things are present or absent. Babies love to play with this impermanence, throwing toys out of sight from the buggy to delight in their return. Adults may know that there are shades of grey and gradations of goodness but this either-or quality is typical of our thought and communication. It's a game that we play too.

Science also presents us with polarities. It sees subjects (us) and objects (the other things and people in our worlds). It presents us with matter and energy, with particles and waves. Our entire world-view is constricted by this mental, philosophical and linguistic habit. It is how the rational mind works.

Link to Spiral Diagram 1

---

[2] www.scienceofpossibility.net/resources

Accepted scientific truth has already gone beyond this limitation. Since Einstein we have been aware of the way in which matter and energy are interchangeable and a whole generation of writers about physics like Fritjof Capra and Gary Zukav has drawn out the parallels with mysticism with book titles such as "Tao of Physics" and "Dancing Wu-li masters". The underlying unity is acknowledged. In some ways though, it remains static. It does not give a picture that includes living, evolution, complexity and time.

To get that picture we have to increase the dimensions. In the spiral, the right-to-left view is broadly representative of the interplay of matter and energy. This is shown at the sub-atomic level where particles and wave functions are interchangeable. It is sustained, on one side through the various manifestations of physical, chemical and biological form – the ways that matter is configured – and on the other via different configurations of energy interchange in electromagnetic fields, chemical reactions and cellular life.

From the bottom to top we have a picture of complexity. This is also a picture of time, since the complexity arose over the vast time-scale of cosmic creation from the "big bang" forward. Its range encompasses the entirety of both non-living material development and evolution into living forms that began with amino-acids and to date has reached the trillion-celled combinations that write and read books. Thus the spiral layers present an imprecise and arbitrary map of this development, from fundamental particles through atoms and complex molecules to living cells, organs and multi-celled organisms. Many of these layers are likely to have their own spirals of internal development; for instance the whole periodic table of elements might be represented in this way. We will see later that there are further spirals above this one which govern the dynamics of social evolution.

None of the two-fold picture so far would challenge orthodox science. Side to side and top to bottom, our picture holds a conventional materialistic view. Matter exists in linear time. There is no spiritual element, nothing beyond the standard picture of random particles interacting in magically and mysteriously wonderful ways. But when we seek to explain a world in which people have intuitive or spiritual experiences, and in which humans find something that they are inclined to call "God" then the picture as described so far is inadequate.

Our intention in this book is to bring the third dimension of this cone to life. This is the hardest to visualise from a flat page and this book doesn't yet come with 3D glasses. The dynamic of existence does not simply swing pendulum-like from one side to the other but rather it loops around, developing, expanding and growing.

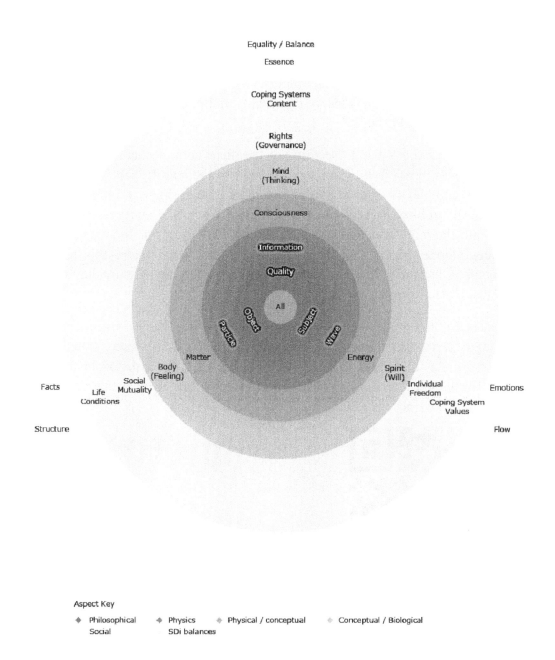

While it grows more complex, it also self-balances. The loop connects each layer of the spiral to the next in a continuous thread. The third dimension, the elements in that balancing process, are represented in the rings of our disc.

Outside the disc we have shown the overall dynamics of structure and flow, which correspond to the matter-energy polarity in physics. The third element is the balancing process itself. Maintaining balance is at the core of an ecological view of the world. The image of an ecosystem is of interactions between multiple species of plants and animals, inhabiting specific conditions of climate and terrain. The question of how this governance of equality and balance comes about in life as a whole is central to this book.[3]

The front right and front left "sectors" of the discs contain labels which broadly correspond with those to the front right and left of the spiral. The same sense of developing complexity should be apparent as those rings progress from the philosophical "subject-object" polarity through waves and particles, matter and energy, into the more conceptual aspects of our biology – our body and feelings, the forces of our spiritual "will" energies. In the social realm this reaches a corresponding dynamic balance that deals with the interplay of individual freedom with social collective mutuality.

This balance is not an accident of random interactions. It is held in a context of information that the universe holds about itself and which manifests as consciousness, as mind and in social governance that are depicted in the upper sector. Throughout our story we will present the picture of what this information is, where it is and how it affects all of the processes that science recognizes but

Link to Trifold Diagram 1

---

[3] Note to Integral Theory enthusiasts: We could have imposed a four-quadrant view on this presentation but on balance feel that this does not work as well for the Physics and Biology as it does for the psychological and societal aspects. The maps are not the territory and our maps are choices anyway which could not possibly be comprehensive.

cannot fully explain. We will also encompass all the manifestations that are not recognised by science but are the core of spiritual experience, religious belief and holistic medical alternatives.

While human habits tend towards polarity, the three-fold view that this diagram presents is not new. It can be found in a Freudian view of the psyche with its ego, super-ego and id. Rudolf Steiner's three-fold social order with its "rights" realm has the social aspect of governance quite explicit. His three-fold view of man presents our interactions with the world through facts, feelings and ideas. It is present in the context that is used by much of alternative medicine as represented by the name of the UK "Mind-Body-Spirit" exhibition. It is also fundamental to Robert Pirsig's presentation of the philosophical significance of "quality" in "Zen and the Art of Motorcycle Maintenance" and to the discussion of "qualia" in other developments within philosophy. Some philosophers see qualia as another word to describe what we will call "consciousness". But despite such pre-figuring, the implications for a fully integrated view of science that encompasses the spiritual nature of the universe have not previously been offered in the way that we present them here.

Every chapter in this book will explore aspects of this integrated universe. We may cover one or more layers of the spiral or explore one or more rings of the disk. As we do so a picture will emerge showing how all the aspects come together as a unified whole.

We could not possibly represent all of the connections in the diagrams. As said at the start of this section, the spiral and disk merely provide a skeleton. We can only paint some of the muscles, nervous systems, organs and circulatory processes which make this living universe. But in the diagram we have drawn attention to three strands of connectivity which are most important in our story. These form the vertical bands through the spiral layers and we draw your attention to them here.

In our chapters exploring the phenomenon of "intuition" we will illustrate the way in which the information that the universe holds about itself is accessible to humans. The "inner sensing" that accesses the information realm has no boundaries. We will give evidence for the ways in which this knowing provides access to every level of existence. We will also provide the evidence for mind-over-matter interaction in this realm, the hard experiments that demonstrate conclusively that there is a complete and active connection down to the smallest levels. We could have used other labels than "intuition" and our descriptions of information and consciousness address the same realm.

Some of the connectivity that runs through layers of the spiral is provided in the realm of body chemistry – in the chemical messengers that deliver one level of cellular communication and in the rhythmic chemistry of cellular life-process that supports coherence and co-ordination. This strand will run through our chapter on the neuro-endocrine system and through the biological picture of life's rhythms. Mind is not the same as brain and at the very least we will show how it is all over the body, though when there is connectivity as described in the preceding paragraph the boundaries to mind itself must be seen as porous.

Another part of the connectedness is provided by a layer of function that we have called "photon and liquid crystal coherence". This strand is present also in the understanding of quantum physics and the orchestrated organisation of our brain function that emerges there. Throughout the text we will be building background understanding for the nature of this connectivity and giving scientific proof for it.

In all of this we will show, illustrate and provide evidence for a world-view that makes sense of spiritual experience, psychic knowledge, energy healing and for a vision of the world as a connected whole. To offer a comprehensive depiction of the world, we have to deal with many different scales of existence and to encompass a hugely complex picture. The small diagrams accompanying most chapters attempt to position each piece within the overall map. Our goal is that some simplicity will emerge on the far side of the complexity. At the same time, the fact that we are including a spiritual world that is invisible by its very nature requires us all to engage our inner senses and our poetic side, to operate alongside our intellectual and analytical forms of understanding.

One aspect of our presentation will call for an expansion in science itself. As indicated earlier, the scientific flaws that we have described derive from an attempt to achieve an impossible level of "objective" proof and a choice to exclude data which might derive from human experience and therefore by definition be "subjective". Our early chapters will provide the evidence that this complementary viewpoint is essential. We are required to place the "subjective" alongside the "objective" in order to achieve a complete view of what "reality" might be. This is the perspective from Goethe and others who tell us that to see something new sometimes demands a new way of seeing. As a Sufi parable about coffee would have it "He who tastes, knows; he who tastes not, knows not." You cannot know all there is about coffee (and by extension about the world) from its objective or physical properties. Of course we all know this and even the most ardent advocate of scientism is not truly detached from the world. Experience shapes their world too and there is art to creating a great coffee blend.

The result of this exploration is an overarching map of creation which we believe is more complete than most views on offer from philosophical, spiritual and scientific perspectives and which combines all three in a coherent whole. As said, we don't hype our message with a claim that "this book will change your life" but thinking of the world this way has certainly changed ours – and greatly for the better.

Review

We have sketched the outlines of the journey that this book will take, a review of what constitutes "scientific truth" and how that excludes other possible "truths". Our path through perceptions of non-ordinary realities, through the mis-presentation of biology and the challenges that physics has created for itself will culminate in a new framing of what is "real" and provide an added dimension to the way humans view and even create that reality.

2. Opening up the scientific viewpoint

Science and metaphysics therefore come together in intuition. A truly intuitive philosophy would realize the much-desired union of science and metaphysics.

Henri Bergson

We don't see things as they are: We see things as we are.

Anais Nin

Theme

This chapter explores what science is, what it is meant to be and what it bases its presentation of "Truth" on. We use Richard Dawkins' "The God Delusion" as a reference point and explore definitions of "God" and claims of objectivity. We open up questions regarding what is "reality" and how human viewpoints may be subject to significant perceptual constraints. This begins the exploration of a new way of seeing.

Even if you are not all that interested in science, you will not have escaped its influence. Doodling your way through chemistry lessons didn't protect you from a perspective on "reality" that pervades Western thought. Millions of people with no appetite for science would like to know how the world works; they just want to be spared the details. Yet more millions have known in their bones that science as currently framed does not work for them. You may be one of the many who know that who you are is different from what they were telling you. That is a disempowering experience, rather like being told that you are not sane. We are here to tell you that sanity is over-rated.

What follows is not simply a technical, esoteric or philosophical head-trip. We are entering the realm of the "just aint so's" which will have affected how you see yourself and your place in the world. So we need to take the blinders off and set out the nature of the flaws that scientism promotes. Describing them is simple and we will use Professor Richard Dawkins' book "The God Delusion" as a foil. While we disagree with his views quite strongly, he is one of the most lucid exponents of scientism and we are grateful for the assistance his clarity gives us.

This also obliges us to bring the word "God" into our narrative. We hesitate because the word means so many different things. However, as described above, we are exploring the territory of human experience which is typically associated with "God" or with "Gods". The Judeo-Christian Bible would have us believe that

God made man in his own image. In later chapters we will show how this is another inverted view, and how humans have fashioned Gods in our own image. In order to present a new understanding of the non-ordinary reality we will need to pick apart the thinking that wraps God, spirituality and consciousness into a confused and confusing muddle. And in confronting scientism we must deal with the way that this belief system creates false demigods of its own – the god of objectivity, the god of physical matter and the god of proof. Scientism has replaced religion and these are its gods; or perhaps we should invent the word "godmas".

1. A meaningful definition for the Divine. There are several definitions given in the first chapter of "The God Delusion" for different classifications of "God" – personal God, pantheistic God, God as metaphor etc. All of these are demolished. This method produces a set of "straw men" which others have used to define the spiritual world and then, with great relish, knocks them down. This demolition creates the illusion that there is no room left for any form of spiritual reality. Yet something inhabits this space which is not covered well by religion and not addressed fully in the choices people are offered. We intend to fill this gap with a substantial, meaningful and coherent definition.

On page 13, Richard Dawkins quotes Steven Weinberg's comment that if you want to say that "God is energy," then you can find God in a lump of coal. He goes on to say that if the word God is not to become completely useless, then the word should be used in its traditional sense to denote a supernatural creator appropriate for worship. Unfortunately this narrowed definition prevents recognition of the real nature of anything "Divine" and confines the relationship to "worship" rather than "knowing", "experiencing" or "engaging with". If God is omnipresent, perhaps we need a definition which meaningfully finds the Divine in a lump of coal. Dr. Dawkins suggests (P.50) that the existence of God is a testable scientific hypothesis like any other. We agree. The hypothesis we present for a meaningful "transcendent spirit", for the presence throughout the universe of a creative consciousness is backed by evidence. It is however far more subtle than the definitions that he offers us.

2. The spiritual world has physical reality. There is a polarisation in our language between "spiritual" and "physical". This is an important and sometimes useful distinction but it also misleads us by seeming to say that there is no connection between them. It implies that spiritual events are without physical mechanisms. We will define later what we mean by "spiritual" but for now please understand that we are not referring to a psychological phenomenon nor to a metaphorical one.

Our version of "spiritual" is a part of the physical, scientifically describable universe. "It's life Jim, but not as you know it." The concepts of Red and Violet are useful, but not to the exclusion of the orange, yellow and blue which lie between them or the over-arching concept of light and an electromagnetic spectrum. The polarisation itself is a self-limiting definition of the physical world. We will connect matter to spirit.

On page 72 Dr. Dawkins quotes Arthur C. Clarke's "Third Law" that "any sufficiently advanced technology is indistinguishable from magic." This is true in the same sense as Patience Strong's homily that "a stranger is a friend you haven't met yet". The false separation between spiritual and physical causes us to regard as "supernatural" the phenomena which we cannot yet explain. In a sense there is nothing in reality that is super-natural or it could not exist at all. What science means by supernatural is either "imaginary" or "we don't have an explanation yet". We will apply more helpful distinctions which view the spiritual world as a non-ordinary reality located beyond our ordinary thinking and perceptions. We will also provide an explanation. The Divine is "ultra-natural" or even "intra-natural"; it is present around or within ordinary reality. We need to expand our frame of reference to recognise the technology that exists in this realm. This calls for a wider bandwidth of receptivity.

3. <u>Spurious objectivity and inconsistent attitudes to proof.</u> The scientific world claims to work to a principle of "objectivity" and asserts that this objectivity is maintained through the standards it sets regarding what may or may not be regarded as "evidence" or as "proof". The standards have been debated deeply in the past and form an academic subject in their own right, the Philosophy of Science. Some of the roots lie even deeper in philosophy itself, in its examination of what may or may not be "known". We will show that these supposed standards are flawed in their reasoning and are inconsistently and arbitrarily applied in practice. Indeed it will be apparent that spiritual knowing cannot find a place in science because it is methodologically excluded. It wouldn't matter how true that knowing is; science as currently constructed could not see it. So it doesn't!

On page 89 of the Delusion we are offered information about the brain and about its susceptibility to optical illusions. We will use similar images. Where Richard Dawkins draws the conclusion that human perception of a spiritual reality is like those illusions, we will suggest that we need to recognise and work constructively with our sensory limitations as well as validating and using our "additional senses". The conventional scientific view is that such limitations make any knowledge from personal experience invalid. But this is a self-fulfilling argument. It is used to deny experiences that are common to millions of humans. Under such conditions what could possibly count as validation for our perceptions? Presented his way, it's an impossible task. But we will present sufficient objective evidence to show that there are very strong grounds to trust that people are not as dim as science thinks that they are and that there is a layer of non-ordinary reality that is both eminently accessible and objectively demonstrable. You are not delusional for already knowing this.

4.   <u>The subjectivity of the choice to be objective</u>. This is a sweet paradox. The choice philosophers and scientists make to pursue this supposed objectivity is a subjective one. The belief that you can avoid all subjectivity is itself a delusion. There has been huge value in attempting to ensure that we have some means of determining and validating what classes as "knowledge" or "truth", and science has benefited greatly from this. We do not advocate mindless gullibility. Belief in an ultimate objectivity is however quite mythological and it leads to the spurious claims of objectivity in scientific methodology noted above. Human beings cannot understand the world they inhabit by excluding themselves and their experience from it. This inconvenient truth also leads to the next mis-perception.

5.   <u>The paradoxical nature of reality.</u>   Quantum physics, mathematics and philosophy have all shown the world to be full of paradoxes. This affects the most fundamental aspects of what may be considered "real" or "true" and is among the reasons why complete objectivity is a myth. Physicists and mathematicians generally grasp this even though they may not see all its implications or try to ignore it in their work. Biologists and others tend not to grasp it, both because they are not necessarily skilled in quantum physics and because it would make their picture more complex. We will see that it is provably impossible for any single view or system of rules to ever provide us with absolute truth. The delicious paradox is that the only absolute truth is that there is no absolute truth. We will also see how wonderful and empowering this is; it opens up a truly creative science, describing a truly creative universe.

6.   <u>The significance of information</u>.   The scientific model is geared towards that which is material and directly observable. There are good reasons for this approach and it too has deep historical roots. Nevertheless it excludes aspects of the world which are crucial to the relationships between living organisms, aspects which lie in the realm of information and by which those organisms regulate their biology and their behaviour. In humans that information is at the heart of understanding the nature of the real world. Information is the connection between consciousness and matter and forms the bloodstream of spiritual perspectives.

Some of these six issues can be traced back through millennia of Western philosophy, at least to the time of the Greeks. It is clear though that these viewpoints have come to inhabit the thinking in our culture, to such an extent that it is not surprising they are now as unquestioned a part of our world as water is for the fish. Nor is it surprising that they are embedded in our language, the common currency of science or of its representatives. We do not define ourselves as philosophers, but the change we are promoting is also a philosophical change and we will inevitably refer to this from time to time.

Nor do we wish to be unkind to scientists as individual human beings. We live in the same world as they do and are similarly subject to its blind-spots. We also have sympathy where the goal of pursuing science as a career is subject to intense competition with strong pressure on funding, tenure, academic publication processes and teaching curriculum that sometimes make true independence a heroic choice. It seems conspicuous that most of the scientists who speak outside of orthodoxy are retired. There are political issues here too. This is neither in essence a political book nor one about market economics but we would be naïve if we did not recognise that there is a political and economic reality that affects the scientific world. In the end, the shift that is required is a cultural shift which will affect all of these areas.

The religion of numerical truth

The fundamental demand from a scientific approach is that we should provide evidence to support our claims for a non-ordinary reality. We should allow the spiritual agenda to be subjected to the same scrutiny as any statement or claim that might be made in other contexts. We are delighted that the question is being asked and eager to rise to this challenge. In return we will make suggestions for proper scientific investigation of various phenomena and hope that they will be taken equally seriously.

Some of our evidence will be based on human experience. Science requires that such evidence should be consistent and that it can be validated by repetition. We will provide an abundance of such evidence and refer to more. We will also illustrate the evidence with personal stories but the evidence should not be dismissed as anecdotal. There is too much of it for that.

At the same time a potential trap is set for us by the inconsistency which is applied to the definition of what may be offered as scientific evidence. The "hardest" of hard science happens in laboratories under conditions which are not particularly compatible with many of the aspects of life that we depict. Many "hard" scientists dismiss "softer" sciences such as those concerned with psychology or social phenomena. A friend, one-time CERN engineer Dr. Vinicio Sergo uses the apt phrase "the Religion of Numerical Truth" and describes our slavery to it as a refined Orwellian situation. We would entrap ourselves if we attempt always to meet laboratory standards of proof. Readers who will think we have failed if we cannot do this are warned that this would be an impossible task. We cannot overcome diehard skepticism and regard the stance described as anti-scientific; it leads to loss of knowledge. The same friend refers to this as "like the loss of biodiversity – a loss for life".

Our approach is consistent with and justified by many other scientific endeavours which live outside the laboratory. Professor Dawkins' own discipline of ethology, or the piecing together of evolution through fossil evidence are examples of studies which rely strongly on careful observation of limited information, and on systematic analysis to determine meaning and relationships. We ask only for the same standards to be applied to our evidence.

We see a parallel in meteorology. "Weather" is a term that you could see as vague and hard to define. You can quantify rain, measure temperature and air pressure and come to conclusions about patterns that enable a degree of understanding of climate and prediction of weather events. But you still might struggle to say what weather is. But we all know it's there and talk about it all the time. It is meaningful, the first commonality for small-talk. You might hardly believe how much dialogue Arizonans can find about the fact that yesterday was 109 degrees and today is 110 when both parties know it will be friggin' hot for months. Today might be ideal for me and too cold for you but subjectively we would agree together that weather exists. Sometimes spirituality exists in a similar frame. We are examining phenomena that live in the realm of human experience. Some of them will be measurable and some less so. If God exists, this is the realm She exists in. You might say that we are engaged in a new science of "deology".

The first title for this book was "God's Ecology". Ernst Haeckel, who coined the term, defined ecology as "the comprehensive science of the relationship of the organism to the environment". That title offers another parallel, the metaphor of treating "God" as an organism and examining the environment in which it lives. Much of today's environmental agenda is concerned with the destruction of habitat and diversity, with an upset to the balance of nature that threatens our present and future lives. Those who inhabit a universe which is in any sense spiritual also find their habitat to be threatened. "The God Delusion" is only one part of a consistent attack on the world of spirituality, holistic health and the validity of belief. In this sense we are eco-warriors for the spiritual environment.

Our purpose is not to tell people what to think. It is however, to encourage a fresh way of thinking that increases our choices of what to think. It doesn't matter to us whether people believe in God and we don't have any new religion to promote. God forbid! Some things do matter to us though. It matters to us that human beings are empowered to improve life on this planet. It matters that we are in a better position to deal effectively with today's problems of existence. It matters that we increase our collective access to health, well-being, ease and joy. We hope and believe that the journey we are about to take will lead in this direction.

A New Way of Seeing

A story is told of scientist and poet Johann Wolfgang von Goethe as a young man. On returning from an educational journey to Italy, he was asked by his father if he had seen anything new. He replied "No father, but I have learned a new way of seeing." We have been brought up in a world which sees itself in a particular way. Scientism's limitations are built in to that mode of seeing so our task is to present a new way of seeing. We can offer a few images of this, and it is one theme that will run throughout our journey.

A simple dialogue with an advocate of scientism will not accomplish this task. Meeting their challenge to provide a science of spirituality will be impossible in a "they say black, we say white" way. The facts are generally not wrong, but equally crucial facts are missing, leading to mistakes of interpretation. We must make a shift of perspective, to see their black and our white at the same time. And then we must paint in the colours to see the whole in its richness.

As a teenager Jon was fascinated by the drawing here, which he repeatedly

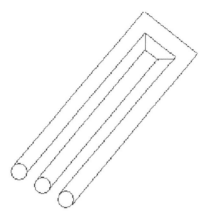

reproduced during dull lessons at school. Those who are familiar with the drawings of Maurits Escher will recognise its central dilemma of our human perspective. In the drawing it is possible to see a squared-off, two-legged figure and a rounded three-legged one, depending on where you focus your attention. Escher's endless staircases and confusing buildings are a brilliant elaboration on the theme.

We do not see the world as it is. The back of our eye is a curved surface only a few millimetres in depth on which rays of light draw a picture that represents hundreds of metres of height and depth. Our inner conceptions of the world are subject to similar limitations. We cannot change our eyes, but we can change our concepts. This is an exercise not in vision, but in visioning. We have to use imagination to reframe our images.

The facts are not new. Many of them will be familiar. It is like the type of movie where the whole story is in front of you, but you do not understand it until the "MacGuffin" moment when the partially hidden motivator is made obvious. A MacGuffin is like a secret hidden in plain sight. It's the territory of Hitchcock's "Charade" and Shyalaman's "Sixth Sense", and as with the latter all the clues were there, but you just didn't see them the right way. It's obvious when you watch the film twice, or when you review the facts. It's the film "Usual Suspects" once you know who Keyser Söze is. Our MacGuffin is not news either. Most of the "clues" we describe have been there all along. Goethe knew it. Proust knew it too when he said "The true voyage of discovery consists not of seeking new landscapes, but having new eyes." You have been watching this movie all your life. If you don't already know the MacGuffin, here are your new eyes.

The MacGuffin

The usual story is that human beings are creatures who inhabit reality. The things of the world – whether the inanimate chairs and tables or the scuttling beetles – are objects that are separate from you. They are as they are and exist independently. You see them, touch them. You the subject make decisions about them, the objects. It is a subject-object world. You make decisions about your experience. There is an external reality and you "know" things about it based on your interpretation of the information that your senses give you about it. That is how most of us were trained to think.

Science seeks to give complete definition to this external "reality" by getting beyond those sensory experiences and the decisions we make about them into a realm where all aspects of reality can be measured and understood independently of our personal experience. It seeks laws that define the way reality works – laws which hold true regardless of any of our experiences. This works very well within certain boundaries and has brought humankind a long way in mastering its physical universe. This knowledge put men on the moon and mapped the chemical sequences of our chromosomes, no small achievements.

This story does not work for an intuitive world, for a prayerful world, for a world of energy healing. There is no science of human experience because human experience is outside the boundaries of science. There would be no problem with this if science were to accept that it has placed an examination of "God" outside of its methodology. It would be quite acceptable for science to say "we have no knowledge of this, it's not our remit, none of our business." The scientific attack on spirituality and holistic medicine steps across this boundary, making it their business and ours. The challenge to provide a science in which prayer and energy healing are intelligent and meaningful can only be met by giving new eyes to the scientific realm as Goethe sought to do and as we do here.

Our view is that the "subject-object" divide is a human invention and has nothing to do with what is "real". It implies separateness between us and the rest of the universe which is quite illusory. In consequence the view of what constitutes scientific objectivity and the presentation of our subjective experience as unreliable are both equally misguided. If you choose to stand in a valley you have only one possible perspective on the terrain. A helicopter view is quite different. The terrain has not changed. When you look at life from within a definition that assumes everything to be separate, you cannot see connectedness. Once you know that the connectedness is there, everything looks different. You can detect the information flow. Reality lives not in subjects or objects, but through and in-between all of creation. Informational connectedness is our MacGuffin.

The hypothesis which this generates is that all "spiritual" experience takes place in the realm of this connectedness. The connection is not amorphous. It operates via information which is retained and exchanged throughout material creation, a universe-wide-web, the Wiki that begins and ends all Wikis. This information actually informs and defines the shape and nature of all material things. Information and our consciousness of it are inseparable. We would see it as what physicist Sir Arthur Eddington meant 80 years ago when he said "The universe is of the nature of a thought or sensation in a universal Mind... To put the conclusion crudely — the stuff of the world is mind-stuff". Throughout this book we will present evidence that demonstrates the presence of this information, its universality, and the way in which humans in particular interact with it. This is the realm in which spiritual experience takes place. It is also a realm in which creative possibility exists beyond what is usually conceived of even in the spiritual context. There is a new slant on the expression "you ain't seen nothing yet."

Review

We have seen that science has aspirations to define what is True and Real and that it regards its approach as objective. We have seen aspects of that thinking which are in fact subjective and which result in unsustainable viewpoints which result in prejudice and in the rejection of evidence. We have seen that once we step beyond the religion of Numerical Truth we have the possibility to open up new ways of seeing. These new ways are at the heart of this book.

3. The Sight Within

A man born and bred to the so-called exact sciences, and at the height of his ability to reason empirically, finds it hard to believe that an exact sensory imagination might also exist.

Johann Wolfgang von Goethe

There may be no absolute division of the energetic Universe into isolated or non-communicable parts.

Buckminster Fuller

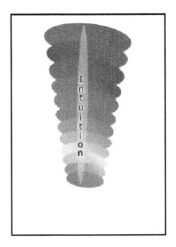

Theme

Our first task is to illustrate aspects of the conscious connectedness we are describing. In this chapter explore core features of the "top-to-bottom" nature of this connection as revealed by human experience. These features are:-

There is a thread of connectedness through all levels of material reality

This connectedness allows human access to information about the world

The information is specific and externally verifiable

Access is both open and "targeted" as well as being received spontaneously or by deliberate enquiry

Information arrives via pathways which resemble our external sensing.

The thread of connectedness does not acknowledge time boundaries either past or future

The connectedness is not simply a matter of passive or active receiving. There is also very strong experimental evidence that humans can "transmit" and that transmission has physical consequences

Flat-earthers and the global reality.

How would it be if you, and all humans, were much more intuitive than you have ever imagined yourself to be? We have to deal with a major challenge that humans face as soon as a subject like intuition is raised. What is it, and where does it fit in "reality"? Most of us see ourselves as biological creatures whose existence runs from birth through life toward death and our concepts of time and space fully reflect this view. Scientifically speaking, reality is not so simple. Mathematicians and physicists work with a model of the real world which has more than our three dimensions of space (up-down, side-to-side, front-to-back) and which does not view time as running in one direction. What's missing here?

Throughout biological creation it is obvious that organisms rely for their survival on understanding space and time very well. You know this each time you cross a busy street. It is a fact which appears to have driven the evolution of sensory apparatus in predator and prey species and we are used to relying on these senses. The physicist's perspective is also psychologically threatening for humans. Potentially it challenges our sense of self to imagine a world of ten dimensions in which time's arrow does not apply.

Perhaps it will help to imagine fewer dimensions and to see ourselves like "Mr. Square", the being imagined by Victorian-era mathematician Edwin Abbott, who created a world whose inhabitants could only see in two dimensions. They could see forward and back. They could see left and right. Up and down did not exist for them. Imagine being such a person and walking straight forward on the globe; you would eventually arrive back at your starting point with no notion of how this was possible. They resemble the humans who expected sailors to fall off the edge of an Earth that was presumed to be flat. Mr. Square could see the third dimension (after all, he had a body) but had trouble conveying this to others.

When the medieval flat-earthers were told that the Earth was a globe they at least had a concept of spherical objects to relate to. They had pebbles, apples and oranges. The information that we present here requires a bigger shift. This is one of Mark Twain's "aint so's". We have to let go of what we think we know. The metaphor of thinking "outside of the box" is very strongly applicable here and as Twain also said "you can't depend on your eyes when your imagination is out of focus".

While remembering that scientifically the box didn't exist anyway, we want to tell some other stories which illustrate how the linear "cause and effect" chain is broken. We want to make it easier to grasp. But we also need to follow this up with some experimental evidence. Stories don't help if they are then treated as myths, mistakes and coincidences. The research evidence is limited because very little research gets funded – the catch-22 consequence of not being seen as scientifically possible. But there is some quality that makes up for the lack of quantity. It is good enough to deserve being taken seriously.

Skilled medical intuitives

The next two examples concern medical intuitives – in both cases people who can be regarded as having marked natural abilities, honed by years of practice. Where Jon's story is of a two-fingered pianist plonking out "chopsticks", these are virtuosi playing Rachmaninov. The first example comes in her own words from Dr Caroline Myss who has progressed from being perhaps the best-known modern medical intuitive to being one of the world's most inspired and clear-thinking spiritual teachers.

By way of introduction, this is what Dr Norman Shealey, M.D, Ph.D says of her:-

> "There have been, throughout the ages, talented intuitives and mystics who have sensed the power centres of the human body. Alice Bailey, Charles Leadbetter and Rudolf Steiner have all written in this field, but no-one has captured the breadth and depth of our electromagnetic spiritual framework as well as Caroline. Never before has the anatomy of the spirit been so powerfully revealed. Herein lies the foundation for medicine of the twenty-first century. ...in my twenty-five years of work with intuitives throughout the world, none has been as clear or accurate as Caroline."

What follows is our précis of Caroline's own description of her experience, which starts in 1982 when she has recently started a publishing company and is still a smoker, coffee-drinker and "not at all primed for a mystical experience". She describes how she started to offer responses to people about their problems, when insights into the cause would pop into her head. She could not figure out how she was getting these impressions, describing them then and now as

"like impersonal daydreams that start to flow as soon as I receive the person's permission, name and age. Their impersonality, the nonfeeling sensation of the impressions, is extremely significant because it is my indicator that I am not manufacturing or projecting these impressions. They are clear but completely unemotional."

She describes the sensation of wondering with each consultation whether "it" would work together with the fear of failing people by being inaccurate and the tension this led to, fearing the judgement of others, or that she would have to live as a mystic. But she was fascinated and compelled to keep on evaluating people's health.

She could also SEE the energy surrounding that person's body, filled with information about the person's history. She says that our spirit is very much a part of our daily lives; it embodies our thoughts and emotions, and it records every one of them from the most mundane to the visionary. "It participates in every second of our lives".

Caroline continues with her story describing a deep experience in which she was "guided" from within, entirely without warning, through the explanation to a woman facing terminal cancer of all the events in her life which had led to her current illness – in "every detail". It is a dramatic and revealing story which we regret having no space to reproduce. We deeply recommend "Anatomy of the Spirit" and all of Caroline's considerable body of work.

As Dr Shealey pointed out, Caroline is not the only medical intuitive but there are several special features about her material. Her partnership with Dr Shealey is one significant element in this. Over many years he would telephone her about his patients. As a result the accuracy of her work was confirmed by true medical diagnosis from a practitioner who could follow cases in detail. Also significant is the nature of Caroline's perceptions and the way in which it captures not just immediate information in the "here and now" but past information.

At the risk of taking too much time with "medical intuitive" stories we also wish to bring in the perspective of Dr Mona Lisa Shulz who is an M.D. with a Ph.D. in neuroanatomy and behavioural neuroscience as well as a practicing intuitive. Where Caroline Myss presents the Spirit-to-mind connection with such clarity, Mona Lisa is able to apply her deep knowledge of the mind-body chemistry in order to map the mind-body relationship in terms of its operational biology.

It is not possible to convey the complexity of this map in a single quotation. It forms a tapestry that weaves through her book "Awakening Intuition". At the centre is her intricate understanding of the way in which the development of disease reflects the nature of the life conditions the subject is dealing with. The disease pattern is mediated physically and biochemically through the way that particular functions become blocked and subsequently break down. Here is a typical example. At the beginning of every reading Mona Lisa has only the subject's name and age obtained over the phone to work with.

"The reading. As soon as I began reading Violet, age 38, I saw a life full of joy and exuberance. She seemed to me like a little sparrow sitting on a fence, warbling away. She didn't like to complain and masked everything with her cheerfulness. But underneath her heart, I saw a difficulty, a person who was getting in the way of something Violet wanted. I saw disappointment in her relationship with this person, which was making Violet's healing a challenge. I saw something Violet wanted to do, some life's path she needed to follow, but I saw she was being held back by this person who had authority over her. This person appeared to be extremely goal-oriented and focused on his own needs to the detriment of the feelings of those around him. In fact he reminded me of a cat who's just killed a bird. He's sitting on the porch, surrounded by feathers. And the feathers are the petals of Violet's heart.

I saw that Violet was feeling less and less joy and was slowly becoming angry and resentful. Her partnership was not one of equality; her partner was the authority and she was submissive.

In her body I saw a little scarring in her lungs from past smoking, and occasional bladder infections in the distant past. But in her head I saw a kind of fury, what I call a tornado in the scalp. This whirlwind of anger and hostility was buffeting her heart and causing it to race and skip a beat.

The facts. Violet was a musician who had an intense desire to become an orchestral conductor. She had the opportunity to work and train toward this end with a leading male conductor. Her husband, however, was opposed to this idea. A non-musician himself, he was threatened by the prospect of his wife's working closely with another man at something in which he could have no part, and he refused to allow her to pursue her plan. Violet was inwardly furious at this, but out of deference to her husband's feelings and jealousy she had put off her training. However she hadn't given up on her determination to become a conductor. She continued privately to plan to do it at a later date, thinking that she could present the idea to her husband in another way so that he would eventually approve it. For the time being, Violet had put a lid on her own feelings to keep her husband happy, but her resentment simmered.   Meanwhile she confirmed that she had developed an arrhythmia in her heart.

Explanation.   Violet had given over authority in the partnership to her husband, but the results of this were causing her great resentment, and this resentment was sitting on her chest. And this in turn was setting the stage for serious heart problems. Sudden death from a heart attack is believed to be due to arrhythmia, a change in the rhythm of your heart. Stress often causes an increase in the production of brain chemicals called catecholamines, which can speed up the heart rate, overstress the heart muscle, wear down its reserves, and enhance the person's chance of developing heart disease."

Dr Shulz's book combines the intricate analysis of different types of illness and their roots in body chemistry with exhaustive investigation into the nature of the intuitive world, drawing on her own experience and that of others.   She quotes a study by Patricia Brenner into the intuitive experience of nurses, who are noted for their sensitivity to the condition of patients undergoing close monitoring, detecting features not obviously present in the data they are reading.

"In an extensive study of nurses and intuition, nurse and researcher Patricia Brenner ascribed the intuitive process of nurses in clinical situations to "skilled pattern recognition"….this theory concludes that preciously acquired knowledge, an expertise based on memory and prior experience, is the basis and source of intuition. A nurse detects something in a patient that rings a tiny bell in her mind and reminds her of a previous similar case that leads to her hunch about the current patient's condition.

One of Brenner's own nurse cases however, contradicts this neat theory. It involves a case of pulmonary embolism. As it happens, pulmonary embolism, a blood clot in the lungs that's nearly always fatal if undetected, is one of the hardest things in Western medicine to diagnose. There are virtually no common symptoms and very often no discernible signs that a patient is in danger of a PE. …Fundamentally it's one of the most fatal medical conditions and tragically, one of the easiest to miss.

The nurse whom Brenner observed saw a patient with cerebral edema, or fluid on the brain. His fluid intake had been restricted, and he was resting quietly. But the nurse was concerned. "Somehow I knew he was going to have a rough time," she reported. There it was, the intuitive hunch. "Somehow I knew he was on the highway to a pulmonary embolism." But what a hunch!   How did she make the unbelievable leap to that extraordinary conclusion?   It was a case of pole-vaulter cognition. This patient didn't have a problem with a clot in his lungs, he had cerebral edema. It wasn't even a case of right church, wrong pew; this nurse didn't even seem to be in the right state! The only possible symptom she could relate to pulmonary embolism was having overheard the patient's wife say earlier in the day that he was anxious. That night the nurse couldn't stay away from the patient's room, even though he was assigned to someone else's care... She found him "sort of pale and anxious,) and even though he was still conscious she called the doctors, and sure enough, just as the doctor's arrived, the patient began to die. The doctors coded, or resuscitated, him.   The pulmonary embolism was caught and the patient was saved."

From these examples we would like to draw out several features of the intuitive experience and begin to highlight some threads which run through the tapestry of the connected world we are describing.   These threads will recur and be developed as this book unfolds and we will see them from a variety of perspectives.

- There is a connection between one human being and another that enables the transfer of information

- That information is capable of being highly specific

- That information is capable of being accessed without an explicit "address" protocol. That is, Caroline and Mona Lisa can tune in to a person simply by hearing a name even when that name is spoken by a third person.

- There is a connection between the information content and the physical events which take place in the body

- Individuals (e.g. the nurse in the example above) can be aware of this information content even when not explicitly seeking it through a deliberate intuitive intention

- The intuitive sensory mechanisms in the receiver vary. They may deliver through pictures, words, or feelings and sensations. Alternatively the receiver may "just know", like Brenner's nurse. But all of these mechanisms use existing sensory pathways.

There are passages in Mona Lisa Shulz's book which describe how a person who is learning to be intuitive must recognise the pathway which is most active for them, and develop their understanding of the symbolic ways in which information presents. This matches Jon's experience and forms an important part of the process he has used for teaching intuition.

The book also explores dream symbols as a communication path for inner knowing about the body-mind.   The symbolic significance of dreams is hardly a new discovery, but it has a link with our next example, written by J.W. Dunne in his 1927 book:-

"An Experiment with Time".

J.W. Dunne (1875-1949) was a British aviation pioneer. As an engineer trained in the classical pragmatism of the late 19th century, he held a number of early British aeronautical patents. He was successful and respected with a comfortable income, to all appearances a normal member of society.

Dunne's book begins by chronicling an experiential journey. Over a period of time he had repetitive incidences of dreams which would in the days following be mirrored back in newspaper stories and headlines of dramatic events. His initial scientific skepticism is documented in great detail, as is the process of analysis by which he sought to test his experiences and to eliminate the possibilities of chance and coincidence or faulty memory and suggestibility. He describes the process with the true engineer's love for precision.

Over an extended period he compiled lists in which he systematically noted correspondences between dreams and events, both when those events were in the past (i.e. before the dream) and when they came after, as if the dream had been predictive. In some cases the dream would be close to the event. In others they were significantly ahead of the time they corresponded to. His eventual conclusion was that dreams were occurring in approximately equal numbers relating to the past and to the future.

Here is one example of a predictive dream with a six-month time-span.

> The dream occurred in the autumn of 1913. The scene I saw was a high railway embankment. I knew in that dream – knew without questioning, as anyone acquainted with the locality would have known – that the place was just north of the Firth of Forth Bridge, in Scotland. The terrain below the embankment was open grassland, with people walking in small groups thereon. The scene came and went several times, but the last time I saw that a train going north had just fallen over the embankment. I saw several carriages lying towards the bottom of the slope, and I saw large blocks of stone rolling and sliding down. Realizing that this was probably one of those odd dreams of mine, I tried to ascertain if I could "get" the date of the real occurrence. All that I could gather was that this date was somewhere in the following spring. My own recollection is that I pitched finally upon the middle of April but my sister thinks I mentioned March when I told her the dream next morning. We agreed, jokingly, that we must warn our friends against travelling north in Scotland at any time in the succeeding spring.

On April 14th of that spring, the "Flying Scotsman", one of the most famous mail trains of the period, jumped the parapet near Burnisland station, about fifteen miles north of the Forth Bridge, and fell onto the golf links twenty feet below.

The experiments described in the book extended over a period of years and were widened to include numbers of other subjects. Examples and analyses fill the first half of the book. Unless one treats the text as either fraud or fantasy the accumulation of evidence is very convincing when viewed in the context of the care with which it is acquired and the dry delivery of a Victorian / Edwardian gentleman. He progresses to the following conclusion regarding what he had originally thought might simply be his own "abnormality" in the experience of pre-cognitive dreaming.

> "...in the light of the experiment, I did not appear to possess even a specially well-developed faculty for observing the effect. Those other people had got their decisive results more quickly than I, and in most cases, those results had been clearer.

> The outcome of the experiments suggests that the number of persons who would be able to perceive the effect for themselves would be, at least, so large as to render any idea of abnormality absurd"

The book's second half is devoted to an enquiry into the meaning of these events. What do they tell us about the nature of consciousness and its relationship with linear time? Much of this is presented in mathematical terms, some of it related to the theory of Relativity, which was quite new at the time. Hopefully we can present the core conclusions in a simple and understandable way. While we could short-circuit the discussion by simply saying that, from physics' point of view there simply is no linear time of the kind that we experience, we feel that it is useful to understand more about the nature of our conscious relationship with time and our perceptions of it.

How we perceive time

Have you experienced something like this? You have taken your seat in a train at a station. There is another train on the neighbouring track. You notice that your train is beginning quietly to move and you see the neighbouring carriages passing. It is only when the last coach one has passed that you realise that the station is still there. Your train hasn't moved at all. You have also not noticed that your assumed motion was backwards not forwards, as if you were reversing out of the station. (You may have experienced something similar as a passenger in a car watching lines of traffic).

Our experience with travelling is that we are usually the ones in motion and that is what we expect. Our minds process the signals from our eyes with this as the in-built assumption. We don't think about it. This tells us something about our position as "observers". We tend to see what we expect to see. The above experiences can include confusion about who is moving forwards and who backwards.

Our view of time also contains expectations. Our everyday waking observer adopts the same view of time that the passenger takes towards scenery. The scenery moves past us and we see one scene after another. Likewise, we experience one moment, then another and then another as time flows. With time we do not think there is any possibility of reversing our journey. We cannot make time go backwards. Still less can we teleport back and forward across the landscape, frog-hopping from past to future.

A corresponding shift of perspective occurs in the dream-experiences described by J.W. Dunne. Just as when we find it was the other train that was moving, his story requires us to turn our view of time around. Instead of time flowing irreversibly with only the present moment being in our view, the whole vista is made available to us and we can indeed do the equivalent of teleporting ourselves into past or future. We can imagine teleporting because that's like an aeroplane but quicker. We can likewise imagine an H.G. Wells-type time machine where the jump is from now to the past or future.

But dreams are also different because we don't physically move. There's no machine, no Time Tunnel, no Tardis. What there is, is an observer that moves. The conscious awareness can roam in time or space, but our bodies stay where they are. You can see how this resembles your experience of memory where you are familiar with an internal journey to images that you have stored, and where your consciousness no longer resides in the present but instead has its focus in that past experience. We all know what it is like, replaying the car accident or the embarrassing moment, recalling the moment of joy at a birth, or when we fell in love. There is a resemblance to imagination or fantasy, where we visualise ourselves in the future, shaking hands as we receive the hoped-for promotion, or that moment as children where we scored the winning goal in the cup final.

In pre-cognitive dreaming however, it is neither memory nor imagination. The observer moves not in his / her own personal world but in the landscape of all worlds. The observer, the dreamer, is not part of the experience but is at a distance, seeing the train fall off the track, watching the stones tumble down the embankment.

In J.W. Dunne's presentation there are two crucial features which distinguish the experience. His investigations do not deal much with the similar realms of intuition or telepathy described earlier, but we suggest that the conclusions would be the same.

Firstly, we are called to reframe our view of reality. In the dream experience we have stepped off time's arrow and have options not available to our physical selves. Our consciousness can move. In a similar way, the intuitive experience takes the consciousness of the observer into realms where there is access to information that is not presented to the senses we call "physical".

Secondly we need to accept that there are two distinct types of observer. Most of us spend all of our waking time in one mode, being "observer A". There is a "right now" in which Jon is sitting at a keyboard, thinking thoughts, typing words, seeing images appear on a monitor. In a future "right now" <u>insert your name here</u> is reading a book and sharing those thoughts. You can be aware of the book, the place you are in, the chair you're sitting on etc. There is also another observer – an "observer B". This other observer takes a back seat in our consciousness. We may never be aware of their observations at all, or we may experience them in dreams, or we may receive them as intuitive experiences. This other observer can roam in space and time; it can see, hear and feel that which is not here and not now. It can know things which are not accessible to observer A.

We do not choose to say who or what this other observer is. If you have a belief system there may be a language you use for it. Words like inner knowing, gut-feel, soul, spirit, higher self, "I am" presence and even "God self" come to mind. But if none of this instantly fits your belief system then we ask you for now to recognise the evidence that is being put forward and leave a space for your own world-view to emerge in a way that will encompass the reality that this evidence is demonstrating.

Who's looking?

Once we accept that Observer B exists, and always has existed, our questions over "God" and spirituality begin to take on a new aspect. The difficulty with "God" is not one of existence or non-existence, it is one of interpretation and understanding. Most historical views of the Divine are too simple to fit the complex reality of a connected world. Many stem from simpler times and thought structures. There is a need for a 21st Century re-evaluation.

For most people in our society Observer B is either not heard, heard and distrusted, heard and ignored (because our science tells us it cannot be real) or simply over-ridden because it does not tell us things that our Observer A personality wants to hear. Whatever Observer B might make it possible for us to know, we have multiple ways of choosing not to know it.

Often, while we may hear the inner voice we do not understand the way in which it communicates. For most people the voice is subtle and understanding it requires some effort. Modern humans have no way of knowing what this experience was like for people 5000 years ago or at the time of Christ or Mohammed. We likewise have little understanding of what it might be for those living now in different ways than we do – the aboriginals, the indigenous peoples who still live close to nature. Perhaps these cultures have been more open and receptive or their members have experienced less competition for their attention from the ordinary reality. Even where a culture treats intuitive relationship to the world as normal it is too much to expect that such a culture might use our modes of analysis or express those views in concepts that are familiar to us. There are a few exceptions however which may help us over this barrier and we will come to them in Chapters 4 and 5.

Much of what now occurs under the banner of religion or in other culture's spiritual beliefs may be easily criticised as irrational and presented as primitive. We prefer to suggest that these apparently soft targets may arise from attempts to make sense of a complex world seen "as through a glass, darkly" and that the over-valuing of the rational mind leads to this potential illusion of superiority. What will emerge through this text is a more coherent way to understand the complexity, the difficulties of interpretation and the nature of what lies within.

As preparation for this we will now present some of the more systematic and explicitly scientific investigations of connectedness. What has been done to prove the ability of minds – of observer B – to roam in time or space and to extend the projection of our sensing mechanisms?

If you were to listen to what scientists say, you could be forgiven for thinking that no-one has done any experiments to determine whether psychic activity is possible. Where there are references to such experimentation, they are almost always dismissive. For example, the famous Prof. Carl Sagan, renowned cosmologist and science-fiction author wrote his final book "The Demon-haunted Universe" to defend scientific knowledge in the face of irrational beliefs. But his chapter on telepathy fails to mention any experimental work. His only concession is to acknowledge elsewhere the existence of three claims in the ESP field which deserve serious study, one of which is the claim that by thought alone, humans can affect random number generators in computers.

Unfortunately he does not say what these deserving claims are, despite the fact that just one such proof of even a minimal effect would demand fundamental revision to conventional scientific theory. In the following pages, we describe some of the more convincing work. Our first choice may well be the one he is referring to.

Psychokinesis

One particularly interesting series of experiments was conducted under the supervision of the Professor of Aerospace Sciences, and Dean Emeritus of the School of Engineering and Applied Science at Princeton University. Robert Jahn is the author of "The Physics of Electric Propulsion", a consultant for NASA and the US Defense Department, and had no special interest or belief in the paranormal. However, he was approached by a student of clinical psychology to oversee her project on that topic. The results were of such interest, that in 1979 he founded a laboratory under the direction of Brenda Dunne "to study the potential vulnerability of engineering devices and information processing systems to the anomalous influence of the consciousness of their human operators."

Twenty-eight years on, the Princeton Engineering Anomalies Research facility (PEAR) announced its closure in 2007[4]. On its website it states

> "Jahn and his colleague, Brenda Dunne, a developmental psychologist from the University of Chicago who has served throughout as PEAR's laboratory manager, together with other members of their interdisciplinary research staff, have focused on two major areas of study: anomalous human/machine interactions, which addresses the effects of consciousness on random physical systems and processes; and remote perception, wherein people attempt to acquire information about distant locations and events. The enormous databases produced by PEAR provide clear evidence that human thought and emotion can produce measurable influences on physical reality. The researchers have also developed several theoretical models that attempt to accommodate the empirical results, which cannot be explained by any currently recognized scientific model."

In one series of experiments, Jahn and Dunne employed a device called a Random Event Generator (REG). The REG relies on an unpredictable natural process such

---

[4] Brenda Dunne is continuing the work through ICRL (www.icrl.org) and has produced a book documenting the PEAR work "Consciousness and the Source of Reality: The PEAR Odyssey"

as radioactive decay, and produces a string of random binary numbers. REG is in principle like a fast coin-flipper which can produce a high number of tosses in a very short time. The statistical expectation is that a large number of throws will tend towards a 50/50 split between heads and tails.

Jahn and Dunne had volunteers sit in front of the REG, and concentrate on having it produce more heads, or more tails. Over the course of hundreds of thousands of trials they discovered that the volunteers did have a small but statistically significant effect on the output of the REG. They discovered that the ability was not limited to a few "gifted" individuals. Rather they discovered that the majority of the individuals they tested could produce some effect. In addition to that, they observed that different volunteers produced results that were consistently distinctive. In certain cases these results were so marked, that the experimenters started to regard them as "signatures".

In a further series of experiments, Jahn and Dunne went from the sub-molecular level, to the visibly physical. They used a sort of pinball-like apparatus that allows 9,000 three-quarter inch (2cm) marbles to circulate around 330 nylon pegs, distributing themselves into 19 different collecting bins at the bottom. The device is contained in a vertical frame 3M by 2M (10' by 6') with a clear glass front. Left un-influenced, the marbles would normally fall more in the centre than the outside bins, and the overall distribution would be a bell-shaped curve.

The experimenters had the volunteer subjects sit in front of the machine, and try to cause more balls to land in the outside bins than in the centre ones, thus flattening the shape of the bell. Once again, over a very large number of trials, the volunteers were able to create a small shift, but a measurable one, in the landing-pattern of the balls. In this they showed that a psychokinetic effect could be achieved, not just at the level of microscopic processes, but on events in the everyday, material world. Furthermore, where characteristic individual "signatures" had been observed in certain subjects during the REG tests, those characteristics were repeated in the pinball-type experiments. This suggested that the psychokinetic abilities of any individual remain consistent from experiment to experiment. There was still clear variation between individual capabilities. The conclusion drawn by Jahn and Dunne was this.

> "While small segments of these results might reasonably be discounted as falling too close to chance behaviour to justify revision of prevailing scientific tenets, taken in concert the entire ensemble establishes an incontrovertible aberration of substantial proportions."

These experiments have several particularly appealing features. Firstly, they did not use "gifted" or "special" individuals. They showed that the results could be achieved even on a wide range of people who had taken no steps to develop such a capability. While it is scientifically significant even if only one being can achieve such results (because the theories of science would still have to allow its possibility and explain its mechanics) it is much more exciting to think of this as part of a generalised reality.

Secondly, these experiments are very hard to "rubbish". In both cases, there is little potential for the experimenters wittingly or unwittingly to influence the outcome. It would be particularly hard for the experimenters to produce the consistent "signature" pattern. The procedures were designed in such a way as to be watertight, by a scientist of considerable repute and no apparent bias. Unless one is willing to accuse the experimenters of deliberate fraud, there is not much room for question.

The only question that might be asked is over what is statistically significant. Perhaps, these minute fluctuations do not seem very exciting. But it is normal in many areas of scientific research to treat small variations in large numbers of samples as significant. The likelihood of the effects occurring by chance vary. Of their overall body of work, including that on Precognitive Remote Perception, Jahn and Dunne state "The composite formal human / machine results are of the order of 10^{-12}; the formal PRP results to the order of 10^{-8}". In normal speech these are chances of 1 in a trillion to 1 in 100 million, hugely less likely than a millionaire lottery ticket. Any reasonable assessment would be that the results are enough to show that the behaviour being observed is not random. In the absence of any other plausible explanation, the one proposed by the experimenters should be accepted.

Please don't underestimate the significance of small-scale effects of this kind. Their mere existence is in antithesis to scientific orthodoxy and blows a hole in scientism. In addition we will show later how small-scale effects can be of great practical significance.

We should also note, as we will in other areas of our scientific material, that Jahn and Dunne have experienced huge difficulty in achieving publication of their results in science journals. Science is not a level playing-field. Those interested in more detail can access the PEAR history and the subsequent progress through the International Consciousness Research Laboratory (www.icrl.org).

Psychic information-transfer

The second experiment we want to report was the subject of an article in New Scientist (15/5/93), by John McCrone. This is how he describes it.

> "Isolated inside a steel-lined cubicle with walls a foot thick, the subject lay back in a chair. Two halves of a ping-pong ball were taped over his eyes, and headphones filled his ears with white noise. Three metres away, in a second padded and shielded cubicle, a "sender" was concentrating on a TV film of an eagle, and trying to transmit the image telepathically. Something seemed to be coming through in the receiver's chamber: "I see a dark shape of a black bird with a very pointed beak with his wings down ... almost a needle-like beak ... Something that would fly or is flying ... like a parrot with long feathers on a perch. Lots of feathers, tail feathers, long, long, long ... Flying, a big huge, huge eagle. The wings of an eagle spread out."

Chuck Honorton, who died in 1992, had set himself the mission of putting the previously disreputable field of parapsychology on a scientific footing, by designing telepathy experiments that would be regarded as rigorous. Maybe he had succeeded. Susan Blackmore, a psychologist at the University of Western England in Bristol, who is a noted debunker of psychic claims, is quoted as saying "He has pushed the skeptics like myself into the position of having to say that it is either some kind of extraordinary flaw that no-one has thought of, or that it is some kind of fraud - or that it is genuine ESP."

Honorton's methodology emerged from a long history of other experiments which were either shown to be frauds, or were criticised for flaws in their methodology. Initially the "Ganzfeld" technique, as it was known, was subject to as much criticism as any others. From a dialogue between Honorton and one of his principal critics, came the modifications to methodology that resulted in the most recent and most convincing experiments.

The principle of the Ganzfeld (from German, meaning "whole field") was the knowledge that when people are subjected to perceptual deprivation, the elimination of outside stimuli to the senses, their internal visual imagery increases. Since mental imagery is a common way to convey psychic impressions to people it was thought that these impressions would be more easily detected under such conditions, making research easier.

As well as sensory deprivation it is necessary to shield the receiving subject from all normal forms of communication. The room has double-doors and walls. The ping-pong balls mentioned in John McCrone's description have red light shining on them, creating a plain visual field. The subject is encouraged to keep his/her eyes open. White noise creates a uniform background, and an additional acoustic barrier. Honorton's experiment also included copper shielding against radio-wave communication.

A computer was used to randomly select from a selection of four videotapes to be shown to the "transmitter" subject. The experimenter is kept "blind", unaware of the selection made. The receiver spends 15 - 30 minutes in relaxation / deprivation, describing sensations to the experimenter, before the transmission session starts. After the receiving session, the receiver is shown four video clips - the one seen by the transmitter, plus three others, and has to attempt to identify which one was transmitted. In the latest versions of the trials, over 240 subjects were used. These were not known to be "gifted" and most only made one trial. Overall, where random chance would have produced only 25% of correct responses, the trials yielded an average value of 34% correct identifications of the randomly selected transmission.

That might not sound high, and you have to understand statistics well to argue about how meaningful it is. Suffice it to say, that the more tests that are done, the less likely it is for any result other than the random 25%. So, if there are enough tests, and the result differs markedly from random, then the result is significant. It indicates that some effect is taking place in an area where no effect is believed possible. In short, it is enough to prove that psychic communication is possible, even if it is that communication is unreliable and low-key.

In general, telepathy experiments have been criticised for leaving too many possibilities for inadvertent communication. If the experimenter knows what the receiver is meant to pick up, he may unwittingly influence his choice. In recent experiments, great care was taken to avoid this possibility. Even the VCR is acoustically shielded from the experimenter, so that the amount of time spent fast-forwarding should not be known. A different copy of the tape is used for the playback and selection to the receiver, so that no differences of tape wear should be involved. The sender cannot communicate with any other person, by any method, during the experiment.

In Honorton's experiment, all sessions were recorded, providing a means of corroborating any results. Although this method made fraud difficult, rather than impossible, one has to be very determined to reject the experiment to make such an accusation. So far even hardened sceptics have assumed some "hidden flaw", rather than deceit to be the cause of the successful result. However, no flaw has subsequently been found.

There was another interesting feature of the experiment. The overall hit rate was above chance. But the hit rate among receivers who did not believe in the possibility of telepathy was as low as 8%. This indicates that certain "believer" subjects had achieved even **higher** success rates than 34% and also opens the possibility that the "non-believers" may have had some intuitive knowledge that enabled them (semi-intentionally) to select correctly **less than** 25% of the time

There have been many attempts at Ganzfeld experiments. Not all have been as tight in their methodology, and many have been challenged. Some have been unsuccessful. Evaluation of this evidence is very tricky. There is no scientific acceptance that telepathy is a proven fact. Nevertheless, there is a weight of evidence building that is making it harder than ever to dismiss the possibility.

We want to draw attention to two aspects of these results. One is that whereas these effects are usually called "paranormal", you should be seeing them as normal. The reason you may not think them normal, is only because science has told you they are not and our culture has built that belief in to our development and education. Secondly, for them to even be possible requires mechanisms of information or energy exchange that are not recognised within the terms of accepted scientific theory. We can hardly say this too strongly. Science acknowledges no way - no mechanism at all, by which such events can happen. Psychic communication takes place without any known wave or particle to facilitate it. These results and any other evidence of psychic activity expose physics as incomplete. When the science does not match the reality, it is science that must be revised - it is fruitless and unscientific to challenge reality. Later when we look at the world of physics we will discover that it already reveals these gaps, even in its own terms.

It is also important to be aware that small effects could be extremely important. The reason for this will emerge later, after we have looked at physical laws, and what is commonly known as "chaos theory" - the study of how dynamic systems change over time. Sci-fi movies and TV series tend to show "psychic" powers in a dramatic way and it might seem that they are not important when their effects are so small and subtle. This is far from true, but it will take a while to show why. We will come to see these small influences as the "butterfly's wing that causes the hurricane". At the personal level we also need to guard that the "Hollywood effect" does not prevent us from recognising our own intuition. Intuitive knowledge comes less with technicolour special effects and more as a still, quiet voice.

We have not reproduced the classic remote viewing experiments of Targ and

Puthoff or the detailed Jahn and Dunne precognitive perception material here. This is purely for reasons of space. The experiential aspects are covered by other material in this book. The experimental results can be found in the source texts and are briefly referred to in the PRP probabilities given from the PEAR proposition earlier.

What is Knowledge? What is Science?

The term "Science" has its linguistic root in the latin verb "Scire", which itself means to know or understand. In consequence it is quite tricky to distinguish the two questions posed in the heading above. It is even more difficult when we bring other expressions into play – words such as belief or faith. We may even have multiple shades to a word like "know". I know that the sun will rise tomorrow. I know that I am hungry. I know that Paris is the capital city of Texas (just kidding).

There are people – maybe you are one such - who would claim to "know" God. They would not say "I believe in God". They would not say they have faith, or not merely that. Knowledge has all of these flavours and more. Page 50 of the God Delusion lists seven and these do not include our own view. Science attempts to be something distinct from all of these and to establish a body of facts which are verifiable and true for all people under all conditions or at least for a clearly defined range of conditions.

In order to do this, science establishes principles of measurement. These principles are most comfortably satisfied when the measurement is performed by apparatus. Science likes it best when the only thing for the human being to do is to read the temperature on the thermometer, to weigh the object on a scale. Provided the apparatus is constructed to the correct standards, the measurements can be repeated in other times and other places by other people and the results will be the same.

Relatively little of science conforms to this standard. Readers of Dr. Dawkins' earlier works will be very familiar with the way in which genetics and evolution offer a great deal of data, and that data is not entirely measurable. Great intellectual effort is required in order to analyse that data and there are deep divisions about what it means. Darwin's theory of evolution through natural selection is a scientific theory but it is not an experiment which can be replicated elsewhere by other observers under controlled conditions[5].

In other areas of science the processes required to measure or detect that which is

---

5 Readers of Douglas Adams will be familiar with an alternative to this statement, which is that the Earth is an experiment constructed by white mice for such a purpose. This hypothesis may not be entirely helpful.

not visible require increasing complexity. Things too small to see require microscopes. The composition of the Sun cannot be determined by grabbing a handful – it is established by indirect means. Brain cells, neurons and dendrites are examined through scanners or by means of thin tissue slivers under electron microscopy. The picture of scientific truth is built piece by piece using many different methodologies, layers of technical intermediaries and multiple human observers, and frequently via deep dispute over what a set of results might mean.

Determining whether knowledge is scientific is therefore not a black and white, yes-no matter. This wide range of data-gathering and disputation involves a systematic attempt to eliminate anything personal, subjective or psychological but it has few absolutes. The truths it uncovers are often relative, often temporary and waiting to be superceded or subsumed, as Newtonian mechanics was by Relativity.

In practice, science is an approach, an attitude to evidence. It's like a police investigation. If you are seen in the neighbourhood of one murder that might mean nothing. If you are placed at the scene of another that might just be coincidence. To be at a third will make you suspect number one.

Science can only measure "things". It might expand the category of "things" on a continual basis to include discoveries like supernovae and quantum particles. It might allow the category of measurement to include the collection of data regarding species and habitats and to enable inference regarding ecological pressure, food-sources, predators, competition etc. Inevitably though, science is troubled by "information". With murder fresh in our minds you might like to think of Iago, whose manipulation of information supply is such as to induce Shakespeare's Othello to jealous murder of the wife he adores. Information has consequences but it doesn't fall into the realm of "things".

When you "know" that you are hungry, that information has the consequence that you are likely to get more closely acquainted with the contents of the fridge / vending machine. Your body is an information processing device. This information is internal and possibly not conscious – certainly based on things that you are not conscious of such as blood-sugar or fat levels. Other information is external – my skin is cold; add clothing or seek shelter. Science can measure some things. It could validate your blood sugar levels or the external temperature. It cannot measure your choice whether to seek shelter or put on clothes. It might have considerable difficulty with your choice when you get to the fridge? Ham? Perhaps you are a vegetarian. Cheese? Maybe a vegan. OK, Hummus. Unless there's chocolate brownies available, making all other choices irrelevant. The choice lies in the realm of information and is not measurable, but the information affects the world. It is yet more significant when the information persuades you to go to war in Iraq.

We could also have asked "what is intuition?" It is a form of knowing – sometimes remarkably accurate – and still it does not quite fit what we would be comfortable to call "knowledge". And yet it is verifiable. In the research cited above, a scientific attitude to evidence has been present to the greatest degree possible, taking into account the limits to that evidence placed by phenomena which are very hard to detect. In the examples of intuition there is external confirmation of the truth of subjective experience. Even so, that experience lay fundamentally in the realm of information, and science has a limited toolkit for this realm.

We are about to delve further into the subjective worlds and readers will each have to reach our own view, find their own answer to the question of "what do I know?" How many coincidences does it take before a pattern is acknowledged? How many reports of similar experience, from how many cultures, are required in order to establish a truth as universal? How much data is required before the scientific model of the physical universe has to change, and incorporate the passage of information from one place, one person, one living being to another into its theoretical frame?

Review

We set out to offer evidence to illustrate the way in which information about the universe is accessible to and accessed by human beings. We have shown that aside from our personal experience of such events, there is validated evidence which shows that humans can seek and locate specific information and that equally it can be presented to them unasked.

We have shown that this information is presented to human consciousness through sensory pathways that make use of the same mechanisms familiar to us in memory and imagination, but providing data which are not necessarily part of our personal world and which can also be accessed in a way which contradicts our habitual experience of the arrow of time.

In addition we have presented strong evidence from rigorous experimentation of both passive information detection and active information transmission. In the case of transmission we have shown the top-to-bottom feature that is in our diagram, connecting humans right through to the atomic particle level. We have also shown that such active transmission has outcomes in the physical world. This is a feature that we will revisit in the context of energy healing and later in relation to the nature of prayer and human participation in creation.

4. I sing the body neuro-endocrine

And all the world is football shaped

It's just for me to kick in space

And I can see, hear, smell, touch taste

And I've got one two three four five senses working overtime

Trying to take this all in

I've got one two three four five senses working overtime

Trying to taste the difference 'tween a lemon and a lime

Pain and pleasure and the church bells softly chime

Andrew Partridge   *XTC 1982*

I SING the Body electric;

The armies of those I love engirth me, and I engirth them;

They will not let me off till I go with them, respond to them,

And discorrupt them, and charge them full with the charge of the Soul.

Walt Whitman   *Leaves of Grass*

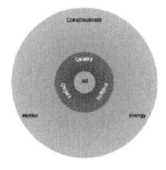

Theme

In this chapter we explore the connectedness that exists at the cellular level in the form of chemical messengers and rhythmic co-ordination. We present a key aspect of the way that "consciousness" shows up the relationship between matter and energy within complex organisms. We show as incorrect the conventional view that "mind" is located in our brain as the seat of our thinking process. Our organs and the cells that compose them have a part to play, as do the cells in our blood. The chemical messengers which circulate our bodies influence our physiological states and the ways in which we experience those states as emotions.

We will also encounter the truth of the age-old perception that the heart is not just a pump, but a coordinator of feelings and decisions. The heart has a particular role in the processes that we describe as "mind" and in our overall intelligence. All of this brings an extra dimension to the connectedness, one that integrates our emotional states with our values and decision-making. It resembles our intuition as an internal, subjective experience carrying personal meaning. We also see the part that biochemistry plays in these inner experiences and in their relationship to the outside world.

Sense, emotion and human experience

The chapter title does not sound as good as the Walt Whitman line quoted, but one suspects he would have absolutely adored the story that this chapter tells, like us finding it rich, subtle, delicate and full of wonder.

The field of science we are about to explore is relatively new. Genetics only began in earnest with the discovery of the double-helix structure by Franklin, Watson and Crick and has taken fifty years to the point of mapping the human genome. But where genetics might be said to be in early adulthood, the discipline which studies the chemical nervous system is barely a teenager. The field which has acquired the name "psychoneuroimmunology" (PNI) started with a paper published by Candace Pert and Sol Snyder in 1979. It was a slow starter. The name was coined in 1984 but even two decades later the full significance is just emerging.

We are about to delve into some intricacies of biology. Let us explain the importance of what is to come. Everything that we are describing in the realm of spirituality is expressed through human experience. Everything that is sensed is sensed by bodies and all that is described comes through a language that derives from a physical process and is reported in terms of our senses. Everything that we see as possible lives in our perceptual frameworks. This is how we experience our relationship with the material world and it is also how we experience the world of unseen energies that can be called "spiritual".

Albert Einstein, when asked to describe how radio works, reportedly (and perhaps mischievously) said

> "You see, the wireless telegraph is a kind of a very, very long cat. You pull his tail in New York and his head is meowing in Los Angeles. Do you understand this? And radio operates exactly the same way: you send signals here, they receive them there. The only difference is that there is no cat."

It is tempting to try and describe spiritual experience as if it is a radio or TV program. It might seem easier and perhaps more "scientific" to say "there is no person". The TV is seen as inert, as a window through which the program is viewed. You can trust the technology, take it for granted when watching that we are seeing what the director filmed and that the TV itself has not altered the original content. In this analogy, with intuition, spirituality and all the related phenomena we describe, the receiver (TV) is an active part of the process. There is a human where the cat..... isn't.

Even more than that, when we describe Candace Pert's work we are in an arena of both biology and communications. The content of the communication, the ways in which it is mediated are at the heart of human experience – subtle, intricate and powerful. The internal relationships and the external ones are intertwined. We will not understand them adequately unless we understand them in their connectedness. This also relates to homeopathy and to other complementary healing approaches including "spiritual" healing that we will discuss later. The connectedness is indispensable.

It's not just a computer

Many of us, even younger readers, will have been educated in an old model which viewed the brain and nervous system as being like an electrical circuit or a computer. There is a system of that kind present within you. It is critical to instantaneous actions such as motor functions. If you touch a hot object you need an instant response through specific muscles if you are to avoid a burn. The kind of co-ordination that is required for sports needs rapid decision-making based on visual clues and a direct pathway between decision-making and muscular action. We know that damage to nerve fibres can be catastrophic. We know that memory and experience combine in a soft-wired component that is like a reprogrammable machine, learning by its own trial and error. For this reason the computer-like model is powerful and persuasive.

That model doesn't give the whole picture though. There is much more happening in the body and a great deal of it is not instantaneous and is not happening in what we think of as "the brain". Living alongside the electrical network is a parallel system composed of molecules and sensors. The sensors are not just in the brain. Every cell is a little part of this network of intelligence.

While it was the second to be discovered, the chemical system was first in evolutionary terms. It exists in creatures without brains and is therefore more basic – more fundamental to life. There is a primitive single-celled creature, the tetrahymena, which is widely used in laboratories for experimental study. This creature makes many of the same substances, including insulin and the endorphins as we are about to describe in humans. On its single-cell surface are opiate receptors just like the ones in human brains.

At the operational heart of this chemical system are peptides, complex molecules which travel throughout the body and act as messengers. Francis Schmitt of MIT coined the broad term "informational substances" because they distribute information throughout the organism. The responses elicited by these molecules can be multiple and complex and they are classified under a wide variety of categories, including hormones, neurotransmitters, neuro-modulators, growth (and growth-inhibiting) factors, gut peptides, interleukins, cytokines and chemokines. Fortunately we don't need to know precisely what each of these does but it is valuable for us to acknowledge the diversity. Some of these messenger names may be familiar, such as insulin for its part in blood-sugar regulation or oestrogen and testosterone for their part in sexual differentiation and fertility regulation.

For a long time it was expected that the brain would be central to communications and control, even if there was a new agent in the process. Any messenger is only of use if there is something to hear the message and take action. There was reason to expect that the receptor mechanism would be found on the surface of a cell and that the brain was the place to look for such receptors and indeed this was where they first began to be found.

Over time the story became more and more complex. It is now recognized that cells in every part of the body have receptors, usually for many different messengers, sometimes acting in combination. Eventually it became apparent that all these cells, all over the body, were not only receiving messages; they were also sending them. Peptides, messenger molecules, were being produced everywhere.

The brain is not the centre of this process; it is but one among many organs, all of which play a part in regulating physical processes.

To give just a little flavour of what this interaction means, here is an extract from Candace Pert's book "Molecules of Emotion".

Molecules of Emotion

"To cite just one recent example, Rita Valentino of the University of Pennsylvania has shown that the nucleus of Barrington in the hindbrain, formerly believed to control merely micturition (bladder-emptying), sends axons containing the neuropeptide CRF down through the vagus nerve all the way to the most distant part of the large intestine, near the anus. Rita has proved that sensations of colonic distension (i.e. the feeling of needing to poop) as well as those of genital arousal are carried back to the nucleus of Barrington. From there, there is a short neuronal pathway (called a "projection") that hooks up to the *locus coreolus*, the nor-epinephrine-containing source of the "pleasure pathway", which is also very high in opiate receptors. The pleasure pathway hooks up to the control area of these bathroom functions, which is located in the front of the brain. Goodness, is it any wonder, based on Rita's neuro-anatomical discoveries, that toilet training is loaded with emotional stuff! Or that people get in to some unusual sexual practices involving bathroom behaviours! Clearly the classical psychologists had grossly underestimated the complexity and scope of the neurochemistry and neuroanatomy of the autonomic nervous system. But the limitations of the past are now giving way before our newfound ability to track these fascinating connections.

If we accept the idea that peptides and other informational substances are the biochemicals of emotion, their distribution in the body's nerves has all kinds of significance, which Sigmund Freud, were he alive today, would gleefully point out as the molecular confirmation of his theories. The body is the unconscious mind! Repressed traumas caused by overwhelming emotion can be stored in a body part, thereafter affecting our ability to feel that part or even move it. The new work suggests there are almost infinite pathways for the conscious mind to access – and modify – the unconscious mind and the body, and also provides an explanation for a number of phenomena that the emotional theorists have been considering."

Dr Pert explores the connection between emotions and memory further, going on to note the work on state-dependent memory by Donald Overton, who has shown that memories stored under the influence of an emotion are recalled better when that same emotion is present. This is a feature used by Neuro-linguistic Programming, Emotional Freedom Technique ("tapping") and other methodologies for personal development and healing.

The perception that the whole body is active, that it behaves like a hologram in which any part holds information about the whole, that the body is a holistic system in which all organic processes are interrelated, is fundamental to all complementary health and natural-tradition medical systems. The relevance to our spiritual theme is that all of these systems share a core perception that the process is also connected with the spirit - in all senses of that word. They would include the spiritual well-being of the individual, their overall emotional tone. They would include the notion that a spiritual entity exists – spirit in the sense of what might be called a soul. They would include spirit in the sense of a relationship with a wider energy world, a relationship which might be felt with nature, with ancestors, with the land or with the "Divine" whether that be expressed as a supreme being or one of the other formulations that describe that realm.

This consistent linking by healers of all kinds between the holism of bodymind and a greater connectivity is not accidental; it deserves to be taken seriously that all cultures worked with these relationships before the dawn of the scientific age – not out of primitive ignorance, but through observation and deep understanding. The gift Candace Pert brings is to describe some of the mechanics which mediate these relationships. Her sentence in the quote above that "The body is the subconscious mind!" encapsulates what we mean by the word "bodymind". Put another way, the body is mind all over.

Mind all over

When we said earlier that the messenger chemicals are produced everywhere in the body, this does not just mean our organs. For example these chemicals are also found on the cells of the immune system. The cells which are carried in the bloodstream and which detect and destroy disease-producing organisms are as much a part of the communications process as any other. Maybe even more significantly, many of the messengers which regulate our processes are integral to the reactions we label as "emotions".

The traditional view was that emotions are some sort of perceptual artifact created by the brain – nothing more than a viewpoint towards the world generated by the mental computer in our head. However we know that our bodies also display these emotional states – we flush with embarrassment; our hearts beat faster with fear and panic. It was assumed that these were nervous system "decisions" but this is quite misleading and leads to a misunderstanding of emotion and its place in our relationship with the world.

We must understand that science finds emotions awkward as they are in the realm of described experience. The physiological reactions can be measured but the emotional labels that we give to them are subjective, hard to isolate and even harder to measure. What one person calls passion may be perceived as anger by his neighbour. You could conjugate these relationships. "I am firm, you are opinionated, he is bloody-minded." Despite this challenge, Charles Darwin was one of the first to be interested in studying emotions. As he observed "...the young and the old of widely different races, both with man and animals, express the same state of mind by the same movements." And elsewhere, "Even cows, when they frisk about from pleasure, throw up their tails in a ridiculous fashion." He interested himself deeply in theorizing how humans and animals display such emotions as fear, anger, disdain, and pleasure and published a book specifically on that subject[6]. It is work that has in most respects been sustained by later scientific research.

In general it is a subject neglected by science since then, except in its more psychological aspects. Pert's work (remember, the title of her book is "Molecules of Emotion") brings this back to the centre of the bodymind. The following passages convey the technical core of this message.

> "In order for the brain not to be overwhelmed by the constant deluge of sensory input, some sort of filtering system must enable us to pay attention to what our bodymind deems the most important pieces of information and to ignore others....Emotions are constantly regulating what we experience as "reality". The decision about what sensory information travels to your brain and what gets filtered out depends on what signals the receptors are receiving from the peptides. There is a plethora of elegant neurophysiological data suggesting that the nervous system is not capable of taking in everything, but can only scan the outer world for material that it is prepared to find by virtue of its wiring hookups, its own internal hookups and its past experience.....
>
> I knew from my brain mapping over the years that the communicating chemicals were most dense in certain areas of the brain and along sensory pathways...And when we focus on emotions, it suddenly becomes very interesting that the parts of the brain where receptors and peptides are richest are also the parts of the brain that have been implicated in the expression of emotion. I don't remember whether it was Michael or I who said the words first, but both of us had the gut feeling that we were right: 'Maybe these peptides and their receptors are the biochemical basis of emotion"

---

[6] Expression of the Emotions in Man and Animals (1972)

As their 1985 paper in the Journal of Immunology stated:-

> "A major conceptual shift in Neuroscience has been wrought by the realization that brain function is modulated by numerous chemicals in addition to classical neurotransmitters. Many of these informational substances are neuropeptides, originally studied in other contexts as hormones, gut peptides or growth factors. Their number presently exceeds 50 and most, if not all, alter behaviour and mood states, although only endogenous analogs (i.e. man-made equivalents) of psychoactive drugs like morphine, valium and phencyclidine have been well appreciated in this context. We now realize that their signal specificity resides in receptors rather than the close juxtaposition occurring at classical synapses. Precise brain distribution patterns for many neuropeptide receptors have been determined. A number of brain loci, many within emotion-mediating brain areas, are enriched with many types of neuropeptide receptors, suggesting a convergence of information processing at these nodes. Additionally, neuropeptide receptors occur on mobile cells of the immune system; monocytes can chemotax to numerous neuropeptides via processes shown by structure-activity analysis to be mediated by distinct receptors indistinguishable from those found in the brain. Neuropeptides and their receptors thus join the brain, glands and immune system in a network of communication between brain and body, probably representing the biochemical substrate of emotion."

Monocytes are immune cells, the ones in the blood-stream which identify foreign substances and destroy them and which also play a part in destroying infected cells in the body. They will "chemotax", moving through chemical influence to sites where neuropeptides are signaling the presence of physiological stress that requires intervention. The biological event and our "feelings" – the way that the biological event affects us emotionally – are linked through the flow of neuropeptides and neurotransmitters and by the way that these chemicals are selectively received by various parts of the brain and body.

In summary then, three previously separated areas of study – neuroscience, endocrinology and immunology, together with their various organs - the brain, the glands and the spleen, bone marrow and lymph nodes – are actually joined together in a multidirectional network of communication, linked by information carriers known as neuropeptides. The emphasis here should be placed on the words "network of communication" as it is the flow of information through the cells, organs and systems of the body which mediates this complex process of regulation, balance and health maintenance.

Regarding the organism as an information network takes us away from old, mechanical ways of looking at the body. To the old view of hardwired reflexes and electrical stimulation, with their limited capacity for flexibility and change is added a form of higher intelligence that is running systems and creating behaviour.

The intelligent heart

To give another flavour of this, let's talk about the heart. Traditionally the heart has been written about with deep poetic understanding as the vessel of human love, the seat of courage and a source of wisdom. We feel emotions as much in our heart as anywhere and it was once regarded by many as the seat of the soul. However a female friend recently complained bitterly about the dismissive attitude a science-minded ex-boyfriend had taken to her discussing the heart in these terms. "It's just a pump" he said.

He couldn't be more wrong! The following information is taken from a lecture by Howard Martin and is derived from the research undertaken by Doc Childre's HeartMath Institute. It presents a very different and once again non-mechanical, subtle and complex view of the human body. Of course the heart is a pump but the intuitive truth down the ages is borne out by recognizing all that it does besides.

The heart beat is present in the foetus even before the brain is formed. This contradicts the idea that the heart is governed by the brain as does the knowledge that it will beat (as it does when transplanted) without a nerve connection to the brain. It is auto-rhythmic, having the source of its beat within itself. It responds to stimuli from the body but is not controlled and particularly not by the brain.

In fact the heart is what people have always known it is – a sensory organ which makes functional decisions and communicates with the brain. This happens in four ways.

1. There is neurological influence. The heart has 40,000 of its own neurons and a distinct nerve pathway through the medulla, the amygdala and limbic system to the neo-cortex. That is, it passes directly through the emotional memory systems to the thalamus and synchronises cortical activity. It sends more information to the brain than it receives – unique among the organs of the body.

2. The pulse of the heart is a wave of energy through the body. The brain synchronises its electrical activity to the pulse of the heart. In effect, the heart rhythm becomes the beat of the body[7].)

---

[7] It is worth mentioning here that acupuncturists and other practitioners use the subtleties of these pulses as their primary diagnostic tool as they provide a primary readout on the status and balance of the 12 governing energy meridians

3. It influences biochemically, producing hormones such as atrio-peptide which regulates the stress hormone cortisone. It regulates oxytocin, the chemical messenger that stimulates behaviours of love and maternal caring. It produces the neurotransmitters dopamine and noradrenaline which are neurotransmitters regulating many aspects of brain function including motor activity, sleep, mood, attention and flight-or-fight response.

4. It is the strongest source of bio-electricity, producing 40 times more than the brain (which is the next strongest). This bio-electrical field radiates beyond the body in a toroidal (ring-donut shaped) field that influences approximately 8-10 feet (3 metres) around. The frequencies of these radiations change with the emotional state of the body. When the body is experiencing what most would describe as negative emotions, this field becomes incoherent. Positive emotional states produce coherence and stability. Aside from anything that we do with our face and gestures, we all broadcast our emotions to those who are close to us. While we cannot cite proof here, we regard it as certain that we can detect each other's fields and that many of us instinctively recognize this additional form of intuitive knowing. In addition it can be shown that the incoherent (negative emotional) states inhibit cortical (thinking) function because they trigger the production of cortisol, while coherent states facilitate it because they trigger DHEA, the pre-hormonal natural steroid which is regarded as an anti-aging substance. Put simply, we think more clearly when we are not under stress. That fact may be obvious, but what is less apparent is that it is regulated by the heart through its production of neurotransmitters (i.e. cortisol and DHEA).

The heart rate is highly variable and the changes are happening continuously, second-by second, thus being a further example of a subtle and responsive process. The prompt of many spiritual wisdoms to live with our minds in our hearts is not just a metaphor. It has a real and physiological root.

The types of interactivity we are describing here go beyond the "power of mind over body" concept that is sometimes promoted. We have used the compound noun "bodymind" deliberately. The world has been influenced for almost 400 years by Rene Descartes and his theory of mind-body dualism. In fact we are seeing and will see more as this story unfolds that the body is mind all over, they are one seamless entity. When we describe either aspect in isolation we are simply changing our angle of view. The power of mind over body enters the whole, influencing at one point in the cycle. Working on the body influences at another point, but it is the same cycle.

How many senses do you have anyway?

In an earlier chapter we discussed the way in which our perceptions are geared to our physical needs and to the history of predators and prey in the way they structure our sense of time. We are accustomed to talk of these physical senses in the way that the song quoted at the head of this chapter does, distinguishing the five described as sight, sound, smell touch and taste.

These five are arbitrary definitions linked more to the physical appearance of our sense organs than to distinct sensory processes. You might as easily say that you have three sensory systems. One is for detecting vibrations in the electromagnetic spectrum, picking up sound frequencies and some bands of light. Another is a network of chemical sensors for odor and flavour. The third is kinesthetic, detecting movement or variations in temperature against the skin.

Our perceptions are narrow and incomplete. We all know that dogs hear sounds at frequencies that mean nothing to us. The same dog might also distinguish chemically between two complex molecules with one atom of difference between them, and which we would not detect at all. In the light spectrum bees see flowers quite differently than we do with senses that extend further into the ultra-violet range. Soldiers use night-vision goggles that translate heat radiation (infra-red) onto a screen to make bodies visible in a dark forest. Probably there are creatures which use this frequency range including many night-hunters like cats[8].

Sensory stimuli are filtered by the processes in the perceptual systems so that we only see what we are accustomed to find relevant. When you are driving there may be lots of information that you are tuning out. Was the person who crossed the road in front of you just now happy or sad? You are unlikely to have noticed.

That filtration is unconscious and happens anyway. But Daniel Simon's selective attention experiments show how a person who is asked to focus on one aspect of a scene will completely miss quite startling events. We have witnessed this in an audience of 250 people with a pre-instructed agenda watching his video clip in which something totally incongruous happens and only about five people laughed. The incongruity was only seen when the clip was repeated and our attention was allowed to open and simply "see" without the instruction toward a specific focus[9].

These filter processes clearly influence our perception of what is "true" or "real". But both phenomena come after the threshold of which information enters our

---

[8] We note that our cat sometimes reacts to the TV remote

[9] http://www.youtube.com/watch?v=vJG698U2Mvo or search "Selective Attention Test"

perception. They pale into insignificance when compared with huge gaps like not seeing infra-red, not picking up radio waves, lacking the visual acuity of a buzzard, not hearing the echo-location soundings of a bat.

In consequence, when we talk of a "sixth sense", this is a misleading term of convenience. It is simply a form of information supply which occurs beyond the ones we are focusing on or are normally inclined to be aware of. It is also the case as noted earlier, that the receiving of this information is processed by the same brain and articulated through the same language channels.

This presents a number of challenges. Intuitive information is a challenge to the person receiving it because it often comes into the consciousness with an appearance which resembles a regular "5-sense" experience but which that individual knows is not coming in the normal way. It is a challenge to that person because it may not arrive clearly and may be more "dream-like" in its symbology, and require some interpretative skill to decode. As Jon has discovered, it is a challenge to teach others to be intuitive because the teacher does not know how the student will function, nor how their internal representation is to be decoded. They have to develop their own knowledge of their unique code.

Not all intuitive information arrives through the regular sensing processes though. Even harder for some people, the information may not come as either pictures or words – it may just be "feelings" or complete "knowings" which then have to be formed into language. This was Jon's experience when detecting the brain tumour (P.6). What came first was a sensation, a physiological experience which included screwing up the face as if pain was being felt, but without the actual pain. The emotional body is included in the mediators for intuitive information.

The use of the same brain and language channels, and the even less well-defined kinaesthetic mode just described also challenges the scientific method. Even those with prodigious gifts like our quoted medical intuitives cannot tell you anything about the ways in which the information enters their bodymind. If science wanted to construct apparatus to detect and validate the experience, it could not. It's a "black-box system" where we know only what little is visible from the outside and what is reported from "inside". Not only is it impossible to measure the content, science cannot even prove that the communication is (or is not) taking place. It does not even know where on any kind of electro-magnetic spectrum to begin looking[10]. Even worse, if it did know, it is in the nature of the experiences we have described that intuitives pick up information which is "out there", but that

---

[10] We will see later that it almost certainly is not to be found on that spectrum

ALL the information is present at the same time. When Mona Lisa Shulz tuned in to Violet, the information for Jon or Juliana or (insert your name here) was equally available to her, but she only got Violet's. What is science able to do with that? It has no apparatus and never will. It just has Mona Lisa.

The intuitive worlds we described in the first two chapters and the neuro-endocrine reality that this chapter portrays share several striking characteristics.

1. They are both occurring inside living humans and are experienced subjectively.

1. They are not susceptible in any meaningful degree to measurement and only in the crudest sense can some aspects be detected by apparatus (for example electro-encephalograph or brain-scan data).

2. They are clearly complex, subtle and deeply meaningful to the person having the experience.

3. They all have to do with an information flow which is part of the relationship between one human and another, and of the connection between one human and the rest of the world.

For all these reasons, it is essential to accept that the only way that we will study these phenomena meaningfully or create any sort of systematic understanding is through working with the experiences as they are reported, finding the consistencies that reveal themselves across generations and cultures. We will explore this world next.

Review

In the spiral diagram we have drawn a theme of connection that runs from the level of sub-components of cells to that of the overall human being. This strand shows chemical communication and rhythmic co-ordination. In this chapter we have described some forms of each of these. Chemical communication is mediated by messenger molecules. Messages are "read" by the cells, cellular activity is modified and other messages may be sent as a result. These messages can affect almost any aspect of behaviour and some of the resultant changes are experienced by humans as what we label "emotions". This chemical activity brings a means by which all cells in the body can be "aware" of the states of others and of the body as a whole. The emotions affect what we view as "reality" and influence our behaviour and decision-making. In this way, the chemical messengers are one element in our relationship with the world beyond our bodies.

A layer of rhythmic connectedness is provided by the heart, which sends more information to the brain than it receives, and which acts to provide a signal which synchronises electrical activity. The heart also produces chemical messengers and generates electromagnetic fields that radiate beyond the body. These fields reflect the extent of coherence within the body's emotional states and influence cortical activity.

Together, these chemical and rhythmic connections form a significant part of the activity that humans experience as consciousness. As depicted on our rings, they contribute to the internal mediation between matter and energy, and to our information about the world - both ordinary reality and the non-ordinary intuitive experience.

5. I talk to the trees.....

The characteristics of conduction in the plant nerve are in every way similar to those in the animal nerve.

Jagadis Chunder Bose

In many respects, plant nervous systems are nearly as sophisticated as our own, and in some plants, nearly as rapid in their actions.

Stephen Harrod Buhner   *Secret teachings of Plants*

I talk to the trees, That's why they locked me away

Spike Milligan

Theme

In this chapter we explore the theme of direct relationship with the world of nature. This means taking the concept of intuitive connection from Chapter 3 a little further and looking at it more specifically.   In particular we will look at the practical applications of this engagement and see something of what can be delivered.

In this and the following chapter we take our exploration beyond the realm of soliciting information about the world into one of active participation in it. Powerful as the passive mode of intuitive awareness may be, there is greater potential emerging here. We first discover how "aware" plants can be and present scientific evidence that plants can demonstrate individual consciousness and connection.   We also discover that the nature spirits are not just general but specific. They have their own tasks and their own form of identity and can engage with us co-operatively.

The "Paint your Wagon" song parody above from Spike Milligan is more than the simple joke that it appears to be. What else is possible once we open up new forms of connection to our world? When our intuition can take our sensing and knowing beyond the limits that we have been used to regarding as "normal", where are the boundaries to who we are capable of being? Let's take our next exploratory step.

The secret life of plants

This is a story that begins in 1966 concerning Cleve Backster, America's foremost lie detector examiner, and stars a houseplant called *Dracaena massangeana.*

You may know that a lie detector works by measuring how much electricity the skin will conduct, using a galvanometer. Backster decided on impulse to attach this device to his pot plant, and see how it responded to being watered. Now if you know anything about electricity you would expect that more water means more current will pass. But the trace on the paper showed a fluctuation that was in the opposite direction - downwards, with a kind of saw-tooth motion that resembles what happens when a human being experiences an emotional stimulus.

He decided to investigate further. When working with humans, police examiners like Backster watch the responses to stress under questioning. One of the most effective ways to get a response in humans is to threaten their well-being. Backster decided to try this on the plant, and dipped its leaf in hot coffee.

Nothing happened. So he thought of a worse threat. He decided to burn the leaf to which the electrodes were attached. The very instant he got the picture of the flame and the action in his mind, the plant responded with a strong upward sweep.

He left the room to get some matches, and found that while he was away, the plant had responded with another upward surge. Reluctantly, he set about burning the leaf, and got a reaction, but less than before. Later he went through the physical motions of pretending he would burn the leaf, and got no reaction at all. The plant seemed to be able to tell the difference between real and pretended intention!

Work continued as he had other collaborators check his findings on other plants, fruit and vegetables. They confirmed his findings, and he decided to set up a lab for more extensive study. Here is just a small selection of further examples.

One time, he was demonstrating his work for a journalist. He hooked a galvanometer to a philodendron, and then interrogated the journalist about his year of birth. Backster named seven years in succession, from 1925 to 1931 to which the reporter was instructed to answer "no" in each case. From the galvanometer chart of plant responses, Backster then selected the correct year of the reporter's birth. That is, the philodendron detected when the "no" was a lie.

To see if the plant would show memory he set up an experiment where six volunteers from among his police students drew a piece of paper from a hat. Five were blank, but the other told one of them to totally destroy one of two plants in the room. Each in turn entered the room. No-one but the person with that piece of paper knew who was responsible for the destruction of the plant. This was followed with a kind of identity parade, in which each of the volunteers went in to the room again, with the remaining plant now wired to the galvanometer. The plant showed no reaction to five of them, but the reading went wild when the "culprit" entered the room.

In their book "The Secret Life of Plants" Peter Tompkins and Christopher Bird describe further explorations in which Backster found that the reactions would persist even when a leaf was disconnected from the plant, trimmed to the size of the electrodes or even shredded. As they describe it:-

> "The plants reacted not only to threats from human beings, but also to unformulated threats such as the sudden appearance of a dog in the room, or of a person who did not wish them well. Backster was able to demonstrate that the movements of a spider in the same room with a plant wired to his equipment could cause dramatic changes in the recorded pattern generated by the plan just BEFORE the spider started to scuttle away from a human attempting to restrict its movement. "It seems," said Backster" as if each spider's decision to escape was being picked up by the plant, causing a reaction in the leaf."

> "At first he considered that his plant's capacity for picking up his intentions must be some form of extrasensory perception; then he quarreled with the term. ESP is held to mean perception above and beyond varieties of the established five sensory perceptions of touch, sight, sound, smell and taste. As plants give no evidence of eyes, ears, nose or mouth, and as botanists since Darwin's time have never credited then with a nervous system, Backster concluded that the perceiving sense must be more basic. This led him to hypothesise that the five senses in humans might be limiting factors overlaying some kind of primary perception, possibly common to all nature." (Our emphasis)

The plant subsequently showed a reaction to simple cellular organisms. On one occasion it reacted to Backster mixing jam with his yoghurt, which he believed to be due to the preservative in the jam killing some of the live cells. This belief was later supported by witnessing a similar reaction when boiling water was running down the waste-pipe in the sink and killing bacteria.

Because Backster was interested in this reaction, he invented ways to attach his electrodes to single celled creatures such as amoeba, yeast, blood cells and sperm. All were capable of producing similar results to plants. Sperm cells, for instance, would respond to the presence of their donor. In more recent work, reported in Erwin Laszlo's book "Science and the Akashic field", Backster took cheek-cell swabs from various subjects and took them several miles from their donors. In one of his tests, he showed his subject, a former navy gunner who had been present there, a television program depicting the Japanese attack on Pearl Harbor in 1941. When the face of a navy gunner appeared on the screen, the man's face showed an emotional reaction—and at that precise moment, the lie detector's needle seven and a half miles away jumped, just as it would have had it been attached to the man himself.

Pierre Sauvin, another researcher and electronics hobbyist inspired by Backster's work, was able to construct more subtle detector systems and to connect a philodendron as a control system. He was able to demonstrate in front of live audiences and even under the lighting of an early TV studio that he could make an electric toy train start, stop and reverse at will by using the plant to detect his emotional changes.

There are three clear and simple conclusions to be drawn from these experiments:-

1.   There is a consciousness that connects living things, that is constantly active, and engaged in (or at least open for) continuous communication

2.   That consciousness - down to the level of individual cells - is capable of picking up on that connection and identifying other consciousness at an individual level.

3.   Individual plants can have a form of conscious presence.

Note that because the tests were done on plants, there is no need for the double-blind approach to be used – at least for the first conclusion. In theory it is possible that the plant could be responding to the experimenter rather than to the source of the signal - the destructive student policeman or the yoghurt. However in the yoghurt example the effect was not expected. In practice it seems much more likely that conclusion no. 2 is also correct. This is consistent with our observations in other areas and is borne out by all that follows below. Conclusion 3 is particularly significant in regard to the stories that will follow.

These conclusions will come as no great shock if you have been following our thread so far. But note that this is the first time we have been talking explicitly about non-human organisms. The core of this chapter concerns the plant kingdoms, though the reactions with yoghurt and other single-celled organisms indicates that we need to keep our boundaries very open.

Additional supportive evidence has been supplied very recently by research published in the June 2007 issue of Biology letters, by Susan Dudley and Amanda File. They have shown that in the annual plant *Cakile edentula*, allocation of effort to root development increased when groups of strangers shared a common pot, but not when groups of siblings shared a pot. Their results demonstrate that plants can discriminate kin in competitive interactions and indicate that the root interactions may provide the cues for this. It is not stated how they achieve this recognition. Some chemical sensing may be involved (there are other examples where this is believed to be the case). However, chemical sensing does not seem adequate as an explanation for how plants would tell one human from another, and the kin recognition may well operate through other forms of connection.

The above is our gentle entry into the realm of "plant consciousness", and an introduction to the idea that plants can have a "spirit". We recognise that we have not so far even attempted to define what it means to describe a human as having a "spirit" – or even a spirit as having a human. We would prefer not to do so and to allow you to draw your own conclusion and frame it within your own experience. Just be aware that the language used by others may contain their interpretations. It may not be the only way to describe the experience and our language may be unclear or even lack the terminology to describe a realm which our culture does not acknowledge. For example, to say as many people do that the soul leaves the body implies an entity which moves its geographical location. As we move through our narrative, you may wish to consider the implications at each stage for who you are, for what your spirit or soul might be, and where it may be located.

Shamanic gardeners

Among the most well-known explorations of this reality, particularly to Europeans, is that which has taken place over the past five decades at Findhorn in the extreme North of Scotland. Their full story is to be found on their website (www.Findhorn.org) from which we quote the following section, which refers to two of the founders, Dorothy MacLean and Peter Caddy.

> Peter decided to grow vegetables. The land in the caravan park was sandy and dry but he persevered. Dorothy discovered she was able to intuitively contact the overlighting spirits of plants - which she called angels, and then devas - who gave her instructions on how to make the most of their fledgling garden. She and Peter translated this guidance into action, and with amazing results. From the barren sandy soil of the Findhorn Bay Caravan Park grew huge plants, herbs and flowers of dozens of kinds, most famously the now-legendary 40-pound cabbages. Word

spread, horticultural experts came and were stunned, and the garden at Findhorn became famous.

The following is Dorothy MacLean's description of the communication that she received when during a meditation, she attempted to make contact with the spirit of a pea plant that she wished to grow.

> "I can speak to you, human. I am entirely directed by my work which is set out and moulded and which I merely bring to fruition, yet you have come straight to my awareness. My work is clear before me – to bring force fields into manifestation regardless of obstacles, and there are many in this man-infested world. While the vegetable kingdom holds no grudge against those it feeds, man takes what he can as a matter of course, giving no thanks. This makes us strangely hostile.

> What I would tell you is that as we forge ahead, never deviating from our course for one moment's thought, feeling or action, so could you. Humans generally seem not to know where they are going or why. If they did, what a powerhouse they would be. If they were on the straight course of what is to be done, we could cooperate with them! I have put across my meaning and bid you farewell."

She describes the Devas as follows:-

> The devas hold the archetypal pattern and plan for all forms around us, and they direct the energy needed for materialising them. The physical bodies of minerals, vegetables, animals and humans are all energy brought into form through the work of the devic kingdom…. While the devas might be considered the 'architects' of plant forms, the nature spirits or elementals, such as gnomes and fairies, may be seen as the 'craftsmen', using the blueprint and energy channelled to them by the devas to build up the plant form.

Our modern concept of gnomes and fairies may make it hard to hear the use of these terms in an appropriate way, but the indication is that these energies have more resemblance to Malidoma Somé's "kontomblé" that we will encounter in Chapter 6 than to the lawn ornament or sentimental Victorian painting we might first think of.

Some readers, particularly in the United States, might have heard more of the Perelandra garden in Virginia, where Machaelle Small Wright has pursued a parallel exploration. Inspired by her reading of Dorothy MacLean and Eileen Caddy, she went in to the woods surrounding her property, announced to them her intention to do in Perelandra what they had done at Findhorn, and invited the nature spirits to help her.

She describes the response as immediate, resulting in a "crowd of voices" speaking to her as soon as she was back at the house and started to meditate. She endorses Dorothy Maclean's view of the Deva's role and in her book "Behaving as if the God in all life mattered" describes it as follows:-

> "The word architect has been used by others when describing what devas do - and I also find this to be the most appropriate word. It is the Devic level that draws together the various energies that make up, for example, the carrot. The carrot deva "pulls together" the various energies that determine the size, colour, texture, growing season, nutritional needs, shape, flower and seed processes of the carrot. In essence, the carrot deva is responsible for the carrot's entire physical package. It holds the vision of the carrot in perfection and holds that collection of energy together as it passes from one vibratory level to another on its route to becoming physical. Everything about the carrot on a practical level as well as on the more expanded, universal level, is known by the carrot deva."

> Each day I would go into meditation and open my connection with devic level. A deva would come into my awareness and identify itself. I was then given instructions. I was told what seeds to buy. What fertiliser to use. How far apart to plant the seeds. When to thin the plants and how much space to leave between them. Spacing between the rows. Desired amount of sunlight, and so on.

> As each Deva came into my awareness. I noticed that there was a slight shift in vibration, but each had its own vibration. After a while I could recognize which Deva was entering my awareness. This led me to develop the ability to call upon specific Devas by aiming my awareness for the Deva's own vibratory pattern. It was as if I was faced with a gigantic telephone system, and I had to learn how to make all the different connections. Then I was able to make calls in as well as receive calls.

> (Later in the summer I discovered that all I had to do to connect with a specific deva was to simply request the connection. For example, to connect with the Deva of the Carrot, I only needed to say "I'd like to be connected with the Deva of the Carrot." Immediately I would feel the familiar vibration of the Carrot Deva in my awareness. I knew the connection had been made an we were ready to work together. It couldn't have been more simple.)

> One day I felt a very different vibration and found myself connecting to the over-lighting Deva of the garden. This Deva talked about such things as the overall layout of the garden, its timing, its progression and its shape. From it, I was also told to change my gardening methods to the mulch method – a method whereby 6 inches of hay, grass clippings and leaves are kept on the garden at all times. Two years later I was told to switch from the traditional straight rows to a garden of concentric circles.

Machaelle Small Wright also takes a similar view to Dorothy MacLean of what happens after the architecting deva has done its work and the nature spirits take over. This is her description:-

> "The job description for nature spirits is "blue-collar workers". The Devas call together the different components of a plant. Once the "package" of energy is formed, Devas then hold the package together as it travels from one level to another, changing and adjusting its vibration as it acclimates to the Earth's density. Once the package reaches the etheric level around Earth, the nature spirits take over. It is the responsibility of the nature spirits, not only to receive the package of energy but to fuse it into its proper form as well. They fuse to a plant its light, its essence, its life pattern.
>
> Devas are universal in dynamic. My Deva of the carrot is the same as your Deva of the carrot. Nature spirits are regional. My nature spirits of Perelandra are not the same as those working around you in your area.
>
> Nature spirits, like Devas, are also bodies of light energy. But because they operate in close proximity to earth and its density, their energy vibration has a denser sensation than the devic vibration does. It is still a very high experience for us to feel. Nature spirits don't feel like energy rocks. They simply feel more dense than that devic energy."

This is just a brief look at the world of nature spirits and devas, which we give in order that you have a feeling for the form and shape that consciousness takes when it is working in nature. Note that there is more than one level involved in this creation. We are trying to amplify the sense of what it can mean for consciousness to be involved in the creation of form and provide a glimpse of the experience which those who have a direct relationship with such energies are likely to describe. Our scientific explanations will be called upon to explain these features.

The picture of this world as given by both writers is rich and detailed. They each relate communications from different plant spirits, and the sense of individuality and distinctiveness between one and another. Both of them give a deep sense of the powerful co-operative relationship that is offered to humans when we begin to listen to and work with the forces of nature. There are many others who have documented similar experiences and the examples we give are simply some of the best known. They are not isolated cases and there are many systems for working with them. The bio-dynamic farming methods which come out of Rudolf Steiner's work in this realm are among the most widely practiced.

Western shamans

In our next chapter we will attempt to convey the depth and strangeness of a shamanic relationship with the world. Step by step we are opening the door to "non-ordinary" reality. Our next illustrations are provided through the experience of some Westerners who have explored these boundaries and first we would like to mention briefly the experience of Dr Alberto Villoldo. In his book "Dance of the Four Winds" he captures with great clarity the experience of a westerner with scientific background as he comes to terms with this alternate view of reality. Having completed his doctorate, studying psychology, neuronatomy and much besides, he is dissatisfied with the clinical models he has been offered and chooses in 1974 to take some time out to discover the ways in which Peruvian healers understand consciousness.

His account switches back and forth between his time in the West and his encounters with a series of healers, shamans and guides who little by little draw him increasingly deeper in to their world. As such it is a revealing depiction of the transformation that is required in order to release western academe and science and find a way into the shamanistic experience. The book covers ten years of exploration as Dr Villoldo learns such aspects as taking on the consciousness of an animal and the healing capabilities of those who learn to see directly into the energy flows of others. He describes working with a healer who has never heard of acupuncture, but when asked to describe the energy pathways he simply "sees" and unblocks on the body of a patient, draws the lines of the Chinese medicine meridians on Villoldo's skin.

In all the descriptions of the journey into shamanism, and of many traditions of esoteric intiation, it seems that it will require the individual to face the fear of death, to overcome and transcend pain and to confront his own inner demons. The world that Villoldo describes seems to be typical of the Latin American view and as such has a different flavour to Malidoma Somé's Africa which we will encounter later. But underneath the cultural differences lie strong senses of correspondence and shared viewpoints.

Alberto Villoldo's experience with an independent perception of the energy meridians links well

to the realm of Plant Spirit Medicine as presented by Eliot Cowan in his book of that name. Eliot Cowan started practising herbal medicine from a book when a vet told him that a sick goat on the farm where he was working, was incurable. The herbs the book told him he needed grew on the farm and produced a cure within a few days. Further experiences convinced him that natural healing was his calling, and he found his way towards perhaps the most respected western teacher of

Chinese medicine and Acupuncture, J.R. Worsley. Some time after qualifying as an acupuncturist he returned to his interest in herbs, following Worsley's advice that anything that needles could do, herbs could also do, but especially to use local herbs because they would be vastly stronger.

In search of information about the plants local to his Californian home, Eliot found that little was available. European texts didn't cover them and the Native American knowledge was effectively extinct from what he describes as "cultural genocide". Because of this he started to study shamanism and to become acquainted with the Huichol people of Mexico, one of the rare cultures which have survived intact. In subsequent decades this journey has led to Eliot becoming the first non-native to be fully accredited as a fully initiated Huichol shaman. His contribution in his own culture has been the development of Plant Spirit Medicine as a healing practice, and he trains others in the core elements of all that he knows.

Essentially this system uses the principles of Chinese medicine, with their several millennia of wisdom about the body as a diagnostic base. However, rather than use needles or herbal preparations as the means of applying treatment, it is the spirit of the plant itself which acts as the healing agent. Embedded in this approach is a relationship between plant, healer and client which is at the heart of this chapter.

This is the opening passage of Eliot Cowan's book:-

> "My friend Peter Gorman is walking down a trail in the Amazon jungle. He is on his way back to the village after watching his Matses Indian friend set a trap for wild boar. The Indian takes advantage of the walk to show Peter some medicinal plants growing along the trail. Within a few minutes he has pointed out several dozen species and pantomimed their healing virtues.
>
> Arriving at the village, Peter summons his interpreter and returns to the hunter's hut. He didn't have his notebook on the walk, he explains, and he couldn't possibly remember all that he had been shown. Would the hunter be kind enough to say once again how the herbs were prepared and used.
>
> The hunter-shaman smiles at Peter and then begins to laugh. He invites all his wives and children over to have a good laugh too. When they have all laughed themselves out, he explains, "That was just to introduce you to some of the plants. If you want to actually use a plant yourself, the spirit of the plant must come to you in your dreams. If the spirit of the plant tells you how to prepare it and what it will cure, you can use it. Otherwise it won't work for you.   Boy, that was a good one! I've got to remember what you just said!"   He laughs again."

This story neatly encapsulates the difference between the shamanistic world and that of pharmaceutical companies, isolating "active ingredients" and processing them into pills. As Eliot puts it, "There is only one active ingredient to plant medicines – friendship. A plant spirit heals a patient as a favour to its friend-in-dreaming, the doctor." Here is one illustration of that relationship.

> "People have returned after treatment to tell me stories of "falling in love with the Earth" or "feeling like I'm not alone" or "seeing fairies in my backyard". One of my favourite such stories involves Karen, a woman in her twenties who was suffering from depression as well as a number of spiritual complaints. I had chosen to treat her with the spirit of hummingbird sage, a beautiful plant that grows in the coastal ranges of Southern California, where I was living at the time. In my dream work with the hummingbird sage, the spirit appeared to me as a jolly, muscular little man full of fun and kindness. He was dressed in a pointed cap, a medieval tunic, leggings and shoes with pointy turned-up toes. This was Karen's report after her treatment:

> "After I left here I felt so tired that I went home and lay down. I was half asleep and had a dream or a daydream or something. It was totally vivid and lifelike. In this dream I felt that someone else was entering my body. I wasn't frightened because I felt he was a very good person, kind and fun-loving. I could see him very clearly. He was short and strong and was wearing funny old-fashioned clothing and shoes with pointy, turned-up toes. I felt he was there to give me something I needed.

> That afternoon I felt an urge to go to my special spot in the mountains. There is a certain place that I go; the smell there reminds me of the smell of the sage that grows in Colorado, in the Rockies. I lived in Colorado until my mother died. I don't know why, I guess I am trying to recapture the feeling about life that I used to have when my mother was alive, so I go to this place. The problem is, I never quite manage to get the feeling back. I get a little glimpse of it but then it fades away. But this time, after the treatment, I went to my spot and it worked! I got that wonderful feeling back! In fact it still hasn't left me!"

> I asked Karen to draw me a detailed map of her special spot in the mountains. After work I drove up there and hiked to the exact location. There I found one of the largest stands I'd ever seen of the fragrant hummingbird sage."

The relationship being described here is magical, the connectivity of viewpoint and experience sublime.

Very early in this book we introduced Jon and described the path that brought him to the alternate reality we are exploring. Some readers may have felt that Juliana was conspicuously absent at that point. We have saved her story until now, because

Juliana works with and trains others in the , expansion of consciousness and spirit-mind-body connection with self and all of life. She qualified as a plant spirit medicine practitioner through training with Eliot Cowan, and all that she does is rooted in extensive relationships with the unseen worlds of Nature and Consciousness. She also works directly with the Nature realms in creating gardens and in healing imbalances in the environment. This is a good moment in our narrative for her to describe the path that she has taken, and to hear her personal experience of these kinds of relationships.

"I have always been sensitive to the unseen worlds; to the energy currents of unspoken words and emotions, to the imprints of what has happened in a building, to the fairies and elemental lives in Nature and to those humans who no longer have a body. I quickly realised that this was not the experience of those around me and I spent many years wondering how to make sense of it all. I was what could be described as a psychic sponge, which was a fairly confusing and uncomfortable experience a lot of the time.

The Catholic church didn't have any answers for me to this or to any of my other questions and I became increasingly disillusioned by their approach and more interested in whether other religions would come any closer to my experience of the world. I had a deep interest in what draws people to religion and in finding a meaningful relationship to the world around me, seen and unseen.

This search led me far beyond the conventional religions and to an esoteric group in the late '70's. Here was the education I had longed for. We studied the natural energies of the earth, the stars and astrology, numbers and language, sacred geometry, the qualities of crystals, the importance of clearing a space of old, stagnant energies and so much more. I became aware of the energies within the human aura and how to bring healing to others, how to feel the energy fields of plants and trees. I re-connected to nature spirits, angels and other beneficial beings and developed the ability to see the energy fields of trees and people – colours and shimmering energies. Finally, the sensitivity to atmospheres became an asset.

I studied many healing modalities and worked as a Colour Consultant, Interior Designer and then for many years as a Feng Shui/Space Clearing consultant – releasing stuck and stagnant energies in buildings to increase life-flow for people, their health, well-being, relationships and finances. It showed me that we are deeply connected to our environment, whether we know it or not, and that it feeds back to us our internal state. If we choose a challenging house it will reflect an inner challenge we are facing and we can resolve the issue either from working on ourselves or on the space around us. I also came to understand the power of the land on which we live. Most of us don't give any thought to the land we live on but there were several instances where I was called to a house that had been built on a cemetery or other 'troubled' land.

One experience of many was when I was called to clear the energies of a house that would not sell.. It was a very well kept house on a newish estate and had been on the market for two years without any prospect of a sale. The family were desperate to move and there was no rational reason why they were stuck. However, when I walked into the house I could hardly believe the feeling of whirring in my head and chest and the chaotic and muddled feeling all around the house. The lady of the house told me that things had rarely come together for any of the family ever since they had lived there. I discovered that the housing estate had been built on the grounds of an old asylum and that this house would have been very close to the sleeping quarters. After the space clearing, which included moving on some souls who had been stuck in the ethers, the atmosphere calmed down, the house sold very quickly and the family were able to move on.

This and many other experiences led me to become more and more aware and more respectful of the land we live on. I began to connect more with the trees and plants in the gardens and to realise how much more connection was possible between these other species and us as humans. I worked for years with flower essences and the positive healing benefits they bring and then was drawn to the work of Eliot Cowan and plant spirit medicine.

In plant spirit medicine we develop deep and respectful relationships with certain plants and pass on their healing qualities to our clients. The levels of possible benefits from plants far surpass the physical level. There is a spiritual level to plants that reflects their connectedness to the earth and to the heavenly forces that flood through them constantly to come into our bodies and to help realign us to the natural world. If you think of a tree – it has its roots in the earth where it draws on the minerals and energies of earth and its branches in the heavens, where it is bombarded with the energies of the stars, the planets and the cosmos. For millennia our ancestors revered this earth as their mother, and in truth there is no life on this earth without the food we eat that is grown in the soil. Our bodies are formed out of what our mothers eat that comes also from the land. Again there is the balance of earth and heaven that combine to give us everything we need for life to grow and sustain. We humans have developed our minds to such an extent that we barely acknowledge the interconnectedness of all things and we have forfeited the awareness of the need for respectful relationships with all the forms of life we find around us.

Herbs and plants have been used for thousands of years and most indigenous cultures will have a form of plant medicine. Many times this will be in the form of a tea or tincture but in plant spirit healing no physical substance is required to be given. When I call upon the plant spirit through my personal connection to it, it comes as a friend would come if asked. I choose the plant that most closely resembles the deepest need of my client and the plant spirit will work to deliver the adjustments that are required to bring that person into balance and harmony. There is generally a huge increase in well-being as well as a sense of connectedness to life, and although physical symptoms sometimes take time to disappear they become less significant when the person's life force is strengthened and their connectedness to life is restored.

One plant I work with is nettle, that most maligned of plants. This is a tough cookie as we all know when we have tried to get it out of the garden! It will grow anywhere, is a wonderful food for wildlife and supports the overall ecology of a garden. It provides nutrients and alkaloids that enable cross-fertilisation to occur and is known in herbal medicine to be a great tonic and blood cleanser. However, at another level this plant is fantastic to help a person face challenges, to withstand knocks in life and bounce back. I would give this plant to those who at a deep level cannot face challenges, who are fearful and timid and who need to develop inner strength and resilience to go forward in their lives. I had one client who was unable to express his anger and was being trampled on by others. He would have a periodic explosion but most of the time seemed to be unable to let people know what he felt. After a few treatments with nettle he has become adept at speaking his mind and is now letting his emotions be known at the time and not needing to store up an explosion about things that happened months before!

Another area of early fascination for me was why people have the symptoms they have. So, if there is a dis-ease why is it that particular one and not another?. What are our bodies asking for when we get sick? We do have predispositions to certain conditions through our genetics but it is often the emotional or mental patterns that we inherit or develop ourselves which result in physical ill-health. I have studied and worked with many healing modalities that incorporate an awareness of the importance of inner balance and which address the energetic blueprint of disease. When we go to the source of the problem, the body will resolve itself quite naturally. Again, this is an example of how everything is connected. Western science has seen our bodies as separate to our minds, emotions or spirits. For me, these are inseparable aspects of a whole being that we call human, which in turn is completely inseparable from the rest of creation.

Everything has an energetic imprint or frequency and we impact each other and our environment with every thought or emotion that we generate. We can choose to generate fear and separation or we can choose to generate a vision of unity that is far closer to the truth of our existence than we have ever acknowledged as a species.

Another example of our connectedness shows up in family constellations. This has grown out of work initially developed by Bert Hellinger and explores the dynamics of our families and how they affect us in the here and now. In constellation work we step into a 'field' of a family, nation or group or whatever we are choosing to represent. This field is completely unseen and unspoken amongst the participants and yet movements, words and actions take place spontaneously to reveal a picture or story previously untold. It is a humbling experience to be part of the unfolding of a constellation. To be a 'representative' of another person – maybe a mother or father or ancestor reveals a level and degree of connectedness that goes beyond logical understanding and yet, can bring a depth of healing to an entire family as a result of this story being told and resolved. How can it be that someone can step in the 'field' of someone's family as a grandfather and begin to limp with the left leg, echoing the war wound of that person, without any cognitive awareness, unless they are tapping into the unseen patterning of that person and family situation.

Many of us carry the wounds of an ancestor on our backs, whether we know it or not. We are not separate from the histories of those who gave us life and on whose shoulders we stand. I have been inspired and motivated by a desire to bring healing and resolution to those areas of conflict that have gone on for hundreds of years between nations, families, religions and races.

The more we are willing to recognise and acknowledge our oneness with humanity and with all life, seen or unseen, the closer we come to being able to live in balance and harmony with the earth and all those who share this world.

We live in a magical world; a world of unseen energies and possibilities that most of us have become blind to. The rational, scientific approach to life cuts us off from awarenesses which are our birthright and which re-connect us to the magical worlds of Nature, of our relationships to earth and sky, to sun and planets, to plants and trees, to animals and other creatures. When we see ourselves as one with all of life, there is no need to fight or hate.. We have been taught to judge – others and also ourselves, believing ultimately that God will judge us; a separation from our source that exists within us all. When we are able to move beyond those judgements and know that there is no punitive God judging us, and recognise how that belief was born out of fear and separation then we can make another choice; one that allows us to move into the loving and deeply benevolent force that manifests as the God within all life, whether it be a flower, a horse or a human-being. What if we have never done anything wrong and there was nothing for us to be judged for? How could that change our world view of ourselves and of others?

What else is possible in how we connect to life? "

We have progressed thus far through an attempt to frame the world of intuition and to show the scientific proof of its fundamental realness. Now we are seeing that our intuition is not purely a passive process. In this chapter we have begun to discover more evidence of consciousness, this time in the plant kingdom and getting a glimpse of the direct and two-way relationships that become possible between us and our world. The natural world has something to say to us if can learn how to hear, and really listen. The depth of that relationship will become more vivid in the next chapter.

Review

In this chapter we took the earlier discussion of intuition into the more specific arena of connection to nature and the environment and extended beyond the realm of soliciting information about the world towards one of active participation in it. Powerful as the passive mode of intuitive awareness may be, there is greater potential emerging here. We saw the evidence for plant awareness and the more particular possibility of connection and co-operation with individual species. Juliana's direct experience also took us further towards the connection between humans and the influence of the past in causing disease. We will explore these aspects further in the coming chapters.

6. The shaman's world

We can find nature outside us only if we have first learned to know her within us

Rudolf Steiner   *Philosophy of spiritual activity*

Sensorially disconnected from their theoretically evolved information, scientists discern no need on their part to suggest any educational reforms to correct the misconceiving that science has tolerated for half a millennium.

Buckminster Fuller – Synergetics

Theme

There is a consistent view of the world which is present in the stories both from indigenous cultures and from western individuals who work through a direct internal cognition of information regarding the natural world.

Our presentation of intuition, precognition and psycho-kinesis showed that information passes between humans and their world. In the preceding chapter we opened the door to more specific forms of connection. The shamanistic perspective takes this relationship deeper. According to most spiritual and religious views, human beings have individual spirits. The shaman's relationship engages with a natural world in which there are many kinds of such spiritual entities inhabiting a layer of reality that is unfamiliar to most of us.

This chapter discusses the characteristics of shamanism and seeks the generic qualities that are described across human cultures. It is largely given over to a major narrative of the experience which comes from someone who was raised, not in the West, but in a culture where shamanistic perspectives are part of their "normal" reality.

The nature of shamanism

In even attempting to describe the world of the shaman we immediately encounter problems. Definitions such as that in the Oxford English Dictionary describe a shaman as a priest or witch-doctor who is a member of a class of people claiming to have sole contact with the gods – a definition which might not entirely distinguish a shaman from a Catholic priest. The use of the term "witch-doctor" also carries some very unhelpful images which we may have derived from B-movies which portray hostile and even cannibalistic savages. We need to get well away from such crude depictions. The truth is so much more complex and subtle.

If we are to present the Divine habitat with any completeness, the shamanistic view is crucial. Shamanistic cultures live in the reality we are attempting to describe. This way of thinking and the approach to life that it engenders is so profoundly different from the one that science frames that it challenges our whole way of thought. Most of us find it very hard to treat these views as having anything to do with reality. Words like "primitive" and "animism" are used; the indigenous people who live this way are presented as child-like, superstitious and naïve.

These descriptions can seem justified when they are applied to cultures which retain superficial remnants of belief systems which have been largely destroyed through contact with the western world and where the true power and deep relationships are no longer living. Where such cultures are surviving, it is difficult for westerners who contact them to cross into their world and very few have made a real attempt since this typically requires years of work as well as openness of mind and spirit. Those who live in these cultures typically do not know our languages and even if they did would not even see the value of communicating a mere description of a deep experience. We are aware of one exception to this, and small glimpses of Malidoma Somé's articulate and powerful account appear below. We invite you, when you read them, to suspend your disbelief and to make the undoubted effort that is required for most Western readers if we are to see them as equally "real" when compared with our own experience.

Wikipedia is more helpful than the OED and defines a shaman as "A member of certain tribal societies who acts as a religious medium between the concrete and spirit worlds". This is a more helpful starting point and is amplified by their definition of the culture of shamanism.

Shamanism refers to a range of traditional beliefs and practices concerned with communication with the spirit world, mostly animal spirits. There are many variations in shamanism throughout the world, though there are some beliefs that are shared by all forms of shamanism:

- The spirits can play important roles in people's lives.

- The shamans work with the spirits, seeking assistance or advice for the benefit of the community, or individuals in it.

- The spirits can be either good or bad.

- Most shamans get into a trance by singing, dancing, meditating and drumming. A minority use entheogens (mind-expanding chemicals)

- The songs and dances describe the spirit's journey or the shaman's own personal journey to the other world.

- Many shamans imitate many animals and bird spirits. This happens when the shaman's spirit leaves the body and enters into the supernatural world.

- The shamans can treat illnesses or sickness. The main purpose of shamanism is to understand nature, work with it and heal the sick.

Michael Harner, a leading expert in this field tells us that the word "shaman" in the original Tungus language refers to a person who makes journeys to non-ordinary reality in an altered state of consciousness. Although the term is from Siberia, the practice of shamanism existed on all inhabited continents. He quotes researcher Mircea Eliade who concluded that shamanism underlies all the other spiritual traditions on the planet, and that the most distinctive feature of shamanism, but by no means the only one, was the journey to other worlds in an altered state of consciousness[11].

These definitions bring us closer to a realistic frame of reference, but are still highly challenging to the scientific mindset that inhabits our western cultures. The Wikipedia definition assumes the existence of animal spirits though it leaves out plants, which we will be talking about extensively. It is inevitably vague about what the "other world" it refers to might be, and as we have said we are opposed to the view that there is "another world" which is "supernatural". We are seeking to depict an alternative sensing and description of the single world we inhabit. The definition also refers to the spirit "leaving the body". None of these are comfortable for our cultural frame of reference. Any of them might be regarded in our system as a defining feature of mental illness.

In consequence, the remaining narrative in this chapter must be read with a fresh mind-set. We hope that by now we have established some of the features of the world which shamans are relating to but we will certainly be pushing the boundaries further with what is to follow. With Cleve Backster's work we saw evidence of consciousness in plants. With the shamanistic gardeners and healers

---

11 More information can be found on www.shamanism.org

we found how some people have worked co-operatively with plant consciousness. Now we enter into the deeper experiential relationship with that reality.

The world of the shaman

Malidoma Patrice Somé is a remarkable individual in more ways than one. That he has survived the cultural journey he has undergone speaks of considerable internal resources. His story is extraordinary in presenting a very rare example of a shamanistic world described by one who has been completely immersed in the experience, but who also understands the Western view of reality and can speak directly to us. In his own culture he is initiated as a medicine man and a diviner. In the West he holds three master's degrees and doctorates from the Sorbonne and Brandeis University. His book "Of Water and the Spirit" is a journey of revelation. For Malidoma it is the revelation of his own traditions. For us it is a doorway into a profoundly different way of understanding, experiencing and thinking about the world.

Malidoma's culture – the Dagara people of West Africa – is a shamanistic culture in which many of the tribe (most of the males at least) would be initiated into the experience of the spiritual dimension and in which all accept and understand that frame of reference. It is in that sense quite different from the model of a single priest with unique access that was indicated by the OED. The close relationship with the natural world is embedded in their way of life. As a very young boy he experienced this very directly, but then at the age of four was removed under duress by a Jesuit priest and taken to a Catholic boarding school, which indoctrinated him for the following fifteen years along with other children whom they hoped to turn into a "native" missionary force. As he describes

> "At the age of twenty I escaped and went back to my people, but found that I no longer fit into the tribal community. I risked my life to undergo the Dagara initiation and thereby return to my own people. During that month-long ritual, I was integrated back into my own reality as well as I could be. But I never lost my Western education. So I am a man of two worlds, trying to be at home in both of them – a difficult task at best.
>
> When I was twenty-two, my elders came to me and asked me to return to the white man's world, to share with him what I had learned about my own spiritual tradition through my initiation. For me, initiation had eliminated my confusion, helplessness and pain, and opened the door to a powerful understanding of the link between my own life purpose and the will of my ancestors."

Malidoma Somé describes the Dagara world-view as "only one of the endless versions of reality." He goes on to say:-

"in the culture of my people, the Dagara, we have no word for the supernatural. The closest we come to this concept is "Yeilbongura", the "thing that knowledge can't eat". This word suggests that the life and power of certain things depend on their resistance to the kind of categorising knowledge that human beings apply to everything. In Western reality there is a clear split between the spiritual and the material, between religious life and secular life. This concept is alien to the Dagara. For us, as for many indigenous cultures, the supernatural is part of our everyday lives. To a Dagara man or woman, the material is just the spiritual taking on form."

The initial encounter with the spiritual world came quite unexpectedly for Malidoma when he was as a child of less than 4 years old.

"One day something very odd happened. As I was running around madly, I stepped on a rabbit. It dashed out of its hiding place and a wild race ensued. Looking for a place to hide, the rabbit ran straight towards a small forested area in the bush. I rejoiced when I saw the rabbit run in that direction because I often picked the fruits there and knew every corner of that little bush. The rabbit disappeared into the bush like an arrow shot into a pot of butter.

I followed with caution, trying to guess where the rabbit might be hiding. The tall grass put me at a disadvantage. I had to beat my way through while the rabbit slipped along easily. When I turned over the first clump of grass, the rabbit was not there. I checked another part of the bush where I knew there was an animal nest. This nest was an earthen hole dug in a little hill, its opening covered with grass and its inside filled with soft straw. I removed the grass and was ready to leap headlong onto the miserable rabbit, but I never completed the action. All my movements were suspended as if by an electric shock.

Where I had thought there would be a rabbit there was instead a tiny old man as small as the rabbit itself. He sat on an almost invisible chair and held a minuscule cane in his right hand. His head was covered with hair so white and shiny that it seemed unnatural. His beard was long and white too, reaching almost to his chest, and he wore a traditional Dagara mantle, also white.

All around him there was a glow, a shiny rainbow ring, like a round window or a portal into another reality. Although his body filled most of that portal, I could just see there was an immense world inside it.

But what surprised me most was that the laws of nature in that world did not seem to operate like anything I had seen before. The little man's chair was sitting on a steep slope, yet he did not fall over backwards. I noticed there was something like a thin wall that sustained him. He was not leaning against the chair he was sitting on, but against that thin wall even though he still appeared upright in the window.

As my eyes moved from that wall and the world behind it back to the man, I say that his thin legs were bare. His toes were so tiny I could barely see them. Petrified by something that was neither fear not mirth, but felt like a tickling all over my body, I forgot to scream as the man said, "I have been watching you for a long time, ever since your mother started bringing you here. Why do you want to hurt the rabbit, your little brother? What did he do to you, little one?" His tiny mouth was barely moving as he spoke and his voice was very thin.

Confused, I tried to reply "I . . .I …don't know."

"Then be friendly to him from now on. He too likes the freshness of this place. He too has a mother who cares for him. What would his mother say if you hurt him? Now go because your own mother is worried."

While the little man was speaking, I spotted the rabbit, which had been hidden behind him in the magic circle all that time. I moved further into that steep marvellous place, and then disappeared behind a tree. Meanwhile, I heard a cracking sound as if the earth itself were splitting open. No sooner had I heard this than the old man stood up, slung his chair over his shoulder, and walked into the opening as if he had commanded it. The earth closed up on him leaving a gust of fresh breeze in his place. At the same time I hear my mother's voice calling me."

Malidoma tells the story to his mother, who says

"Oh, my dear ancestors, my child has just seen a Kontomblé. What else can it be? Don't talk anymore. Let's get out of here. I'll never take you out again."

She was concerned because Malidoma was very young for such an experience and it had significant implications for their relationship if it became known to others. Her reference to a "kontomblé" denotes a spirit who lives in the underworld.

Smart chickens and tree spirits

As an infant, Malidoma had been very close to his grandfather, one of the senior shamans in the tribe who had recently died but whose stories come to his mind regularly. At one point shortly before his kidnapping he is watching chickens in the yard and recalls an incident where his grandfather translates the conversation that chickens are having, and explains that they are about to scatter the millet from a basket where one of the women has painstakingly separated and cleaned it. Immediately after, Malidoma watches as one of the hens jumps into a basket and causes a fight with the woman that ends with the basket overturned and the contents beyond collection, except by the feasting chickens.

In case you should still question why, in a book intending to provide a scientific view of the spiritual world we are telling such stories, the reason is this. The reality we are describing is very alien to most of us. The idea of little spirit men and chickens which communicate is quite fantastic. The journey we are taking requires us to suspend large amounts of culture-engendered disbelief. Do you believe that Malidoma's grandfather could really understand what the chickens were communicating with each other? Do you believe that chickens could even make a plan of any kind?

Questions are also raised regarding the true nature of what is being described. Is the Kontomblé something solid that one could touch? Do chickens, when they communicate, use a language that resembles ours? Was grandfather hearing their thoughts? Since we are not shamans, we cannot know for sure. Some elements of Malidoma's story are common features when encounters with the spirit world are described – the sense of an altered surrounding reality and the feeling of heat or tickling to the skin. These are cross-cultural elements. But actual descriptions vary a lot, as do the pictures people draw of what they see. It seems probable that the visions are translated into the perceptual systems of our everyday reality and that messages are received as ideas, to which the recipient applies language. There is some indication that the forms are filtered through a culturally derived perception and that the spirit world communicates in a way that resonates with us, whatever form that requires. Indeed there are other reasons we will discuss in later chapters why this is probably so.

Although these features are not crucial to what we are saying in this book, we need to caution those who might be looking for complete consistency to understand that it may be unrealistic to expect this. We need to look underneath at the nature of the world that is being described, and to understand its significance. Piece by piece, the non-visible aspects which the western world has eliminated from its conceptual framework are emerging as potentially more important than all the material manifestations that we are habituated to.

In earlier chapters both of us have described experiences across the boundaries of this alternate reality. But neither of us has had the deep shamanic initiations that those like Malidoma Somé describe. It is very difficult to have such experiences and live in the world of TV and mobile phones. In consequence it calls for a great deal of trust on the part of any of us to even hear these stories as a form of "truth", to not label them as fantasy, hoax or insanity.

The description that Malidoma Somé gives of his initiation, after returning to his own culture as a twenty-year-old, takes over one hundred pages. It is a rich account, but he indicates that it is nevertheless incomplete, as there is more that his elders do not allow to be shared. The excerpts we are using merely hint at the reality, but we include them because they are the only examples in our book from someone raised in such a culture. All our other stories came from those who have crossed into such a culture from the West. Our final example occurs during his initiation.

The background to the story is that some of the tribal elders have been unwilling to see him initiated because of the impact of his Westernisation, while for him it is his only way back into connection with his people. The exercise he is engaged in requires him to stare at a yila tree until he "sees something". He is struggling to achieve anything and has just tried to "fake it" bringing a mixture of scorn and derision, which has left him feeling a confusion of aloneness, broken pride, anger and alienation.

"Through my tears, I managed to continue keeping an eye on the tree. Then I finally began speaking to it, as if I had finally discovered that it had a life of its own. I told it all about my discontent and my sadness and how I felt that it had abandoned me to the shame of lying and being laughed at.....

I then spoke to the tree again, not angrily but respectfully. I told her that, after all, it was not her fault I could not see, but mine. I simply lacked the ability. What I really needed to do was to come to terms with my own emptiness and lack of sight, because I knew she would always be there when I needed to use her to take a close look at my own shortcomings and inadequacies.

My words were sincere: I felt them while I said them. My pain had receded somewhat, and I found I could now focus better on the tree. It was around mid-afternoon, but I was not really interested in the time. I had something more important to deal with, for suddenly there was a flash in my spirit like mild lightning, and a cool breeze ran down my spine and into the ground where I had been sitting for the past one and a half days. My entire body felt cool. The sun, the forest, and the elders and I, understood I was in another reality, witnessing a miracle. All the trees around my yila were glowing like fires or breathing lights. I felt weightless, as if I were at the centre of a universe where everything was looking at me as if I were naked, weak, and innocent. Indeed I thought I was dead. I thought that something must have happened while I was trying to reconcile myself to the shame of being caught in a lie.

To substantiate my impression, I thought about the hardships of the day – the baking heat of the sun and my sweat falling into my eyes and burning them like pepper. I had lost all sense of chronology. I told myself that this is what the world looked like when one had first expired. I felt as if I were being quite reasonable. I could still think and respond to sensations around me, but I was no longer experiencing the biting heat of the sun or my restless mind trying to keep busy or ignoring my assignment. Where I was now was just plain real.

When I looked once more at the yila, I became aware that it was not a tree at all. How had I ever seen it as such? I do not know how this transformation occurred. Things were not happening logically, but as if this were a dream. Out of nowhere, in the place where the tree had stood, appeared a tall woman dressed in black from head to foot. She resembled a nun, although her outfit did not seem religious. Her tunic was silky and black as the night. She wore a veil over her face, but I could tell that behind this veil was an extremely beautiful and powerful entity. I could sense the intensity emanating from her, and that intensity exercised an irresistible magnetic pull. To give in to that pull was like drinking water after a day of wandering in the desert.

My body felt like it was floating, as if I were a small child being lulled by a nurturing presence that was trying to calm me by singing soothing lullabies and rocking me rhythmically. I felt as if I was floating weightless in a small body of water. My eyes locked on to the lady in the veil, and the feeling of being drawn toward her increased. For a moment I was overcome with shyness, uneasiness and a feeling of inappropriateness. I had to lower my eyes. When I looked again she had lifted her veil, revealing an unearthly face. She was green, light green. Even her eyes were green, though very small and luminescent. She was smiling and her teeth were the colour of violet and had light emanating from them. The greenness in her had nothing to do with the colour of her skin. She was green from the inside out, as if her body were filled with green fluid. I do not know how I knew this, but this green was the expression of immeasurable love.

Never before had I felt so much love. I felt as if I had missed her all my life and was grateful to heaven for having finally released her back to me. We knew each other, but at the time I could not tell why, when or how. I also could not tell the nature of our love. It was not romantic or filial; it was a love that surpassed all known classifications. Like two lovers who had been apart for an unduly long period of time, we dashed towards each other and flung ourselves into each other's arms.

The sensation of embracing her body blew my body into countless pieces, which became millions of conscious cells, all longing to reunite with the whole that was her. If they could not reunite with her it felt as if they could not live. Each one was adrift and in need of her to anchor itself back in place. There are no words to paint what it felt like to be in the hands of the green lady in the black veil. We exploded into each other in a cosmic contact that sent us floating adrift in the ether in countless intertwined forms. In the course of this baffling experience I felt as if I were moving backward in time and forward in space. ...."

It would be understandable if the depiction above should seem unreal or fantastic to you. Yet it is far from being the most other-worldly experience that Malidoma Somé describes. Here he remains to a significant degree in the world we know. In other experiences the sense of dislocation, of travelling beyond the boundaries of our reality, is much stronger. They are journeys into foreign territory and last for much longer. But all the descriptions have the same dreamlike quality and have in this a resonance to "dreamtime", the phrase used by Australian Aboriginals for their otherworld.

We will save discussions of what "reality" is for a later chapter. For now we would merely note that there are strong philosophical grounds for treating these apparently subjective experiences as equally "real". The use of the word "dream" carries an automatic cultural implication for us. We need to step away from that implication because that view is created by the very philosophy we are questioning here. It is a habit based on a false assumption and if we remain locked inside that habit, none of the descriptions in this chapter can ever be seen as other than fantasy.

Very few of the writings on this aspect of philosophy have been written by a Western thinker who is capable of direct knowledge of the intuitive, spiritual world. The one exception known to us is Rudolf Steiner, who explores it in depth in his books "Knowledge of Higher Worlds" and "Intuitive thinking as a Spiritual Path". He points out that the supposed differences between "external observation", by which he means the use of our regular senses, is not as different from "internal observation" (the intuitive acceptance of a mental picture) as it is commonly taken to be. As he says, this "makes the error of characterising one percept as a mental picture while naively accepting the percepts of ones own organism as objectively valid facts." He sees both as equally misleading. In his view

> "for a relationship to exist between my organism and an object outside me, it is not at all necessary for something of the object to slip into me or impress itself on my mind like a signet ring on wax. Thus the question 'How do I learn anything from the tree that stands ten paces from me?' is all wrong". It arises from the view that the boundaries of my body are absolute barriers through which news about things filters into me."

Instead he states that we belong to the same world as the things we think of as external. The segment that you think of as "you" is run through by the stream of the same Universal life process.

We will return to this topic and to the boundaries we are in the habit of perceiving, but introduce it now in order to underpin the notion that the dreamlike, esoteric and internally perceived world is worthy of the same possibilities for validity as the world of sight and touch. We are doing all we can to keep solid ground beneath our feet and to avoid dismissing what is unfamiliar to our cultural definition of what can or cannot be "real". This is essential if we are to reach our goal of a single world, not one split in two.

The earlier references to Alberto Villoldo and Eliot Cowan, as well as Juliana's own working experience, see Malidoma's engagement with a nature spirit not only as real, but as enabling an engagement with the natural world that can support health and healing for us all. This is only one of the areas in which communication and connectedness offer some sort of possibility for "energy medicine". Eliot Cowan's story illustrates the spiritual dimension of the healer, the intuitive nature of the connection, the subtle energetic flow of the healing and the ability of the client also to respond to aspects of the experience that were not directly known by her. It summarises and weaves together several aspects of our text so far.

In the next chapter we will look at another very different modality and see the systematic and scientific depth beneath homeopathy. This will lead us toward further examples of miraculous capabilities in the world of healing before heading into the scientific heart of the biology and the physics through which it all manifests.

Review

This chapter has presented some definitions of the terms "shaman" and "Shamanism" and looked at the characteristic features of belief systems and views of the world which are very different from anything that is normal in the modern scientific and largely Western world. These viewpoints are discussed because they have been widespread across all parts of the world and represent something that is common to human experience.

In order to illustrate what that experience is like we have offered descriptions in our language, but written by an individual who was raised in the culture where such viewpoints are normal. We have invited you to consider that they are no less "real" than anything that we or you would consider to be so.

7. The science of homeopathy

Two women are on a bus, talking.

First woman: "What happened with that homeopath you were seeing?"

Second woman :"Nothing. She gave me a couple of pills but they didn't do anything."

"So who are you seeing now?"

"Oh – nobody.   A few weeks later the problem just went away."

Anon: Joke enjoyed by homeopaths

Any sufficiently advanced technology is indistinguishable from magic

Arthur C Clarke

Theme

In this chapter we use homeopathy as an illustration of the scientific method as applied in alternative healing, and show the way in which its systematic study has revealed various aspects of the connected world that we are describing. Homeopathy shows the effects of informational relationships in a very distinct way.   It also provides evidence that there are aspects of that information flow which should influence our perception of what happens between one generation of humans and the next, that there is more than simple genetics involved.

There has been laboratory proof of the way in which homeopathy works – proof which has been not simply ignored, but suppressed in a quite unscientific way. This has allowed the myth that homeopathy operates through the "placebo effect" to perpetuate. While the placebo effect tells us something important which science also is unable to explain adequately, it has nothing whatsoever to do with homeopathic effectiveness.

How homeopathy developed

Homeopathy is a science. Many other complementary health practices might say the same, but homeopathy may be unique in having developed explicitly according to western scientific methods prior to 1900.   This is one reason we will concentrate on it at the expense of other strong contenders such as Chinese medicine and acupuncture which, regardless of their millennia of practice and huge curative history, are more difficult to frame in western terms and whose origin is more obscure. The other reason for a focus on homeopathy is that it illustrates perfectly the nature of the universe we are describing and exposes the gaps in the scientism paradigm.

We will use the term medical here to denote the core of pharmacologically and surgically based western Medical practice. This is not to deny that complementary therapies are also medical – it is simply a narrative convenience.

A convention has developed among the medical and scientism advocates in their references to homeopathy. That convention is to dismiss its practice in a very simple way. They note that its principle is to use remedies prepared in such a way that potentially there may be not one single molecule of the original substance from which the remedy derives, present in the pill taken by the patient. By this simple logic, Medicine takes the right to attribute all experiences to the "placebo" effect. What they mean by this is that the patient "imagines themselves better" because they are made to feel good. We will show that the notion and implications of the placebo effect are important. Indeed they are more significant than Medical science is able to grasp, but not at all applicable to homeopathic treatment.

This convention has been present for so long and is so entrenched that it is largely taken for granted that no further study of its theory or history is necessary, creating a conspiracy of ignorance. Such trials as are undertaken are not conclusive. Often this is for methodological reasons we will examine below. On Page 167 of the "God Delusion" Professor Dawkins echoes the conventional viewpoint, which he repeats regularly in hostile TV polemics and one suspects, as with all his co-conspirators, that he has not been willing to look any further. In most cases we simply disagree with his views of science. In this instance it is fair to accuse him of totally inadequate research.

At this point let us remind ourselves of what the fundamental principles of science are supposed to be. A simple view is that science is the search for one kind of truth about the universe. Scientists might say that this method is the only way to determine truth. Such truth would be based on a straightforward set of processes.

- Observe the nature of the world with detailed care.

- To the greatest possible extent, undertake such observations in such a way that they can be measured and repeated by others under the same conditions.

- Try, when measuring and compiling the data, to avoid interfering with the subject under examination in such a way as to distort the events observed.

- From these observations attempt to formulate theories which would explain the relationships of cause and effect that determine what is seen.

For example: "I release an object at height. It falls to the ground. Therefore there is a force which operates on objects. I will name it 'gravity'." This is the background to the image of Newton's stroke of genius under the apple tree. Theories can develop to greater complexity such that we can then measure and determine mathematically just how strong earthly gravity is, how quickly a falling object will accelerate towards the ground, and what the effect might be of resistance to that motion from the air such that feathers fall more slowly. Further theories might then develop regarding why it is that the Earth exhibits a property such as gravity and explanations would appear for the observed motions of the planets.

Depending on the phenomena observed, we would then repeat the observations under controlled conditions in order to test the theory and determine as completely as possible the boundaries of the theory. Is it always true? Under what circumstances does that theory vary, and how? Does the theory of gravity still hold true on the surface of the moon?

At the core, this is what science is – an attempt to create systematic processes for describing the observed world in a way which can be labeled as "truth". The process is simple in theory but due to the great complexity of the world the practice is not easy. Science has grown because a theory (e.g. Newtonian gravitational mechanics) was insufficient to explain why light bends and does not always travel in straight lines so that it is distorted when under the influence of gravity from large objects in the universe. A new theory is required and Einstein finds curves in space-time (don't worry – we don't need to understand that here).

The example above illustrates the requirement for science to change when new facts emerge. There is a linguistic misunderstanding over the phrase "the exception proves the rule" because we hear the modern meaning of the word "proves" as "shows it to be true". The original meaning of "prove" was "test". Thus, if a new and exceptional fact emerges (e.g. we notice conditions under which light bends) we test if the rules of Newtonian mechanics still apply. Depending on circumstances, we must either find a bigger theory in which our current theory is seen as a local approximation, or we have to throw the theory out and find a better one.

Forgive us if we are labouring this point. To do so would be insulting to scientists in particular were it not for the tendency through history for them to behave in fallibly human ways or to be influenced by political and economic forces. Ask Galileo. Science has regularly experienced the cycle of influence when a new fact emerges that does not fit with previous accepted theory. The fact in question is then denied by scientific authority. There is then a prolonged period of struggle (sometimes ending with the death of a particular authority figure) after which the fact is admitted and a new theory becomes authoritative. This has been neatly summed up by Schopenhauer who said :-

"All Truth passes through three stages.

It is ridiculed

It is violently opposed

It is seen as self-evident",

which brings us back to homeopathy, whose current scientific credibility, continues to lie between ridicule and opposition. Homeopaths must yearn for the day when the Earth will revolve around the sun.

Homeopathy's Galileo is Dr Samuel Hahnemann who was a physician and expert in pharmacology born in 1755, one of whose medical successes was to live to the age of 88. Hahnemann was by any standards a remarkable man, who was fluent in seven languages, and the initial prompting for his research came when he was translating the work of an English doctor, William Cullen, into German.

He was unconvinced by a statement of Cullen's regarding the reasons behind the effectiveness of Cinchona bark as a treatment for malaria. In order to find out more about its properties, he took repeated doses of the substance, up to the point where its toxic effects began to show. What he noticed was that the toxic effects were fever, chills, and other malaria-like symptoms. From this he theorised that rather than being due to astringent properties, which Cinchona bark shared with other substances that were ineffective against malaria, the reason for its effectiveness was that the symptoms produced by the bark were similar to those of the disease. He set out to test this hypothesis systematically.

His method was scientific and since one observation does not make a science,

Hahnemann spent the next six years, with the assistance of a small group of followers, testing the observation on a wide range of substances and conditions, using his family too as subjects. At the end of the process he published his findings in a medical journal. He immediately met with opposition, both from physicians, and from apothecaries (the pharmacists of that time). The latter were upset at the potential damage to their businesses, because Hahnemann recommended only one medicine at a time, and in small quantities, so they were not disposed to comply with his prescriptions. When Hahnemann found that homeopathic medicines were not being prepared correctly by apothecaries, or that they were taking it on themselves to prescribe different medicines, he began to do the preparation himself. This was unlawful, and resulted in his being charged, and forced to leave his home in Leipzig.

It is perhaps helpful to see homeopathy against the background of its day, when treatments were very coarse - bloodletting and the use of leeches being among the common techniques. It is recorded that in 1833, 41 million leeches were imported into France. Orthodox medicine also used preparations from arsenic, lead and mercury - all poisons, as well as strong herbal purgatives. But despite this it was homeopathy that was labeled as "devilish", "cultish", or "quackery".

Like cures like

The first principle discovered by Hahnemann, and the foundation of homeopathy, stemmed from the initial observation, and can be summarised as follows:-

- Every pharmacologically active substance produces symptoms in a healthy, sensitive individual which are characteristic of that substance.

- Each disease has a characteristic set of symptoms.

- An illness can be cured by administering to a patient a small, homeopathically prepared dose of the same substance which, during trials, produced symptoms similar to the illness in healthy individuals.

His approach to treatment represented a dramatic move away from the established method. Allopaths (conventional medics) establish the existence of a particular disease, clarify its symptoms, and then test the effectiveness of various medicines on it by the use of substances that oppose the symptom(s), a principle of "opposite suffering". An illness accompanied by fever and diarrhoea, for example, would call for the combined use of substances that calm the fever and others that normally constipate, and so in a crude way, a total balance would be found by using a number of appropriate medicines together.

Homeopaths tried the opposite approach: first build a repertory of substances for medicinal use, they said, by giving them to healthy volunteers, and carefully noting the symptoms produced. Then use small quantities of the substance which produces the correct combination of symptoms as the single medicinal agent for those with disease conditions, a principle of "similar suffering".

To use a substance which potentially produces, rather than suppresses the symptoms seems at first counter-intuitive, but was found to work in practice. A simple example of the principle that "like cures like", would be that if you were suffering from the particular type of cold symptoms combining streaming and burning nose, watering eyes and bouts of sneezing, the remedy to be used would be derived from onions, which as any cook knows, produces the same symptoms. Likewise *Allium Cepa*, the remedy referred to, is used to combat hay-fever with similar presenting symptoms. The cause of those symptoms is not necessarily relevant. The fact that one may be caused by a viral or bacterial pathogen, and the other by an airborne irritant does not matter, because the body is seen to mobilise the appropriate curative resources when its energy is stimulated by the remedy to fully engage.

Returning to Hahnemann, in the six years of work before publication many examples of this process were collected. In doing so he set a further fundamental principle in the development of the science of homeopathy; it should be based on detailed observation, extensive trials, and systematic testing of theoretical and philosophical ideas, through careful experimentation. This tradition thus meets the criteria described earlier for a scientific process and continues to the present day.

Homeopathy was successful, and spread rapidly. It was taken to America in 1825, and expanded so rapidly that in 1844 the first national medical association was formed, by Homeopaths. Two years later the American Medical Association was started with the specific intent of slowing the growth of homeopathy. The AMA specifically excluded homeopaths from membership, and expelled members who admitted any contact with it (which in many states was a precursor to loss of license to practice).

Nevertheless, the rapid spread of the science had the beneficial consequence that there were many practitioners, and a vast body of growing validation of its effectiveness. A second effect of this growth in numbers was that many practitioners were formulating, sharing and testing new theories (as too was Hahnemann, who remained active throughout his long life). As a result the body of information and experience as well as the recognised range of effective remedies, increased steadily. We want to stress here, that the point of this chapter is not to compare levels of homeopathic effectiveness with that of modern medicine. It is not a question of which is better. Our point is firstly to show that the principles discovered by Hahnemann and his followers were real and scientific and that there are sound reasons why homeopathy is effective. It is secondly to show the implications that this has for current scientific thinking.

Less is more

Having established the first principle, Hahnemann's second line of investigation was to determine what amount of the "similar" agent would be required to bring about the best curative effect. Some of the patterns in disease resemble the actions of seriously dangerous toxins. For example the sickness and diarrhoea that occur with food poisoning are sometimes like the effects of arsenical poisoning. It is obviously desirable that only the smallest quantities of arsenic would be used in treatment.

This line of research led to the second counter-intuitive finding. Just as it was better not to suppress the symptom with an opposite, Hahnemann discovered that the less of a substance he administered, the more effective would be the cure. His second axiom for treatment is the Law of the Minimal Dose. This states that the effective dose for a disorder is the minimum amount necessary to produce a response. The process by which a homeopathic remedy is prepared is known as potentisation, and involves a sequence of progressive dilution and a rhythmic shaking, termed succussion. In a typical method, one part of the source substance is added to 9 parts of water and shaken rhythmically. This is known as a 1x (decimal) dilution, or 1 part in 10. One part of this is then taken and added to another 9 parts of water, again succussed, to give a 2x dilution, or 1 part in 100.

These dilutions can be repeated a large number of times. A typical UK health-store remedy might be 6x (or 1 part in a million). But it might also be 6c, a centessimal (1 to 100) preparation where the original substance is diluted to one in a billion levels. Practitioners often use dilutions down to one in a trillion and well beyond. While the toxicity of such medicines is obviously very low, the dilutions quickly approach levels where it is questioned whether a single molecule of the original substance remains. This is the cause for the chemical reality which underlies the "placebo effect" dismissal. In one sense it is a valid challenge since the results are not arising through chemical action. It would be more valid if there was an attempt to say "so how is it working?"

To answer that question we will need to reveal much more about the findings of homeopathy and its scientific implications, but it would be good to get the "placebo" issue thoroughly out of the way. While it might be apparent to an open-minded reader that the founding processes are so detailed as to make the placebo theory inadequate, there are many more facts which support the case for homeopathic effectiveness.

The placebo myth

One simple fact giving lie to any imagined or psychosomatic effects is that homeopathy works with infants and animals. The numbers of parents who have experienced the almost miraculous calming effects on their teething infants of chamomile in homeopathic potencies probably amounts by now to millions. This could be enough on its own, but it is far from being an isolated example.

Jon has practiced homeopathy as an amateur for over 20 years now with friends, family and pets. In his early days he visited with friends whose small cat had suffered for days with worsening symptoms of sneezing, runny nose and wheezing. Cat-lovers will know that this can be quite serious – more so than a human cold – and the hosts (you can't "own" a cat) were getting worried. They gave permission that evening for Jon to experiment, despite his never previously having treated any animal. A few remedies were placed in the cat's food and water and by the following morning all symptoms had vanished.

If this was a one-off occurrence it could be regarded as coincidence, but it is not and there are homeopaths who specialize in veterinary practice. What it emphatically cannot be ascribed to is a placebo effect. It is even less appropriate than with infants to apply such a term.

There is strong epidemiological evidence too. In 1830 cholera, a disease that had never been seen before in Europe, was having devastating effects all across it. Nowadays it would be controlled by hygiene and by preventing diarrhoea and dehydration but even now if cholera occurs in an unprepared community, case-fatality rates may be as high as 50% according to World Health Organisation figures -- usually because there are no facilities for treatment, or because treatment is given too late. In 1832, two hundred and fifteen deaths occurred in Sunderland alone, and by the summer of that year the disease had taken toll of some eight hundred lives in nearby Newcastle-upon-Tyne. From North-East England the disease quickly spread to Southern Scotland causing three thousand one hundred and sixty-six deaths in Glasgow. In April the disease appeared in Hull and in

Liverpool where one thousand five hundred and twenty-three deaths occurred. Leeds, Bristol, and Manchester were also soon afflicted as well as many other towns and sea-ports, the disease being especially rampant amid the shacks and hovels of the new industrial districts.

In 1854, cholera broke out again in London. At the London Hospital where homeopathy was used, of the 61 cases of cholera reported 10 died (83% cured), and among 341 cases of choleraic diarrhoea just one patient died (99% cured). In contrast the neighbouring Middlesex Hospital received 231 cases of cholera and 47 cases of choleraic diarrhœa. Of the cholera patients treated conventionally 123 died, a fatality rate of 53.2 per cent.

When doctors could not cure the disease they attempted to treat the wound to their professional pride. The homeopathic cases were excluded from statistics presented to parliament due to the claim that they would give "an unjustifiable sanction to an empirical practice alike opposed to the maintenance of truth and to the progress of science". (Evidence of the Medical committee to the Parliamentary Board of Health).

During this period of Europe-wide epidemic the Russian Consul General reported results from homeopathic treatment practiced at two locations. Of 70 cases, all were cured. In 1849 Cholera had reached America, and an outbreak in Cincinatti was treated similarly, with a 97% cure rate in a sample of 1116 patients. Similarly in Naples, a Dr Rubini treated 225 cholera cases without a single death. At this time the success of allopathic medicine was generally no better than one in three patients cured. Similar effectiveness was reported later in the century, for treatment of a Yellow fever epidemic in the Southern USA.

We said earlier that the effectiveness of homeopathy against cholera was important in itself, and not as a comparison with antibiotics, or any modern medical procedure. The importance is that you cannot conceivably achieve a 90% cure rate for as hostile a disease as Cholera with a placebo effect. If you could, other treatments would have been able to do so as well. It is clearly quite ludicrous to dismiss homeopathic results in this way. There has to be a better explanation and indeed there is one, which takes us back to the theory.

Small doses stimulate

In conventional (scientifically accepted) pharmacology, one of the basic tools is the Dose-Response curve. This graph illustrates one of the rules of thumb in drug use: that an increased dose of drug will give an increased effect. But this applies only at higher dosages. One of the very earliest laws of pharmacology, known as the Arndt-Schulz Law also expresses the homeopathic effect. Formulated by Arndt in 1888, the law states that for every substance, small doses stimulate, moderate doses inhibit, and large doses kill. Allopathic medicines, with their emphasis on moderate drug doses, work in the inhibitory part of the scale, and are used to suppress symptoms. Homeopathic medicine, on the other hand, begins at the stimulatory end of the curve, and moves to smaller and smaller dose ranges. Its emphasis is on the stimulation of the body's natural response mechanisms.

But although the basis is there in pharmacological theory, we must go beyond pharmacological action to understand the homeopathic effect. Since there is often insufficient substance to have a pharmacological effect, the action has to be taking place at some kind of "energetic" or "informational" level. We accept the statement from critics that there may not be a molecule of the source substance present. Therefore the effectiveness of homeopathy does not lie in chemical action or conventional pharmacology. Equally clearly, it has great capability to be selective and specific about the effectiveness of different substances, or their actions would not be so precisely targeted. Homeopathic theory validates the transmission of a healing effect through a medium which is purely "energetic" or "informational". We will add more definition to these terms shortly.

By now you will hopefully be recognizing the fundamental reason for our earlier focus on scientific principle.

- Something is happening in homeopathy that is scientifically backed by volumes of systematic evidence gathering but which does not fit with existing medical theory.

- Since the existing theory is failing some tests of evidence the theoretical model needs either to expand or to be replaced.

- In order to develop a more comprehensive theory further investigation is required into the data and the underlying principles of similarity and minimum dosage

Fortunately there is more evidence for us to look at. Unfortunately medicine and science are in a state of denial (ridicule and opposition) around this too. They have cause to do so because the evidence strikes at the very roots of the scientific challenges and gaps which we listed in our introduction. But the alternative theories which emerge from this evidence are very exciting indeed.

Suppression doesn't work

As we delve more deeply into what homeopathy has discovered, the evidence throws up yet more challenges to scientific theories, and in some very interesting ways. We are going to have to abbreviate a lot more of the theory to get to this. There is so much of value regarding the way in which healing works, and we can only encourage you to read of it elsewhere (the work of Deepak Chopra being one excellent example).

We referred above to the choice allopathic medicine makes to cure by suppression of symptoms. Most alternative and holistic approaches are fundamentally opposite to this. They promote a basic respect for the actions of the body - an assumption that if the body produces fever, it is because it needs to do so. This approach is inclined towards supporting the body in its natural response and views suppression as likely to drive fundamental causes of ill health deeper into the body, bringing worse trouble later on. In homeopathy the use of a substance that shares the underlying symptomatic signature (similar suffering) is designed to support the natural process as it propels the symptom(s) towards completion of the healing.

Examples of the ill effects of suppression form a very strong part of homeopathic case lore. A whole strand of investigation in homeopathy relates to situations where actions that were taken to suppress a disease resulted in the later appearance of another symptom pattern. In such circumstances allopathic medicine set out to treat the new symptoms it has produced in its attempts to treat the original presenting issue. (Now there is a whole area of conventional medicine devoted to "iatrogenic" problems.) However, homeopathy would deal with this situation differently. Instead of treating the latest symptoms it would set out to treat the disease that was originally suppressed. Study after study has shown homeopathy's methods to be effective no matter how long before the current problem it may have occurred, or how absent those original symptoms may be.

A typical example of this which occurred frequently during the 19th century, when the development of homeopathy was at its height and when venereal disease was also very common, arose from the treatments used to suppress Gonorrhea. The case we use to illustrate this dates back to 1875, when the eminent homeopath concerned was treating a man of 60 for an obstinate case of rheumatism. This patient was walking with a cane, wrapped in a muffler, thin, bent and aged in appearance, and his condition had persisted for several months. Unable to shift the problem with remedies conforming to the current symptoms, the homeopath recalled the correspondence that had frequently been observed by him and his colleagues, between arthritic conditions and earlier treatments to suppress gonorrhea. His insight was to treat the patient with the remedy derived from that disease. The case notes describe that the patient returned ten days later feeling well, and that within the month he had ceased use of the cane and muffler. His weight subsequently increased from 140 pounds to his previously healthy 212 pounds.

Although this example illustrates powerfully why homeopaths and other alternative practitioners are so concerned to bring disease out of the body, rather than push it deeper in by suppressing symptoms, that is still not the main point of this story. There is a further and even more remarkable stage to go yet. Homeopaths have consistently observed that the effect we have just described could span the generations. That is, an individual could present the symptoms of rheumatic disease, and that this could be cured by the gonorrheal remedy, even though the case of suppressive treatment had occurred in the patient's parent. What this means is that there is experience in homeopathy that the energy pattern relating to a disease can pass from generation to generation, and that the inheritance can be treated. This observation has huge significance. For the avoidance of any doubt we must stress that according to accepted theories this inheritance cannot occur by way of the genes and has no known alternative scientific explanation. It clearly needs one. We will examine this area extensively when we get to grips with the true mechanics of genetic processes.

Energy sickness down the generations

In fact this understanding is one of the basic strands of homeopathic science, which recognises that there are certain diseases, such as Tuberculosis, Gonorrhea and Syphilis which have widespread influence through inheritance. That is, they are so embedded energetically and informationally in the human race that their influence may be detected several generations beyond the last known experience of the disease in the person's lineage. A modern homeopath might rarely encounter a patient who knows the last time tuberculosis or syphilis occurred in his or her family, but it is nevertheless frequently an element in that individuals "make-up". Homeopaths call such an energetic lineage a "miasm". Treatment of miasms is a strong part of a practitioner's armoury, deeply validated by practical experience.

So now we have a second instance whereby some sort of energy is perceived as being transferred, without any known mechanism. It is not being suggested that the miasm is passed via the DNA, or that it has been incorporated into the genes. So just as in some way, the energy of a substance can be put into the fluid that is used to impregnate a homeopathic tablet, the energy of a disease can be passed - perhaps in the cytoplasm of a sperm or egg cell - from one generation to the next. The implications of this fact (and I repeat that homeopaths have been healing people on this principle for a very long time) are profound. This means of transmission implies the strong influence of a vibrational or information-carrying energy component in the disease, and on its passage between generations. We cannot over-emphasise the significance of this fact. In case minds are beginning to wilt in face of the apparent improbability of the facts being described, we also have to repeat the statement that these facts have been repeatedly validated through systematic observation by multiple practitioners across several continents and over very many decades.

The laboratory evidence

Clearly some aspects of these effects are not suitable for laboratory examination. We cannot deliberately infect patients with gonorrhea and then manipulate their treatments for the purposes of measurement, still less monitor the effects on their children. But there is scientific evidence in the laboratory of a means for this transmission to take place. It was discovered by Jacques Benveniste, when he was director of the French National Institute of Health and Medical Research and specialist in immunology. The evidence demonstrates a phenomenon known as "molecular memory". This is akin to a kind of subtle electromagnetic language, whereby the "sound" of one molecule could be recorded by another, like a tape-recorded sound. Benveniste's research was first reported in the magazine "Nature" (Vol. 333, No. 6176, pp. 816-818, 30th June, 1988).

Benveniste had taken a substance which when mixed with a blood serum preparation in a test-tube, typically produces the chemical activity associated with allergic reactions,. He diluted the substance tenfold, and repeated the experiment. He continued this process repeatedly, and as with homeopathy, progressed way beyond the level where any molecule of the substance remained. In effect his solution was just distilled water, containing in theory, 1 part of the original antibody to 10^{120} parts of water. (10 followed by 120 zeroes - a trillion multiplied by a trillion repeated ten times). The effect on the blood serum persisted regardless.

This experiment was replicated in Jerusalem, Toronto and Milan with the same results, and his paper was signed by twelve other researchers. This should have been hailed as revolutionary and groundbreaking. Even now it should be seen as hugely significant for scientific theory. At the very least its results give considerable credence to all that Hahnemann and his followers had discovered in practice, but the implications go much further – right through medicine and biology. Instead of hailing the research the medical and scientific establishment treated the results as a problem, one reminiscent of an anonymous poem which runs "Last night when walking up the stair, I met a man who wasn't there. He wasn't there again today - I wish, I wish he'd go away." This is very much the response that Benveniste's work has met with.

The poem's last line may also be substituted with "I think he's from the CIA". It can be risky to challenge authority, and the attacks on Benveniste in the years since have been outstanding in their abusiveness, including Nature's choice to send its editor and two "fraudbusters" to Benveniste's laboratories. One of these was the noted stage magician James Randi, an arch-skeptic whose name appears regularly as a "debunker" of alternatives and who was reported to have taped information to the ceiling "to prevent tampering". Benveniste complained that the process was unprofessional, accused the team of poor controls and of using one week's work to wipe out the activities of five years research in his and five other laboratories. The results Benveniste obtained in their presence were mixed, and inconclusive, with the first three trials providing some confirmation, but a further four showing nothing.

There are many subsequent failures to replicate the original work but there was a notable success in 2004 when Madeleine Ennis, who claimed to have begun as a skeptic published a study which stated "it has been shown that high dilutions of histamine may indeed exert an effect on basophil activity". (Inflammation Research 2004: 53; 181-188). Such mixed evidence is unhelpful but who knows what happens with such subtle energies, especially in such an environment as the chaotic Nature investigation and under such stress and hostility. If there is a relationship between thought and energy such as we are suggesting, then confusion is exactly what you would expect from this scenario. One clear result was that Benveniste was hounded from the scientific fraternity. He died in 2004 following heart surgery.

English researcher Cyril Smith has also demonstrated the ability of water to store electromagnetic frequencies, and French physicist Michel Schiff likewise participated in replications of the experiments and in his book "the Memory of Water", acknowledges that the water memory effect does seem to occur. Despite the totally central role of water both biologically, and at a planetary level, there remain many properties of water that are not understood. Those who want to know more might care to look into the work of Viktor Schauberger. A further source of wonder comes from those such as Masaro Emotu, who has photographed the

Benveniste effect. You can see examples at www.hado.net .

As far as homeopathy is concerned, we repeat once again that the effects we are describing are consistent, and have been repeatedly observed for over a century. It is frustrating to continue to read material that speaks of homeopathy as unvalidated. This is completely untrue. The evidence is being ignored or misrepresented, and the mythology passed on in medical schools. The consistency and volume of evidence inside the homeopathic world has been developed over two centuries and could only be denied in this way by those who have not actually read it. People's lives and health are at stake, because we are not making full use of the healing techniques that are at humanity's disposal. We feel it appropriate to issue a challenge to science and medicine: put the same degree of funding and constructive open-minded effort into the investigation of the theories described here, and refrain from attack until an adequate alternative explanation is on offer. Such would be a genuinely scientific response.

Our explanation follows a pattern that will now be familiar to readers. We are giving consistent evidence that there are phenomena occurring which could take place only through some kind of transmission of information. That information has to be specific enough to convey the characteristics of a molecule even when that molecule is chemically absent. It has to be specific enough to convey the characteristics of a disease pattern (or at least of the triggers it provides to the body) such that the "picture" of suppressed gonorrhea, or the "picture" of family tuberculosis can travel down the generations. This requires an informational content that is both complex and subtle. That it persists at all speaks of its power. That it is so hard to detect the mechanism speaks of something that is not visible to our normal processes of investigation. In these features it corresponds deeply with the nature of a spiritual reality, of a pervading consciousness. This is our explanation, it is the one which is consistent with the other phenomena described in this book and the one which corresponds with the varied aspects of human experience usually described as "spiritual".

Why science often can't find the evidence

A typical example of the debate over homeopathy took place on the BBC "Today" program, (23/05/07). On one side was Ray Tallis, a Professor of Geriatric medicine who is attempting to persuade UK NHS Trusts to abandon their co-operation with homeopaths. On the other was Peter Fisher, the Clinical Director of the Royal Homeopathic Hospital. Ray Tallis claimed that the "authoritative reviews" of published studies on the efficacy of homeopathy conclude that there is no proof of any benefit. He further described homeopathy as an example of "magical thinking" because the basis on which remedies operate is "impossible". In contradiction to this, Peter Fisher cited other reviews showing value from Homeopathy, in particular citing the views of the major health insurers in Germany – the "Krankenkasse" – that homeopathy is adding benefit and proving cost-effective.

This debate exemplified the prejudices and lack of understanding we have just examined, but we must deal explicitly with the issue of "experimental evidence" and of adequate design. Scientific experiments aim to isolate phenomena. They therefore work best in simple scenarios and are well suited to a situation such as that pertaining to drug-testing. In such tests a set of people with a single condition are tested against a single pharmacological substance. Usually this is conducted in such a way that some people get the drug, some a placebo, and that neither subject nor experimenter know who is getting what. This is known as double-blind placebo control.

Homeopathic remedies are not drugs. They are informational stimuli which promote reactions of self-healing by the body. They are not selected by homeopaths on the basis of correspondence to a disease diagnosis, but on the symptom pattern. A single disease may manifest with different symptom patterns in different individuals and be treated with different remedies that match those symptoms. Also, homeopaths would often not administer just one remedy because they are treating the individual holistically over a period of time. This means they would be administering other remedies which support the patient constitutionally or which deal with the underlying conditions that are seen, based on personal and family history, as antecedents to the currently presenting symptoms.

This complex and holistic approach does not fit the experimental methodology that medicine currently regards as scientific. In fact it is notable that when discussing the evidence, the anti-homeopathy camp cites those studies which do show positive benefit as being those which have the greatest weaknesses in methodology.

This sets up a situation where it is likely to be very difficult, or even impossible, for homeopathic medicine to be evaluated on anything like a level playing-field. Science is making the rules, and the homeopathic approach does not fit those rules. It goes to the heart of the case we are making – that the scientific model breaks down in these areas.

It is also appropriate to take more note of the issue of "placebo effect". The homeopathic effect is distinct from a placebo result because it is highly specific in the information that it carries. But since the effect is mediated through the bodymind response to the remedy stimulus, there is no way in which to distinguish it from a placebo effect. They look the same from the outside, just as from the ground it appears as if the sun orbits the earth. On superficial examination we might still think that Copernicus and Galileo were wrong. The evidence of effectiveness comes from cumulative health improvement and there is no means by which to demonstrate beyond doubt that the patient could not have healed spontaneously. In the complementary medical world, all healing is self-healing. That's the whole idea! It's a model of health creation, not of disease control.

The mere existence of the placebo effect speaks volumes regarding the relationship of mind and body in health and healing. It goes to the heart of the very relationships which complementary medicines work with and which science would like to convince us do not exist. Conventional medicine cannot explain the placebo effect – it has no model adequate to accomplish this. The perverse consequence of this is that rather than investigating an obviously powerful phenomenon, science works hard to eliminate it from all research. The recent advances in neuro-endocrine immunology have been showing science the way out of this cul-de-sac, but to date these advances are under the same pressures as homeopathy and the medical community as a whole has yet to embrace them fully.

The writer who conceived the idea of communications satellites, Dr Arthur C. Clarke notably said "Any sufficiently advanced technology is indistinguishable from magic". From our point of view, homeopathy is a technology which is in advance of scientific understanding – it is inevitable that they will label it as magic. But the use of the word "magic" in public debate is more corrosive because it is intended to imply naivety and gullibility on the part of homeopaths and patients. It is a subtle but deep insult.

As we have pointed out, there are experiments which show the "molecular memory" effect in water. We believe that these were done reliably by Jacques Benveniste and reproduced by several others, despite being extremely subtle. The work of Candace Pert and others described earlier underpins many of the relationships that are involved within the body (or bodymind). Generations of homeopaths, complementary practitioners and their patients, work daily with the evidence in systematic ways which conform to scientific principles, even if not to scientific "knowledge". They are not gullible or deluded and it is time that they ceased to be treated as such.

But there is one more aspect to the experimental evidence which we must address. You are by now becoming familiar with the connected world we are describing, one in which energy and information are communicated at all levels, through all types of organisms. You know of our view and of the evidence that the human mind has the capability to be a receiver and a transmitter, and that the homeopathic effect, like a hands-on healing, can be passed directly.

The implication of this for scientific experimentation is profound, and for the kinds of trials required to "prove" homeopathy it is catastrophic to scientific methodology. If the remedy can be transmitted to the patient by pure energy means, which includes the fact that the homeopath has it in his or her mind and might transmit it directly, there is no such thing as a placebo or a double-blind trial since all patients are receiving the remedy (at least in some measure). Equally, if the experiment is being run under conditions where actively hostile skeptics are involved, in principle the energy effects from their thinking can also interfere with the outcomes. This could explain why those who initially replicated the Benveniste experiments were successful, and why later researchers with a less open mind-set were unable to achieve results.

Lastly, practitioners know that the engagement of the patient with their healing process contributes to its effectiveness. To describe this as a part of the placebo effect grossly oversimplifies the relationship and goes against the whole thread of neuro-endocrine immunology described earlier. The bodymind relationship is too subtle and complex for this. The patient is not a mechanical object. Any good doctor knows this. It is the pharmacological approach which drives the scientific model. This is not the place and we are not the people to lead discussions about conspiracy theories. But we would be naïve not to recognise that drug companies are there, that they are influential in research funding and wield enormous influence with huge amounts of money at stake. At the very least this has to be seen as creating massive inertia and pressure against the changes we are promoting.

Science is a powerful tool and its experimental methodology is very valuable in sifting truth and understanding underlying mechanisms. But in the area under discussion it is not effective. For sure, we need to apply scientific understanding and analysis and to be systematic about our observations, which homeopathy does, and is. But the standard experimental tool is too primitive. The observer cannot be separated and the variables cannot be controlled. We would not use a chainsaw for brain surgery and we must recognise that there never will be an effective double-blind placebo trial proof for homeopathic healing. It's an inappropriate methodology. The healing processes described in the next chapter present even greater difficulty to science. They also take us back into the heart of the spiritual debate.

Review

The evidence from homeopathy fits centrally with all that we are saying. Homeopathy demonstrates:-

- A scientific and systematic gathering of evidence

- Epidemiological proof of effectiveness of remedies

- Consistent and specific relationship between substances and the conditions that they treat

- Transmission of a healing that is neither pharmacological nor a placebo effect

- Evidence of the passage of disease-related information across generations, not mediated by the genes

- The weakness of experimental approaches when dealing with alternative healing

- Dubious standards of scientific objectivity applied to its evaluation

The implications of these facts support the case that we are making for the types of connection and information carrying that we see as central to the science of spirituality and add another element to this consistent picture.

8. The routine of miraculous healing

In every culture and every medical tradition before ours, healing was accomplished by moving energy

Albert Szent-Gyorgi (1937 Nobel Prize Winner)

The mystery of life isn't a problem to solve but a reality to experience

Frank Herbert

Theme

It is common for our collective view of non-medical healing to be contaminated by incredulity. Many of us have been conditioned by science to regard energy healing, spiritual healing and many kinds of alternative therapy with disbelief. In this chapter we will attempt to create a new perspective on both what might be seen as a miracle and what should be regarded as ordinary. We will use some examples from direct personal knowledge and others from clinical observations in order to illustrate many remarkable features of healing and to complete the foundation of evidence that our science is required to explain.

The miracle of human existence

What if you are vastly more than you have ever conceived yourself to be? How confident are you that you know what your body is really capable of or what you as a human being might encompass? How many limitations have you been fed with and conditioned into which have nothing at all to do with who you truly are? What do we consider to be miraculous when it is truly ordinary, or take for granted when it is genuinely wonderful?

Humans have a patchy perception of the miraculous. Take, for example, the famous physics equation $E=MC^2$. It doesn't look that impressive, and perhaps the demonstration of its raw power – a Hiroshima-style unleashing of the energy locked up in a small ball of matter – has negative connotations that you do not want to think about.

Underneath that fact and the associated image lies the truth that your body encapsulates the raw power of a tiny sun. When, in the early stages of creation of the universe, inconceivable amounts of energy were formed into matter, every atom that was created locked up a tiny portion of it. Each atom was forged in the centre of a star. Eventually some of that matter found its way to be Jon, Juliana or you. If the energy inside your body were to be released, it would free up a larger power than Hiroshima. That portion of creation, that piece of something like a primeval sun, is inside you.

Alternatively, take a look at biology. The human body is the most exquisite miracle of organisation. Everything about it is extraordinary. Take a look at what our biological systems are able to do:-

- Starting from one single cell, we build an organism composed of trillions of cells.

- The new-born arrives with some capability of function and behaviour, e.g. nipple rooting response, suckling mechanism, eliminative functions, breathing.

- The organism will grow over two decades, in a co-ordinated way, at a largely predetermined rate typical for our species.

- Every one of those trillions of cells has a specific function, and is in its correct location.

- Each of those cells has sub-components concerned with individual internal functions

- Every one of those cells will take in nutrients, expel waste materials and regulate its environment

- Every one of those cells, when exhausted, is capable of producing its replacement

- The organism is capable of responding to malfunction of its own replication process, and deal with unhealthy cells. A cell, if damaged, will make it possible for the body to recycle its materials.

- The cells will work together to function as organs.

- Each organ will function within the context of the body, regulating its activity as required and in harmony with other organs.

- The body will consume energy as required, converting it into activity, or new tissue.

- The body will dispose of toxic by-products of its metabolism.

- The body will respond to environmental variability such as heat and cold to maintain temperature stability. It will regulate water balance, salt balance, oxygen supply and more.

- It further regulates these factors in relation to demands for activity, including the substantial changes required by emergency responses which range from peak instant performance to long-term endurance under deprivation.

- It will repair damaged tissue or broken bone, responding to injury with incredible rapidity

- It will resist and repel the attacks of viral, bacterial and fungal parasites, and cleanse the body from a vast range of environmental toxicity

- It will reproduce whole organisms to replace itself and perpetuate the species, carrying its immature offspring within its body and nourishing it there for nine months.

- It will display a variety of skills - intellectual, social and physical, in mastering and ordering its external environment

All the above are scientific facts. They are "everyday miracles" and we will revisit this list later in more depth. The things that we are familiar with no longer seem miraculous. Both the physics and the biology above are very hard to grasp. For most of us it is quite a stretch to visualise oneself as containing the energy of an atom bomb, or hold the image of 50,000,000,000,000 cells. Our day-to-day senses just don't go there. Part of our challenge in this book is to help you see a number of equally miraculous features of our world as likewise equally mundane, to take your senses towards places that they probably don't ever go. We faced this credibility gap in the discussion of shamanism. We face it again here in relation to "miraculous" spiritual healing.

In the previous chapter we described the systematic validation that homeopathic science has given to a particular form of connectedness. That connectedness has characteristics of being able to transfer defined energy signatures, to convey information from a substance to the human body systems and to transfer similar types of information across generations, independent of a genetic mechanism. We also referred to experiments which have shown that these energy signatures can be proved in the laboratory to imprint on a carrying medium such as water.

We focused strongly on homeopathy because of its systematic approach, one which maps as closely to an experimental methodology as is ethical when our subject is the health of a living human. In this chapter we will widen the examination of other modes of healing in order to broaden our understanding of this energy and these relationships. We have used the expression "energy signature" as a term of convenience. It is vague; in much the same way that "magnetic field" is vague. What can we do to create a richer description, and perhaps even have it be equally acceptable as a scientific term?

In keeping with some earlier chapters the first approach is illustrative. We will switch back into personal stories in order to provide some pictures which help the presentation of theory that follows. As previously we recognise that a few anecdotes do not make a science. At the same time we remind readers that to see these as "miracle stories" places them in the realms of the extraordinary when they are in fact examples of widespread and frequently reported phenomena. The limited extent to which these are validated by a body of conventional scientific investigation is a symptom of their wide exclusion from research. It is easy for scientism to present an argument that it is up to us to prove that the phenomena exist, not for science to prove they do not. This is logically correct. However there is also a duty on science not to ignore or dismiss from consideration the evidence that is available. In addition there is an obligation to fund the appropriate research into effects that are reported across all continents and cultures.

The next three stories come from personal experience. In chapter 1 we wrote of the training in intuition and to Jon's first experience of a psychic event. The founder of the training which gave rise to that experience also led courses in hands-on energy healing. This story begins with Jose Silva's 1983 healing course in London.

The timing of the course coincided with the rise in awareness of AIDS and two students in particular were acquainted with people who were ill or who had died from this disease. They called on others present to form a group to use their new skills to help with this rising crisis.

A protocol was designed which involved three visits per day by three group members together to the client. When visits started, "Peter" was in hospital, unable to leave his bed. His family had been summoned from Switzerland with the prognosis that he had about 4 days to live and the priest had given the last rites. His weight was down below 100 lbs, not much for a man of 6ft 2ins – even a naturally lean one.

Over the first week, the visits proceeded regularly with sessions of approximately

40 minutes of "hands-off" energy healing accompanied by meditative music and guided visualisation. Peter was visibly improving. Jon did not participate in the first "wave" of healing sessions. By the time Jon first saw him one week later Peter was already reported as more alert, and had some colour. He was receiving a lot of attention, a total of nearly two hours per day. This improvement continued, and after three weeks, he walked to the bathroom. After five weeks, he left hospital. He was still fit and healthy and apparently pronounced by his specialist to be free from any trace of HIV infection when Jon last saw him about 6 months later.

It is worth noting also, that during the first session that Jon participated in, Peter reported at the end of the treatment that he had felt a physical change take place in his mouth and that all the symptoms of oral thrush he had been experiencing had simply disappeared. It never returned. This specific relief of an individual disease pattern was a unique occurrence, but is mentioned to show both how rapidly a change can occur and that on this particular occasion the connection between healing and the presence of the healing team was quite specific. Because it was Jon's first time, the Candida healing was jokingly attributed to him (he claims no such impact), but for some time the more camp members of the group took pleasure in calling him "Candy".

This case was admittedly an isolated success. Only a couple of other clients were able to receive the same treatment regime, without similar results. Before long the group disbanded despite the success, largely because the time and effort required was too demanding for all concerned. Members were covering two different hospitals and a personal home; travelling from all around London at personal expense was plainly not viable. We would perhaps not tell the story if the healing had been less dramatic and also had it not been documented throughout by dramatic changes in the T-cell count by which immune system damage in HIV patients is monitored. Even though this was too demanding on a volunteer group, how much might be achieved by one which was funded and even resident in a hospital? A case like this begs for scientific research since such treatment would still be hugely cost-effective compared with any currently available alternatives. Note too, that the healers concerned were recently trained and none would have claimed any special gifts. This protocol is eminently capable of replication.

A second story once again has connections with homeopathy, but with a distinctive twist. It concerns a time when our young son was an infant of about 1-year old. He was sitting contentedly on Jon's lap during a meeting. Exciting things were being discussed and at one point in the meeting Jon jumped up to cheer, raising his arms (and son) above his head.

There was no reaction at the time, but sometime after midnight, the baby woke suddenly and began to cry – howling with distress. For ten minutes both parents sat with him, trying to figure out the cause for such a sudden eruption. Jon was scanning for symptoms and thinking about potential homeopathic remedies when a flash of insight took his attention back to the meeting. In the space of two seconds, the thoughts went through his mind – meeting – shock – find shock remedy - and he reached for his kit. By the time he had opened his bag, the crying had stopped, as abruptly as it started and the baby was sleeping again.

Sceptical readers – especially those who have raised children - will know that events like this can sometimes be hard to explain. It would be easy to dismiss what happened and Jon has questioned this himself over the decades since, but the experience was too strong to ignore. The two seconds during which the recognition of the cause and the treatment took place was such that experientially there was no doubt of the sense of connectedness and that the application of the "remedy" had taken place through a direct engagement. The energy required to treat the condition had been applied through direct connectedness and not through a prepared pill.

One more example is appropriate concerning the same infant son earlier in his life before he was on solids and when he had never experienced bottle-feeding. Juliana had contracted a severe gastric upset, possibly from food poisoning, as we were in a hotel attending a convention. The result was that she was mostly too ill to breast-feed and when she did manage it the milk caused our son to vomit. But by chance, a friend at the convention had recently weaned her son and was still producing some milk so was suggested as a wet-nurse.

This worked well, but by the evening our son developed a quite strong fever, which was very unusual for him. We consulted a very experienced homeopath who was also present and he advised that the wet-nurse showed strong indications of carrying the miasm that comes with family history of tuberculosis. As we discussed in the last chapter, such disease "imprints" are carried and maintained through the generations.

At this point things get even more "alternative". No-one at the convention carried the required remedy, which is not a part of any first-aid kit. Since the fever was not going down at all, we adopted a radical strategy. We decided to use hands-on healing to "program" the remedy into water, by consciously attuning to the nature of the remedy – its "signature" and consciously visualising the energy going into the glass container. The drops were then given directly to the child. Within an hour the fever was clearly subsiding and he was asleep. By the morning he was completely normal.

The picture we are attempting to paint with these stories is of a quite seamless connection which runs between individuals whether this be through the physicality of a substance like breast-milk or through the connection of "thought energy", as when the shock remedy was recognised. That connection also runs between individuals and other substances as shown when the energy imprint of a remedy is transferred by thought alone. Even more, it indicates that the miasmic effect referred to in the previous chapter can communicate not just between generations, but between one individual and another. (Note though that there may perhaps be some indication of susceptibility to the energy on our son's part and that without this he might have been unaffected.)

In the case of the AIDS healing it is not possible to be specific about how healing was brought about. However, we would point out that no touch was involved. At the time this was discouraged both because of uncertainties and caution around contagious transmission and because of the risk of dislodging the various tubes keeping Peter alive. The effect is not one of mere soothing. It should also be noted that during the first week he was not very conscious. Even when awake, although he had accepted the offer of this treatment, he was quite negative, resigned and dismissive of any likely success. At that stage this could not be regarded as an example of positive thinking.

From the Silva healer's point of view, the method of delivering healing is through visualisation. Simple techniques are used to achieve an atmosphere of mental and physical relaxation (technically intended to increase the proportion of alpha-wave rhythms in the brain). The mind is then engaged in a way that pictures the existence of healing energy and its transmission from their hands to the client. Each healer uses their own image. As an example, Jon's at that time was to imagine the silver lining in the clouds as a source of energy and to imagine it being pulled down as a pillar that enters the body between the shoulder blades and leaves through the hands. There is an assumption which applies to most such healing practices (such as Reiki) that the client's body will make the best usage of the energy supplied. Unlike homeopathy, where the apparent power derives from the very specific stimulus provided by a very small quantity of remedy, hands-on healing is typically intended to deliver a lot of energy. Typically healers may well experience a lot of tingling or heat in their hands, though this is not essential.

We don't ask or expect anyone to take the above stories as scientific proof. The fact that an experienced homeopath identified the relationship between the constitutional history of the wet-nurse and a disease-pattern in a child is not a scientific validation of anything. It's just an opinion. Since so little study has been performed in the areas we will cover, this is to be expected and we would argue that the cause of this lies in scientific thought-structures and scientific funding processes.

Others have presented volumes of evidence concerning many other healing methodologies. We will have to focus our attention elsewhere. However, we would not want you to think that the examples above are the only personal ones we could have given and we could find countless others of our direct acquaintance who would bear witness to similar events. Such miracles are genuinely everyday events which have been testified to by many people. We also point out their similarity to previous examples like those from Mona Lisa Schulz and Eliot Cowan, and that we have not included any of Juliana's examples with her own clients. Stories are merely stories, included as an encouragement to open-mindedness so that the scientific model we are about to present can be better understood.

Mind over matter

One of the most articulate, knowledgeable and respected writers in this field is Deepak Chopra. An American of Indian extraction, he qualified in conventional medicine. His cultural background also led him into profound exploration of the traditional ayurvedic healing system and into the meditative spiritual practices that are also strongly associated with that tradition. His blend of East and West informs all that he writes. Among his many books, "Quantum Healing" is the one we point to as most directly relevant to our text.

Deepak Chopra goes into great depth regarding the extent to which mind, emotion and physical form are intertwined, and showing much common ground with neuroscientist Mona Lisa Schulz. It is a book full of revelations and insights. All of it deserves to be included here, and in its absence, we are merely going to attempt to represent one of its points.

One thing that he reveals is the extent to which the mental process can influence body chemistry. A simple but graphic example he cites is the case of Timmy, a boy with Multiple Personality Disorder (MPD). MPD is a strange syndrome in which a single body exhibits two or more personalities. In the case given by Psychiatrist Daniel Goleman, one of Timmy's personalities is allergic to orange juice, and breaks into hives when he drinks it. If Timmy's "normal" personality returns while he is in the middle of the allergic reaction, the itching stops right away, and the blisters begin to subside. On the other hand, if "normal" Timmy drinks the juice and the other personality appears while he is still digesting it then, and not until then, the allergic reaction will break out. You probably think of allergies as being a direct biochemical response to a purely physical cause, especially if you are a hay-fever sufferer, but Timmy's response is totally at odds with that way of thinking.

This feature of bodymind can be worked with. In 1971, Dr Carl Simonton was inspired by his experience of the Silva Method to try out a process of visualisation with one of his patients, a 61-year-old man with advanced throat cancer. The man was encouraged to enhance his radiation therapy by visualising his cancer, and choosing an image which appealed to him, showing the attack of his immune system on the cancer cells. The man chose to see the white blood cells of the immune system like a blizzard of snow, covering the black lump of his tumour. This process was repeated a few times per day. He had fewer side-effects than expected from the radiation, and after two months, the tumour was gone.

It is an unfortunate and puzzling truth that although there were other successes from this methodology, and the Silva network had many stories of such reversals in cancer and also in arthritis, the statistical success rates from Simonton's research and others of a similar kind are quite poor. Nevertheless, we include it, because it reveals that a process akin to that which occurs spontaneously in Timmy has the potential to be instigated by conscious choice. In both cases, the mind is shown as having a direct impact on processes otherwise seen as "biological" and "unconscious".

In coming chapters we will encounter the part of body "intelligence" which is continually active, cleansing, balancing, repairing, nourishing and protecting. Many scientists still choose to conceive of this intelligence as no more than the result of a chemical accident (or series of them) that happened to produce life. That point of view is hard to refute - you either choose to believe it or you don't. But when we are dealing with conscious intelligence, we need to be rather more careful in our consideration.

It is ironic, considering that each of us lives with an experience of consciousness all of our lives, that we have almost no idea what consciousness really is. As we discussed before, some scientists view consciousness as purely an effect, a description that we overlay on a biochemical process. It is clear that we share with others the different view that it is a crucial feature of humanity, and one for which a specific explanation is required. As we said in the introduction, far from seeing consciousness as a side-effect, we see it as the prime cause and the simplest explanation that is consistent with the facts.

You might view this choice as being just as deep a personal decision, and one taken with as little hard evidence either way as whether you have chosen to believe in God. It may almost be the same choice. For one person, the glories and intricacies of the natural world are clear evidence of a creative intelligence at work. For others the same data are evidence of the spontaneous, random occurrence of the most miraculous, self-sustaining chemical reaction imaginable. In one, the question "why are we here?" yields simply "chance". In the other, a creative consciousness is involved.

The difficulty in distinguishing between these two viewpoints lies partly in the fact that consciousness cannot be located. It is indistinguishable from the physical activity which generates it, or which provides our evidence of it. As Dr Chopra points out, the implication of Timmy's allergic reaction through one of his personalities, is that the white cells await the arrival of the orange-juice, and then a decision is made whether to react. For this to be the case, the "intelligence" (Dr Chopra's word) has to be in the cell. "Moreover", he says

> "its intelligence is wrapped up in every molecule, not just doled out to a special one like DNA, for the antibody and the orange juice meet end-to-end with very ordinary atoms of Carbon, Hydrogen and Oxygen. To say that molecules make decisions defies current physical science - it is as if salt sometimes feels like being salty and sometimes not."

We will close this chapter on the miraculous with two more stories – just to ensure that we are stretching your credibility to the maximum. This is not done lightly. This book is about "pushing the envelope", a phrase used by Chuck Yeager, the first test pilot to fly faster than the speed of sound. We might hesitate to include the first one had we not both witnessed it ourselves.

We use the sound barrier comparison with purpose. As Yaeger was flying close to the threshold of the "sound barrier", his plane began to vibrate as if it would shake apart. He backed off, and when on the ground asked the design engineers what was going on. They had no idea, since this did not show up in any calculations. It is a mark of Yaeger's bravery that he was willing to take the plane up again and push it through the turbulence, which got worse as the threshold approached. Once beyond, the flight became smoother than ever.

For humanity our envelope is the conventional view of the world. It is as invisible as the sound barrier and as difficult to penetrate. For some of us the resistance may be quite strong. As Gloria Steinem remarked "the truth will set you free, but first it will piss you off." Yet alongside that sits the huge possibility of the world beyond the envelope, a world in which the things we now think of as miracles may turn out to be routine, and become part of the repertoire of normal human activity.

"Miracle" Healings

In Essex, UK there is a healer called Stephen Turoff. He has been widely reported on in recent years, both in print and on television and there is a full autobiography available. Stephen works as a "psychic surgeon". He describes himself as being a channel for a long-dead German physician, and working under his control "like a puppet". ("Kindred Spirit", June 1996). His capabilities too are beyond belief, such as treating a man unable to walk for three years due to a brain tumour, who stepped out of his wheelchair and walked. Another story is of a woman with a heart complaint, whose doctor had taken X-rays and warned that the heart could fail at any time. After treatment by Stephen Turoff, that same Doctor reacted with shock to follow-up X-rays, not just because the patient's heart was fine, but because it appeared to be a quite different heart. We know personally several people who have received treatment from Stephen, and some of these had clearly benefited.

In one case we both observed an operation on Juliana,'s mother, a lady then in her late seventies who walked into the treatment room with the aid of a stick. The operation dealt with long-term severe back pain by removing a sliver of bone from her spine. We were about five feet away with a clear view. We could hear the scrape as the bone was cut. The operation took place at high speed, and the patient was back on her feet ten minutes from the start, with a small line marking where the incision had been, and no bleeding. For the duration of a one-hour journey home she was uncomfortable and recovering from the shock.

When she got home, she had been in the lounge for a few minutes when she noticed a small piece of debris on the carpet. She bent over to pick it up – bending from her back. She was clearly quite unaware that she was doing anything out of the ordinary, but our jaws dropped. She would have been quite unable to do this a few hours earlier. It was very apparent that her mobility was considerably improved, and this sustained in the following months. There is no doubt in our minds that Stephen Turoff is genuine. You could fake the operation with conjuring tricks, but you cannot fake results like that, which is perhaps why his waiting room always has very long queues.

Stephen Turoff was an "ordinary" man in every sense, except that he grew up with a strong sense of a personal connection with God. He had practiced "laying-on-of hands" healing for fifteen years prior to "Dr Kahn's" intervention in his life, and though he called from deep within to make his connection manifest in his work, he had no idea that he would do what he now does. Could something like that happen to you? Can you envisage creating such a connection with the "source"? If you could, what would you like to be able to do? What connections would you like to make with the "resonances" which would enable you to do it?

We do not see it as being of consequence here to know what Dr. Kahn is or whether he is real or imaginary. The fact that Stephen Turoff is able to attune himself, and to use energies and resonances in the way that he does, is all that matters. It is of little importance what he thinks, or feels, or visualises in order to achieve that. We emphasise this point, because in our view the power is with the individual. Stephen Turoff sees it this way too. When asked how he does what he does, he now says "because I am God". Then he says "and so are you". If you ask him "Why can't I......?" he will likewise say "because you do not know yet that you are God". Given his track record, who has the authority to disagree? His successes testify to the potential for great powers in each of us.

"Miracles" for all

It is the power in each of us that is of most significance of all and this chapter would not be complete if we could not give at least one example of a methodology which is showing itself to be teachable and generalisable. This story begins with a single individual, Dr. Joe Dispenza who is a neuroscientist and chiropractic physician. Here, in words taken from his own website is the background to his story.

> "It sounds crazy, but in 1986 I had the privilege of getting run over by a truck in a triathlon. When I received the diagnosis that I broke six vertebrae, that I had bone fragments on my spinal cord, and that I probably would never walk again, I had to make some important decisions. After I opted against a radical surgery recommended by four different experts and facing the prognosis of paralysis, I left the hospital with only one conviction: "The power that made the body, heals the body." My mission was to make contact with this innate intelligence, then give it a template or a design with very specific orders and finally surrender my healing to this unlimited power.
>
> I really had nowhere to go at the time of my accident and I did not have many things to do, so it was the perfect opportunity to experiment with using my mind to

heal my body. For two hours twice a day I went within and I began creating a picture of my intended result: a healthy healed spine. If my mind wandered to any extraneous thoughts, I would start from the beginning and do the whole scheme of imagery over again. I reasoned that the final picture had to be clear, unpolluted, and uninterrupted for this intelligence to take my condition to the next level.

Over the course of ten weeks, I experienced a wonderful and veritable healing. At eleven weeks, I was back in my office seeing patients again without surgery or a body brace (both of which were recommended by the physicians at the time of my injury). As a result of this experience more than 20 years ago, I have spent the remainder of my life investigating and researching the mind-body connection as well as the concept of mind and matter"

In our chapter on intuition we described how Mona Lisa Schulz is able to map the connections between brain chemistry and physiological problems. Dr Dispenza has similar expertise in brain function, and has learned how to use those connections so that instead of creating sickness, they produce cures.

The brain and nervous systems are not fixed. Neuro-scientific research indicates that experience can actually change both the brain's physical structure (anatomy) and functional organization (physiology). Our brains do not distinguish between external experience and experience that we generate from our own thoughts. This principle, known as neuroplasticity has been drawing increased research interest in recent years and also underlies the increasing use of cognitive behaviour therapy. To emphasise, the research shows that it is not just information storage that changes; the alterations are visible in the physiology.

If, ten years ago, you had asked a typical brain surgeon whether such change could happen you would have received the conventional wisdom that this mind-to-body causation was not possible. Our references above to Timmy's hives, to Carl Simonton's patient and to Dr Dispenza's personal healing all indicate that it is and that it goes way beyond changes to brain physiology, beyond the chemistry of allergic response and immune cells, all the way to structural change in the densest and seemingly most fixed parts of the body structure. Not only is there brain plasticity but also body plasticity.

Understandably, Dr. Dispenza does not make any public claims regarding the extent to which his methods are effective in others but we know others who attend his trainings and we hear some remarkable stories. The techniques that he teaches are those of deep meditation and visualisation which he used on himself. At one level they are not miraculous at all since they require deep focus and considerable attention over time. At another level, because the techniques are simple and teachable they are miraculous in comparison to many aspects of technological medicine.

We have indicated that our core purpose with this book is to place a scientific context around stories of this kind and to give corroboration that supports that science. This chapter has strayed as far into personal anecdotes as we are willing to allow, but we would not have wanted to leave out the very experiences which have led us to our viewpoint. Even if you are still unsure about this reality, we invite you to continue the journey. Maybe it will all seem more believable after we have described just how these processes work – that there is a science of this "spiritual" healing. They work because they are an aspect of how the universe itself works.

Although our approach has been scientific up to this point, we have been in the "soft" sciences, assembling the outward evidence. It is at last time to return to our core agenda and describe just what kind of physical organism experiences these things and how it functions biologically. We must gain some understanding of the physics that makes this universe possible and offer a scientific model for how both that universe and we humans come to be. It is a picture by turns exciting, complex, multi-layered, wonderful and rich. Beyond the complexity we will draw out some simpler patterns which make deep sense of it all and supply the mechanisms which all of these healing processes are utilising. Underneath the miracles that happen we will find a deeply miraculous and connected world.

Review

There is much in the world that we are accustomed to, take for granted and do not think of as miraculous. There are other aspects to the world which we are in the habit of treating as extraordinary which might well be regarded as mundane if more of us had the training or the personal beliefs that empower them.

The examples quoted continue to hammer home the thread of connection, that information can pass at the material level (through pills or breast-milk) but equally well without physical contact, as when hands-on healing or visualisation are involved. Energy healing can have almost instant results, and equally be used over a longer period in the most serious of conditions.

Within the body, psychological information at the level of a complete personality is shown to dramatically affect physiological response, to an extent where it appears that each cell is affected by the psychological change, and the consciousness appears to be present in, or influencing the cell directly. The most capable of healers are capable of delving into this reality in order to firstly diagnose a condition and then to "operate" surgically on it, without any blood loss and with instant healing of the wound and no scarring. The healer we witnessed is performing such operations hundreds of times per year, with widely attested results. Other teachers are supporting larger numbers so that they can achieve similar results for themselves.

This chapter closes our wide-ranging exploration of the evidence both old and new, and the experiential stories which illustrate the connected world that we are presenting. From here we move into the territory of science itself where we will see how biology, evolution, cosmology and physics all need to be viewed differently than they are now and are all susceptible to interpretations of the evidence that support our presentation of reality.

Section 2 : Our biology

What was your face before you were born

When the heart bursts into flame

history completely disappears

and lightning strikes the ocean

in each cell

There before origins

when the double helix

is struck like a tuning fork

there is a hum

on which the universe is strung

Stephen and Ondrea Levine   *Embracing the beloved*[12]

---

[12] Information regarding Stephen and Ondrea's work can be found at www.levinetalks.com

9. Simplexity and the taming power of the small

All life is an experiment

Oliver Wendell Holmes

A single group of atoms existing only in one copy produces orderly events, marvelously tuned in with each other and with the environment according to most subtle laws... we are here obviously faced with events whose regular and lawful unfolding is guided by a 'mechanism' entirely different from the 'probability mechanism' of physics

Erwin Schrödinger   *What is life?*

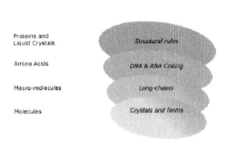

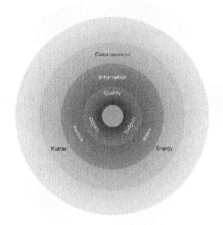

Theme

Questions such as "who are we?" and "Where did we come from?" seem to require answers at the high level, derived from our perceptions of ourselves as whole organisms in relation to our external world.   Throughout this section we explore a contrasting viewpoint, that in order to understand who we are we also need to see what is inside – the origin of life in single cells and how those cells function together in trillion-celled creatures.   It helps if we understand how the apparent simplicity of a single being which perceives itself as one conscious entity emerges from huge underlying complexity.

The section's first chapter explores the concepts we are working with. We look at some of the core ideas and questions, the extremes of scale, the types of evidence available in order to create a high-level map of the territory, a helicopter view of the journey that we are about to take. We have several strands to weave:-

- Knowing who we are requires knowing where we came from (evolution)

- This must be integrated with understanding how we "work" (living functionality)

- Knowing how we work connects to how we were "made" (individual development and growth)

The way Jon is made both resembles and differs from the way that his father was made and this change is a tiny reflection of the way in which planetary life changes over long time-scales. We come full-circle to evolution, and the question of where we came from.

None of this happens in isolation. Every thread in this tapestry occurs in relation to an environment. Each cell relates to what is outside it. Groups of cells such as organs function in relation to the bodies they are part of. Each organism relates to its surroundings. Groups of organisms function together in eco-systems where they are in relation with each other, with the physical surroundings of the planet (geology, climate) and with the change in both. The change may be daily cycles, seasonal changes or alterations in climate or the development and extinction of other species.

The answers to questions such as who we are and how we came to be must incorporate all of these layers and must understand how they form a whole. It is a significant task. It is also an exciting one because it opens up the field of our biological possibility.

What if most of what you have been told about genetics, evolution and human biology was seriously misleading? What if despite decades of brilliant work and quantities of hard-won facts, you have been presented with a mythology and misinterpretation such that your picture of who you are and how you came to be is inaccurate? In this chapter we will shine light into these areas in a way that is likely to transform your understanding of biological reality.

Simplexity

If we are to provide an adequate picture of the biological realm we will need to go further into the depths of the science than in any other section. The detail provides the keys to our understanding and the sharp point which penetrates the "envelope". We have borrowed the word "simplexity" from Jack Stewart and Ian Cohen, whose thinking has influenced this section strongly. We are attempting to show how an entity that we think of as a single part – a human being – emerges from the complex detail which forms it. Whilst is not crucial that you understand every piece of the detail it is essential to be aware of their existence and to have a grasp of the number of levels and relationships involved. So please bear with us, and don't give up if some parts are more technical than you would prefer.

Our chapter heading "The taming power of the small" also borrows the name of a hexagram in the I-Ching. We use the phrase to encapsulate the way in which this big organism is managed through microscopic events. The behaviour of single cells and the co-ordination of information flow via activity beneath the level of those cells are central to the coherence of that big organism and to its way of relating to the environment. Little changes have big consequences. As Anita Roddick observed, the power of the small is only overlooked by someone who has never attempted sleep with a mosquito in their room.

Everything pertaining to the human spiritual experience is mediated through our incarnation. Whatever anyone may imagine or conjecture regarding the existence of a soul independent of the body, the fact that we are beings of flesh, blood and bone and that we have and describe all our experience through nervous system, sensory apparatus and cognitive interpretation puts biology right at the centre of our narrative. We encountered some of this truth in Chapter 4, but the endocrine system is just one among many layers. We will examine several other layers to this relationship in order to show where and how spirit and biological matter meet.

The genetic strands

In the past 160 years the scientific world has been rightly excited, first by Darwin's insights into evolution and more recently by the discovery of how DNA works. This excitement, however justified it may be, has led to exaggeration of what is known and proven. Some of the images have been further distorted by simplistic or inaccurate film or TV plots. The underlying scientific work has been phenomenal and the complexity of the picture is indeed widespread in the scientific literature, but the public presentation has been misleading, or even unconsciously oriented to exclude our view of consciousness and spirituality.

As Richard Dawkins says of himself in his preface to "The Blind Watchmaker" his book is "….not a dispassionate scientific treatise. ….."I may not always be right, but I care passionately about what is true and I never say anything that I do not believe to be right." Although we would not dispute the honesty of his intention there is a risk involved – one that we potentially share. Beliefs are dangerous as soon as they begin to lead people into stating as truth things that are unproven or half proven. They are damaging if they become a spurious certainty which causes perceptual filtering such that evidence which does not fit the belief is ignored or avoided. We are about to show the ways in which science has gone down this route, with Dr Dawkins among those leading the charge and taking the Western world along with his powerful and articulate presentation. We need to open up this narrowed presentation and re-introduce some uncertainties, mindful of our earlier assertion that questions are often more powerful than answers.

Uncertainty 1. In the face of a story which refers to "genes" as if these are always simple discrete units determining identifiable "characteristics" in an organism, we will have to present the full tale of the way that organisms develop, and show how misleading that simple story truly is.

Uncertainty 2. Beyond the supposed relationship between "genes" and "characteristics" we have been led (or at least allowed) to believe that DNA and chromosomes determine everything about the organism. They are presented as a full and complete instruction set which is in total control. There is no flexibility in this model of how an organism develops, nor any responsiveness to the environment. This is simply not accurate.

DNA is represented as a "code", one which decodes itself as it constructs organisms. Just what can codes do, and how are they "decoded"? How do the features of codes really work, and what does this mean for the relationship between developing cells or developing organisms and their environment?

Uncertainty 3. Cells are presented as if they are just blobs of jelly and their internal functioning is easily overlooked. This is far from reality. Since we are composed of huge numbers of cells, the real story of how they interact with each other and how they mediate our interaction with the corporate cellular environment is central to how we function.

What are the processes of life in terms of the second-by-second functioning of a complex organism, and what do these processes tell us about the nature of life on the grand scale and how it is organised? We will need to include in our answers the way in which physics affects biology and show some aspects which have been seriously neglected.

Uncertainty 4. Another function of genetic material, besides building organisms starting with a single cell, is to guide the process of inheritance from one generation to the next. The core of Darwinist fundamentalism is that all life evolved to its current state without any kind of guidance system. This suggests a wonderfully complex but ultimately mechanical relationship. On one hand we have random and spontaneous change through re-combination or mutation of genetic instructions. On the other we have the process of natural selection, by which some of those changes will result in organisms that are more successful than others in reproducing. We are not about to repeat any of the arguments that Richard Dawkins has so eloquently demolished. Nevertheless we will raise some questions around the absolute absence of guidance in the fundamentalist viewpoint. Here the core question is not about whether Darwin was right. It is about the boundaries within which he was right. Evolution is presented as both entirely random and entirely mediated by changes to chromosomal DNA. To what extent are these presentations open to question?

Uncertainty 5. Over the decades Richard Dawkins has extended the argument beyond the pure science of how evolution takes place. His campaign is not purely in favour of fundamentalist Darwinism, it is also against a primary alternative, the existence of a "designer" in any form. His view that no designer is required is a strong platform on which he builds a very different argument – that "God", in the sense of a single creative being, does not exist. We won't challenge this view. He then takes the argument further and seeks to exclude any kind of spiritual component to life. These are so strongly bound in his presentations that they have come to seem almost as if they are the same thing. We need to disconnect the two, because the second contention is the one which we are challenging very deeply. Rejecting the biblical Yahweh, and eliminating all possibility of a non-material, spiritual or "Divine" realm are two different things. The difference is quite crucial.

In our previous chapters we have presented evidence which needs to be encompassed within the scientific model. We have described Cleve Backster's experiments (Ch.5, P.72) which demonstrate the presence of consciousness in plants. We have quoted the understandings from Malidoma Somé, Eliot Cowan and others which depict the presence of some form of spiritual being within the plant kingdom. We have shown how others like the Findhorn gardeners have worked with these beings in a way which influenced the development of plants –

the 8Kg cabbage phenomenon. In our chapter on homeopathy we showed the evidence of disease suppression in one generation being carried somehow to the next. Where does all this evidence find space within the biological model?

To answer that question we must look at other theories which indicate that there is a missing element in the picture of consciousness and attempt to fill crucial gaps in our understanding. We will summarize and draw together the strands in a way which shows the potential for non-visible energies to play a part in all of these biological processes; we will show the ways in which such subtle and small influences have the potential to drive larger and more visible events.

The size of the task

Thus we embark on such a discussion of genetics, development, inheritance and evolution with some trepidation as we attempt to cover scientific evidence which ranges from the microscopic investigation of molecular-level events to the grandest scale of evolutionary evidence, based on several billion years of organisms living on planet Earth. Throughout we will keep in mind our core questions, those which have driven both religious and scientific thinking for millennia.

Who and what are we?

How did we come to be?

What, if anything, lies behind our existence?

The science of genetics is now central to the answers that the scientific model offers and it comes at a particular place and time in the development of thought.

In mathematics, early insights came from the Greeks. Beginning with Pythagoras about 600 years BC, they began to discover the way that space is ordered. Early models of the cosmos, which can be seen in the construction of astrolabes, viewed the sun and planets as moving in circles, because that was the most elegant way they could imagine it to be. It was a model that humanity found helpful for over 1500 years. Similarly, the shapes of platonic solids, the cube, tetrahedron and octahedron could be seen mirrored in the fundamental structure of matter, like a crystal of salt. God's universe was ordered by the harmony of the spheres and the unbreakable relationships of numbers. These discoveries travelled into the Arab world and were central to Moslem culture and art through the European dark ages, as well as to architecture in the use of oriental sacred geometry.

These mathematical relationships are timeless, or perhaps for all time. The art in

Moslem temples, however beautiful and glorious, is regular and patterned. Pictures have no perspective. Music of that time is linear without structural depth. It is only in the centuries leading to the renaissance that this begins to change as Moslem culture and mathematics re-emerges into Northern Europe. Art acquires a sense of three-dimensional perspective and astronomy shifts its viewpoint with Copernicus' revolutionary perception (1543) that the Earth is not at the centre of the universe, but rather revolves around the sun and Kepler's recognition (1609 and 1618) that planets move in ellipses. Mathematically these insights involve a new understanding. Perspective recognises the laws governing the way to map three dimensions on to a two-dimensional surface. It is a new way of seeing. Elliptical motion brings the fourth dimension - time - into the mathematical models of the cosmos and this develops further with the invention by Newton and Leibnitz of what we now call calculus.

Biology lagged behind physics and it took a further 400 years before it made the shift from a static view to the dynamic one. The biblical model of biology is that God made everything and that all creatures were here by the seventh day. Biological science was about cataloguing species and labelling organs, with some recognition of cyclical change where everything comes back round the same. Time only really enters biology properly with Darwin.

It is hard to overstate Darwin's genius. Much of the material now viewed as evidence has only been gathered since his time, and examined in the light of his vision. "On the origin of species by means of natural selection" was published in 1859. His brilliance was to look at a snapshot of nature in time, a small portion of all the species in existence, and recognise some core truths answering the fundamental questions we ask in this arena, questions which mirror those above.

How did all these different living organisms come to be on the planet?

Why is one creature different from another and how do those differences arise?

Where do we (humans) come from and what is our relationship with the rest of creation?

Since there is a creation, is there a creator?

Bear in mind that the mechanics of inheritance were not known until later. At the level of individual characteristics this was revealed by the work of Gregor Mendel, published in 1865 but barely recognised for decades after. His understanding of "hereditary factors" paved the way for modern genetics. The molecular chemistry that underlies this activity begins to be understood with the discovery of chromosomes by Flemming (1882) and the sequences and combinations of nucleic acids in a strand of DNA in 1953. Darwin could not have had any knowledge of how the biological development of the individual organism takes place, nor of the mechanisms by which one generation is related to the next.

So Darwin's insight, which has driven scientific understanding for almost 160 years, was a huge and brilliant step. In place of a static catalogue of living organisms, pre-determined in six days by the Old Testament God and with humankind his finest creation, came a moving picture that unfolded over time. Species change. In time they differentiate, and features are accentuated or lost according to how well they enable the members of that species to survive and reproduce in their environment. That environment is itself changing in relation to climate, geology and alterations in other species.

The question is not whether Darwin is right or wrong; we are not questioning Darwinism. It is about the boundaries to this insight and about some of the slants within it; our question relates to fundamentalism and rigidity in the way others present Darwin. Both Darwin and his contemporary Alfred Russell Wallace, who independently formulated similar views, were strongly influenced by their intellectual culture. We know more now than they did. Darwinian fundamentalism takes the principles that Darwin put forward and argues that they are complete and that nothing more is required to explain everything we see in either the development of an organism or the evolution of life. We intend to show that this fundamentalist view is unsustainable, and is imprisoning biological science as well as wider understanding of life itself.

Echoing the two core questions of who we are and how we came to be here are two strands to our story. These are the biological mechanics of what we are and the evolution over time of who we are. Both are huge subjects to cover and they are linked as closely as the two strands of a DNA double-helix. We will try to keep the strands clear, but we cannot separate them entirely. While doing so we cannot help but meet the even deeper question of the possible existence of a creator. Then with or without a creator we also face the question of an underlying "plan" or inevitable logic that drives the process towards higher complexity in function, behaviour and mental capability? Within all these, and central to the theme of this book is the question of whether there is thread in this tapestry that has any relationship with a "spiritual" view of the world?

The scientific challenges

An anonymous humorist once said of Christopher Columbus that when he left for America he didn't know where he was going, when he arrived he didn't know where he was and when he returned he didn't know where he had been. A voyage into the evolutionary record is not much better. Some of the work done by those who specialise in this area is amazing and brilliant, but their successes come in the face of major evidential challenges. So much has to be inferred, conjectured and tested against this limited information.

A few billion years of organic life have left relatively few visible traces. Those we have are patchy and sporadic, such as fossil remains. Whole categories of creatures have come and gone (e.g. dinosaurs) and it is certain that many species left no trace at all, particularly aquatic ones which end up as mud. Even where evidence exists, we see mainly the hard-tissue remnants (e.g. bones and teeth) and quite obviously the organism is dead. We do not see how it grows, or witness it move, breathe, eat etc. Our picture of the environment in which it lived is also threadbare.

Our picture of the timing of any supposed events is equally fragmented. We know that an organism existed but not necessarily when it first existed, what the true geographic boundaries to its existence were or when it became extinct. Note too that the geography and geology themselves have changed continuously over the millennia involved.

Many of the conclusions reached about genetics and evolution also require considerable understanding of how to classify the remnants. In human taxonomy (the branch of science which catalogues these classifications) debates have ranged over decades regarding the precise significance of a few bones – whether they represent something pre-human, late-ape, or just where in-between they belong. As stated by researcher J Shreeve, 'Fossils are fickle. Bones will sing any song you want to hear'. If you would like an example, web-search the virulent debate over "Toumai", but expect to find more heat than light. Such complexities of classification, affecting the entire historical record, reduce the sharpness of focus on even the patchwork of material that we do have. Investigators are presented with a big challenge if they are to draw reliable conclusions in this area.

But since evolution works through genetics, our first challenge is to understand what is claimed to be known about genes and genetics as distinct from what is really known. Here the challenge is not lack of evidence, but the opposite – vast volumes of intricate detail which we must slice through to reveal the crucial features. Fortunately there is a clear if complex pattern discernible beneath the detail. As this is revealed, so too will be the areas which are relevant to our theme of a relationship with the world of subtle energies and information content that underpin our notions of consciousness and spirituality. We will find the non-physical mind, and meet what Gilbert Ryle called "The Ghost in the Machine" and what we think of as the spirit of the matter.

Life as a lottery

People like Richard Dawkins and Stephen Jay Gould have written whole series of books delving with passion and erudition into aspects of these issues. In attempting to cover our central "spirituality" question in a few chapters we will have to cut to the chase. We will abbreviate, summarise and inevitably over-simplify some parts of the discussion. This is a risky business. Professor Dawkins in particular is ruthlessly scathing when demolishing opponents who over-simplify.

We agree with much of what is said by Gould, Dawkins and other scientists who have developed and extended the understanding of evolution and genetics based on the Darwinist model. We are not about to argue for the view that there is an external creator as literally depicted in Genesis. We acknowledge that it is possible for complexity to emerge through subtle change and selective pressure, and that the existence of wonderful design (e.g. the eye) does not imply the existence of a designer as such. The central questions we address are the nature of the subtle change involved and precisely how, when and through what mechanisms the environment influences that change.

As stated above, the conventional view from these and other campaigners is that random change, together with selective pressures on survival to the next generation for genes (and / or the species that contain them) is a sufficient explanation for all of life's wonder and diversity and that nothing more is required. The image that Dawkins borrows from Paley of the "Blind Watchmaker" encapsulates some of the wonder that such a wonderfully complex tapestry could indeed arise through a purity of chance, chemistry and eons of time. The detail of his writing demolishes, one after another, the arguments that there has to have been a creator, or indeed any natural law that drives gene selection or evolution in a pre-destined direction. If nothing more is required then it is against scientific principle to add anything. This model of Darwinism is made to appear so complete that it is very hard to challenge.

Nevertheless we must challenge it. The evidence does demand that we add something. And even if it is only 0.1% incomplete, the fundamental difference is as significant as when you walk into your living room at night and switch the electric light on. Physically the only thing that has changed is a small amount of moving energy in a section of wire – everything else is as it was. But life with the light on is very different. Our search is for something that lies in between a creator god and utter randomness. Fortunately that is a big space to play in, and it starts in the cell.

The complexity of cells

The science of genetics seeks to explain several related aspects of biological existence. The "what we are" aspect is about the mechanics for controlling and managing a complex organism. The "how do we come to be" has two perspectives. One is the development of individual organisms. This is a lot to explain when you consider that each of us started from one single cell. The other is our evolution and the introduction of changes to that development over the generations.

Let us break into some of that complexity. Multi-celled organisms are a miracle of existence and it helps to be aware of all the things that we and other creatures can do biologically. At the start of the previous chapter we listed the many miraculous things which our biological systems are able to accomplish. We refer you back to this list and ask that you continue to keep it in mind.

Just as miraculous, and maybe more surprising is that when you compare a multi-celled organism with a single-celled creature, or compare it with one of its own constituent cells, the list you get is not much different. As Dr Bruce Lipton puts it in "The Biology of Belief":-

> "Each eukaryote (nucleus-containing cell) possesses the functional equivalent of our nervous system, digestive system, respiratory system, excretory system, endocrine system, muscle and skeletal systems, circulatory system, integument (skin), reproductive system and even a primitive immune system, which utilises a family of antibody-like "ubiquitin" proteins".

As above, so below. Cells are not blobs. We hesitate to use the word "intelligence" here as it can so easily be taken to mean we are saying that individual cells think. But it is hard to find a good substitute. There is deep capability akin to intelligence built in to anything that shows such functional complexity and that manages itself in relation to its environment to such a degree. Perhaps, like the complexity of the eye, it could arise randomly. Or perhaps there are other factors.

We are accustomed to looking at whole organisms, and we need to remember that organisms may co-ordinate many cells together for a purpose, but nothing happens unless it is facilitated by cell-level activity. As throughout our story, it is necessary to grasp many different scales simultaneously. One helpful perspective is to see where cells came from.

Back to the very basics

Here again we ask the basic question "What is life?" but in a more specific sense than before. What we mean here is – at what point does "life" become recognisable as distinct from the background of what we see as non-living material existence? Our normal descriptions don't attribute life to lumps of coal, but they do so to algae and bacteria.

As best we know, this planet started as a ball of molten material which cooled and eventually formed a ball of rock with an atmosphere. This is not the atmosphere we are familiar with, having no oxygen, but quantities of methane and ammonia. An experiment by Stanley L. Miller and Harold Urey in 1953 showed that a flask containing such an atmosphere, and having electric current passed through it in a way that simulated the lightning storms believed to have existed at the time, would start to form carbon-based organic compounds and eventually amino acids of the kind that are required to make proteins and which are the basic prerequisites for cellular life.

So where and how would we define the transition from amino-acid slop to a living organism? The chain of evidence is extremely thin, and the next stage that we believe we are capable of recognising is the presence of fossils in 3-billion year old rocks, which look similar to what we know now as bacteria. To reach this distinguishable state requires the object in question to show itself as having a boundary. That boundary is the cell membrane. It is this which defines an object that is capable of organising itself in distinction from its environment. The most primitive form of organic life, we suggest, is first recognisable when there is some sort of autonomous entity. The cell membrane is more than a marker in space though. It is not like a plastic bag containing the cell's "jelly"; it is an active component of the living process and the means by which the cell maintains its autonomy. This is the boundary at which environmental influences on the cell take place.

The "intelligence" in the cell membrane

What we are saying is that this relationship between organism and environment is not merely as old as life itself, but rather is the definer of life itself at its most basic, and that the relationship is facilitated by the cell membrane. The membrane is so thin that it is barely visible, seven millionths of a millimetre thin, and can only be detected using an electron microscope. But functionally it is highly complex and contains a multitude of molecules called phospholipids. These form a barrier to the passage of molecules in and out of the cell. Within the barrier are a further set of molecules, the Integrated Membrane Proteins (IMPs) which are its gatekeepers, controlling what may pass through the phospholipid barrier. They are classified as being of two kinds. There are receptors, which detect environmental signals – the equivalent of sense organs- and there are effectors which control the behaviour of the cell – such things as motility, shape, synthesis of molecules and supply of cellular energy.

In your school biology you may possibly have studied the amoeba. As one of the simplest creatures it is a favourite topic. It is a single-celled organism which is common in ponds and is an easy study both because of its size – some are as large as half a centimetre and because of its visible "behaviour" as one of the most primitive predators. Typical foods are smaller single-cell organisms such as paramecium. The way the amoeba catches its prey is to alter its shape. Its membrane extends either side of the prey, and then extends further around behind, under and over the prey until they meet up in three dimensions. These extensions are known as *pseudopodia* (fake-feet) though for us an analogy is more like putting our arms around an object and hugging it. However, during the process, the amoeba does not touch the prey until after it has completed its encircling.

To do this, the amoeba first has to sense its prey, detecting it chemically. It then has to respond to this detection by extending – not towards the chemical it senses nor away from it, but at an angle to it. The membrane has to continue to sense the position of the prey as it extends, so the surface has detectors all along such that can always maintain the distance appropriate to the size of the target. If it gets too close, the paramecium could take avoiding action using its cilia (tiny hairs) for propulsion. There is no brain or nervous system involved in this activity yet the sensing and movement are as if guided. The "intelligence" of this primitive cell is built-in to its membrane detection receptors and to the internal effector structures which change its shape. There are good pictures of this process to be found on http://www.microscopy-uk.org.uk .

This is just one example of the capability of a simple one-celled creature and shows what is possible with a combination of receptor and effector proteins. It is typical of cell function in all organisms and all happens independently from the genes. The genes are necessary to build a new cell and for it to reproduce itself, but the cell can function in all other ways even if the nucleus, which contains the instructions for reproduction, is removed. Even when the genetic nucleus is present, its activity is controlled by the effector proteins which determine how to "read" the code according to environmental signals picked up by the membrane's receptors.

Co-operation to overcome single-cell limits

We have no way of knowing how many IMP's were present in the earliest cells. This cannot be discerned from a fossil. But evolution is a process by which the "intelligence" or functional complexity of the cell increases by the incorporation of larger numbers of proteins into the membrane over time, eventually reaching the thousands and even hundreds of thousands. There is a limit to how many integrated proteins can be fitted into one cell membrane. Eventually, life could only improve further by forming units of more than one cell.

In its earliest stages, single cells used messenger molecules to "co-operate". This is the type of process which allows slime mould amoebas to co-ordinate when they live as separate cells, and control when they come together to form a group to produce new spores. It is assumed, in the absence of any evidence, to be the beginning of a process which leads eventually to genuinely multi-celled creatures. That is to say, that the first multi-cellular life forms communicated chemically. You may recognise in this, the likely seed of the neuro-endocrine communication system Candace Pert describes.

Through this history, or something like it, we have reached the point where individual cells exist, where there is life on the planet, and there is potential for cells to group together into more complex multi-celled organisms. Now we have that context we can start in earnest to look at the ways that genes are involved in development and change.

Review

We set out in this chapter to introduce the central issues in our examination of humans as biological beings, and to indicate the areas in which we will explore the potential for a spiritual dimension to the way that life came about, the process of evolution, the growth of form in the developing organism and to the relationship between organisms and their environment.

We indicated that this would need our understanding of small scale detail such as cell function alongside large-scale events such as 4 billion years of evolution. We indicated some of the claimed certainties in the conventional explanation and the areas in which we believe that doubt should be applied. As a first step we have looked at the many functions that cells perform, at how the cell responds to its environment and explored where the first cells came from, as a prelude to exploring the genetic mechanisms that guide these processes.

10.   Genes and characteristics

All of life is an experiment. The more experiments you make, the better

Ralph Waldo Emerson

The most heinous offense a scientist as a scientist can commit is to declare to be true that which is not so; if a scientist cannot interpret the phenomenon he is studying, it is a binding obligation upon him to make it possible for another to do so.

Sir Peter Medawar.   *The limits of Science*

Theme

We have indicated elements of life that genetics are intended to explain:-

What are you? How do we explain similarities and differences from other humans.

How did you develop from that first single cell?

How did it come about that life exists at all and that so many different species exist together on the Earth?

At the centre of the answers, science gives us the gene and provides us with a model of how it delivers what we are asking for. In the last chapter we indicated that we would need to unpick some of the certainties that genetic science claims to have achieved in answering these questions. The first spurious certainty lies in the implication that there is a simple relationship, where one gene determines one identifiable characteristic in the organism. We will show that this does not stack up and that the knowledge that would substantiate it is a long way from being complete.

The second appearance of certainty is given to the developmental process itself and the way that science encourages us to draw over-simplified conclusions about how the genetic coding system delivers outcomes. We will see that the story is much more interesting than that, and that genes simply cannot explain everything – indeed that there is much about the process that has not been adequately explained at all.

These two uncertainties drive significant gaps into the story that genetics has presented to us. We start with a refresher of what that model says about the genetic part of the answer.

The conventional view

The instruction book to make a human being is contained in the first, single egg-cell. It is contained in strands of a chemical known as DNA. The strands are tightly wound, packaged in a set of predefined units known as chromosomes, of which there are 23 pairs. The pairing system enables each new organism to inherit half of its chromosomes from the father through a sperm, and half from the mother via the egg.

The long chains of DNA, coiled up tightly in the chromosome package, consist of two linked strands, and these strands are capable of separating. It is usual to imagine this process as being like cutting a ladder down the middle of each rung. DNA is an unusual chemical, because each side of the "rung" is a particular type of component, called a base. There are four types of bases, and they only connect with each other in specific pairings. This means that by assembling bases, one by one, to match up to the free half-rung, the other half of a ladder could be re-manufactured exactly as before. That is, a strand can split, leaving two halves, each of which can be re-matched, and the result is that you have two DNA strands, each identical to the original. In this way, when a cell divides to create a copy of itself all 46 chromosomes split, attracting new bases to complete themselves. One of the new

sets of chromosomes then becomes the nucleus of the new cell, and one remains with the original. (A chromosome is one tightly coiled, very long DNA strand-pair).

The base-pairs enable genetic material to replicate, and while that is a remarkable achievement it is only the means to an end. There is no point in creating copies of something meaningless. The bases themselves form a code. There are chemical messengers capable of reading that code and using it as an instruction set for the manufacture of other chemicals. The chemicals that are made (proteins) will be all that is required to create a functioning body. There will be proteins to make a transparent lens for the centre of your eye, and proteins to make hard enamel for the surface of your teeth. The instruction set is phenomenally clever, since the same set of instructions – the same original nucleus from a single fertilized egg - will make a liver cell or a brain cell, and put it in the right place in your body. It will wait 6-7 years and produce a new set of front teeth. It will wait another 6 or 7, and then cause the body to change, growing facial hair or breasts, widening hips or shoulders as appropriate for the different sexes. A 3D animation video describing the human genome project can be found on "Youtube"[13] and is much more illuminating than any graphics that we could insert here.

The key elements in this theory are:-

- There is a chemical code which controls the process by which cells replicate and organisms develop

- Sequences of those chemicals can be identified and are called "genes"

- Those sequences are associated with particular outcomes in the body, or characteristics, leading to the idea that you can have "a gene for cancer" or that there is a gene responsible for being left-handed or for Mozart's genius

- Everything about our development is dictated by our genes, by the sequence of chemicals contained in our chromosomes.

We need to unpick this theory and find just how much of it is proven, and how much is a combination of conjecture, scientific over-simplification, wishful thinking, commercial distortion and tabloid tosh.

So what's a gene?

In the above paragraphs we would have liked to avoid the word "gene". Linguistically the word has complex roots, at the core being the Latin genere to

---

[13] http://www.youtube.com/watch?v=VJycRYBNtwY

engender, but being closely related to "general", to "genesis" and to "genus" (stock or race) and possibly to "genuine" (as in authentic to its stock). Francis Galton, a cousin of Darwin, was particularly interested in the inheritance of "genius" (extraordinary inventive capacity), though perhaps had more influence in his idea to use fingerprints in identification of criminals. But what actually IS a gene?

For a word in such common usage it is remarkably hard to find a definition. Many books simply use it without giving one. Wikipedia has the brief "a unit of heredity" and the longer "A gene is a hereditary unit consisting of DNA that occupies a spot on a chromosome and determines a characteristic in an organism", which is similar to that given in the Concise Oxford Dictionary. Part of the trouble with the science we are discussing, is that this pervasive lack of definition causes much confusion. Identifying what is a "unit" is less obvious than it might sound. The simple link between "a unit of DNA" and "a characteristic" is more an exception than a rule.

Science, as well as Hollywood ("Jurassic Park" for example), is responsible for creating, and allowing a myth. The CD that you play on your stereo contains the code for a piece of music, but it requires a CD-player. Genes do not operate in isolation either. So far, children require mothers to grow in (or at least eggs, in the Jurassic instance). Test tube babies are not born out of test tubes, merely fertilised there. Science implies a fixed, deterministic development. That is, you have a set of genes; the codes they contain say how you will develop and who you will be. That supposedly is that, the whole story.

Another image that is often used is "genetic blueprint". But the architects' drawings of your house conveyed little of how it finally looked and said nothing about how it got built. The engineer's drawing of your car has no fuel in it. So even though we used the expression "instruction set" earlier and may use such metaphors elsewhere, we need to make clear that there are limits to their accuracy. The areas where they break down are crucial to our whole understanding of the fit between the scientific building-blocks, and the energy-systems of the complementary world.

The description of replication and development above involves a good deal of "shorthand". The relationship between a chemical sequence of DNA and a characteristic in the organism it produces is complex. Sometimes scientists may pinpoint a unit of heredity, especially with simple organisms, and identify simple processes which they have discovered how to alter (to make tomatoes which are slower to break down and rot, for instance). But more often they won't know, since much of development inevitably involves multiple processes, and many different strands of DNA, possibly on different chromosomes. It may be far from obvious where or what a "gene" is, in any physically definable sense. It is a convenient but misleading abbreviation to designate a notional portion of the DNA which corresponds with a characteristic.

Often what scientists see of gene expression is dictated by what goes wrong, for example what happens if the "clotting" mechanism for blood is absent, as in haemophilia? A few diseases which can be traced to a specific locus in the DNA sequence appear to validate this form of thinking. Cystic Fybrosis is one example. But such examples encourage stupidity and misrepresentation in the press of the "we have found a gene for cancer" kind. If only it could be that simple! As Dr Bruce Lipton puts it:

> "..single-gene disorders affect less than two percent of the population; the vast majority of people come into the world with genes that should enable them to lead a happy and healthy life. The diseases that are today's scourges – diabetes, heart disease and cancer – short-circuit a happy and healthy life. These diseases however are not the result of a single gene, but of complex interactions among multiple genes and environmental factors."

In fact, the suggestion that there is a simple relationship between a "gene" and a "characteristic" is at best a metaphor and at worst a dogmatic assumption that is quite erroneous and always was. We would like to use a childish joke to illustrate this thinking.

> Did you know beetles have their ears on their legs?
>
> No, how do you know that?
>
> When I pulled the legs off mine and told it to jump, it couldn't hear me. (Bada Bing!)

In practice there would be many potential "genes" for deafness - one that fails to form the auditory nerve, one that fails to form a tympanic membrane (ear-drum) or the small bones of the cochlea, one that blocks the ear canal with excessive wax, or another that stops the canal from forming at all. In that sense, there is no gene for "hearing" or equally, there are many genes for it. The relationship between genes and characteristics is opaque.

There are abundant examples to illustrate the complexity of isolating (or failing to isolate) a genetic cause for something observed in the individual. The attempt to understand how homosexuality arises makes a great example for the obvious reason that such a preference appears to operate counter to the expected evolutionary drive towards reproduction. Clearly a feature such as this does not bode well for competition to produce offspring. The Wikipedia article on this subject makes it clear how complex the relationships may possibly be both in the context of genetic pathways and in the potential relationship with both social and maternal factors. Whatever the genetic cause of homosexuality, it would have to be a side-effect of something with adaptive advantage. (Note that the widespread phenomenon of identical twins with differing sexual orientation also indicates that this is only a hypothetical illustration.)

The other side of this question would be to ask "Is homosexuality a characteristic?" Do we have a definition of what a characteristic is? Is that something "simple" like eye-colour, or more complex like sporting prowess. Is the gene for baseball the same as the gene for tennis? Is the liking for olives a characteristic? If so, is there a gene for it? It all sounds rather fuzzy and that is because the question is close to unanswerable except for a minority of quite specific instances where the characteristic is clear and the link has been found.

The "usual story" tells us that the gene package determines what develops in the organism in which it is present. This story is further undermined by some "genes" which do not affect the organism itself. An example of this occurs in snails, whose shells are formed in spirals. Viewed from the rear, they coil either clockwise or counter-clockwise. It might be expected that this is coded for by some gene, and that this could be detected in the organism itself. There is indeed a gene (or genetic factor) and it can be detected, but it is not in the organism. The gene is in the parent. That is, there appears to be a gene which says "this organism's offspring will (or will not) coil clockwise". The effect of the gene is delayed by one generation. You can equally well describe the gene as being "the tendency to produced clockwise children".

The way that this comes about is highly significant. When the snail is developing, it does so according to the "package" of material that it receives from the parent. That is more than the chromosomes in its DNA. Just as a human foetus grows in a whole maternal environment, a bird's egg does not just merely contain the genes for a new generation. It includes a whole growth medium, and in a less obvious way snails' eggs contain other materials besides DNA. In the mother's ovaries, even before they are fertilised, the eggs are accompanied by pre-packed messenger RNA and proteins. Accordingly, when development starts, this predetermined package and not the chromosomal DNA dictates the direction of shell formation.

The origins of pre-determined elements and the balance between such fixed features and the flexibility of environmental influence is critical to our understanding of the subtlety in these relationships. We will return to it several times. Before doing so it will help us to understand some more of the metaphor that is contained in such expressions as "instruction set". This too is not as simple as it seems.

Codes, Blueprints and Builders

Any code is capable of being used in more than one way. You can use an alphabet code to construct the German or Italian language as easily as English. In either of those languages, you could write instructions for baking a Christmas cake, the rules for football or a description of the Taj Mahal. You could write in the present tense, or in a historical past tense, or you could write of imaginary things that will happen in the future. You could also use the alphabet as an encryption for itself. A simple version of this would be where each letter is used to stand for the one prior to it in alphabetical sequence - B stands for A, X instead of W. KVTU MJLF UIJT. (Go on, try it!)

It is entirely possible and even quite likely that the genetic "alphabet" of bases in the DNA strand could be used for more than the encoding of proteins. In Douglas Hofstadter's book, "Gödel Escher Bach", he devotes several pages to the postulation and exploration of a made-up game which he calls Typographical Genetics. In this he shows how it would easily be possible for a further level of code to be in place, which would determine how strings of genetic material could be cut up, moved in sequence, switched around, deleted and reassembled. With this level of coding, it is entirely possible to construct a set of instructions that tell a sentence to copy itself. This is therefore the kind of mechanism that would be necessary to trigger a cell to reproduce itself. It might also be the type of mechanism that would be required to control timing. Simple versions of this type of process can be modelled in computer programs which can be made to produce self-replicating code - and even some which are to a limited degree self-repairing. The human genome project, having completed the basic code sequences, is now investigating this "fifth base" aspect, the higher-level phenomenon of control.

If you look at a strand of chemical it is far from obvious where the clock is, or even could be. So how might control of timing be done? What would it take to construct mechanisms such that a six-year-old loses and replaces her front teeth, but not her molars? What is needed to instruct a 14-year-old boy to grow facial hair but not grow breasts, expand his voice-box, widen his shoulders but not his pelvic girdle?

The process has to stop as well as start - what determines when these events are deemed to be complete? Why do we continue to grow only until we have reached a certain size, and not indefinitely throughout our life-span? Equally, why do we eventually cease to replenish, shrink, age and die? The last question may be different than the rest, since one possibility is that ageing processes happen because of cumulative maintenance failures, rather than as part of an intentional "program". But the others seem to be purposeful, ordered, largely consistent throughout our species and well-controlled.

In simple terms, what would be required is that certain instructions would be dormant until turned on by other sets of instructions. This is rather like ensuring that in the genetics of house building, the roof-truss genes do not operate before the wall-erection genes have completed. This at least would provide a mechanism for establishing an order of events. In the case of sexual differentiation we do know something of how this works, and that the presence or absence of hormones like testosterone is a trigger. But that doesn't mean we know how the genes achieve this. This kind of mechanism could also be used to delay the production of an enzyme, which would then be fed back into the system. Such a chemical device is capable of performing the function of a molecular clock. Moreover, since delayed feedback in systems generally produces oscillations, it would be likely that the body processes that we see would have an element of cyclical operation about them. There are in fact many of these. We have mentioned some of the longer-term ones, but a typical and important short-term cycle, is that which operates in the mitochondrion (a sub-component of the cell) to produce the energy which powers the cell. Called the Krebs cycle, tricarboxylic or citric acid cycle, it lasts approximately four minutes. As we will see later, there are much shorter ones.

We hope that it is becoming apparent that genetics is immensely more complex than a blueprint, or an architects' drawing, or the "instruction set" that we called it earlier. It appears that the very first cell supplies the drawing, but it also supplies the mechanical shovels that dig the foundations, the concrete that is poured in, the bricks, beams, window-panes and roof-tiles, plus bricklayers, hod-carriers, plasterers and electricians. And if that was not enough, they also tell each operative what to do, brick by brick, joist by joist, year by year, under variable environmental conditions.

In the light of what we will discover to be a limited number of genes, it is clear that a lot is dependent on these control and timing sequences. But these too cannot be separated out easily. Maybe the hod-carrier is effectively a part of the roof-tile - an extra piece of chemical designed to ensure that the chemical it is attached to cannot be used until it is in the right place. This makes the decoding of genes potentially very difficult. To distinguish control from function requires almost arbitrary lines to be drawn. And if the hod-carrier is a determinant from outside the DNA strand as it was in our snail example, it is just as hard to show where it is and how the triggering occurs.

This detail is not yet present in scientific knowledge. It is at best an over-simplification and at worst a myth to pretend otherwise, and that is Uncertainty number 1. Four to go!

Codes require code-readers

There is yet another whole area of further complexity in organism development. As stated earlier, a code is of no use without a decoding machine. The compact disc conveys nothing to you when you lick it, however hard you try. Pressing it to your forehead, staring intensely and rubbing it with a finger don't work either. You have to put it in the CD drawer of a computer or player, and activate the software which decodes its bit-patterns.

With the snail example, we indicated that development involved an interaction between the instructions supplied in the cell nucleus and other factors supplied by the mother in the egg's environment. This is far from being the only effect that is environmental. In a laboratory, much effort is expended in establishing and controlling the conditions for a reaction to take place. Any cook knows that the temperature of the oven can often be critical. Too hot, and the outside browns before the inside is cooked. Too cool and the dish dries out, or fails to develop the structure it needs, like a soggy sponge cake. Home wine-makers know that they have to give the right temperature to the yeast in their fermentations, without which they are inert, or killed, or produce flavour-spoiling by-products.

In the living world, organisms may have to cope with much variability. For instance there is good evidence that there are large proportions of additional DNA in egg-laying creatures which enable the egg to respond to differences in temperature that are encountered during the developmental process. But these sequences are believed to be redundant or missing in mammals, where the temperature of the growing-medium is well-regulated by the host-organism. A frog embryo may well be capable of coping with temperatures from zero to 20 centigrade (32 - 80F), the potential variation from dawn to peak daytime in a spring pond. In comparison an in-vitro human embryo will not cope with variations much above 1 degree C.

Interaction with the environment

Thus in some circumstances the genetic process works to reduce the effect of the environment - to make the outcome the same, even though the conditions are widely varying. But the opposite also occurs. There are other areas where the developmental process responds to the environment with variation. One example of this occurs in insects, some of which will breed earlier in their lives when food-surpluses are present. Amazingly, some will even breed while still in the larval stage, rather than wait for adulthood. So there is a degree of flexibility supplied even within the genetic instruction set. Cell activity is not fixed. It is capable of responding to environmental conditions, and will produce some enzymes only if those conditions demand them.

This is not new knowledge. In the work which won Jacques Monod and Francois Jacob the 1965 Nobel prize, they showed that some strains of a bacteria, which in one environment normally lacked the enzyme to metabolise the sugar galactose, would proceed to manufacture the required enzymes when exposed to it. The indications are that this flexibility is carried in the DNA, and that there are mechanisms there for manufacture of repressor proteins and detector proteins. With these the cell can detect the presence of a substance, or in its absence, suppress what would be uneconomical activity. This type of mechanism clearly confers adaptability to a variety of conditions upon its owner, and such mechanisms are found to be widespread among micro-organisms. But it also indicates very clearly that the genes do not simply dictate outcomes. There is interaction with the environment and "choices" are made.

A rather different example of this occurs as trees develop, where root growth will be inhibited if the tree is not subject to wind. In enclosed environments like the Eden project or the Biosphere, growers were obliged to shake the trunks of saplings regularly to ensure that the trees were stimulated to produce adequate root systems. When you think about it, this is quite a subtle mechanism involving not a one-off "switch" but a progressive series of responses over time.

All of this gives us added cause to regard with great caution the notion of a blueprint, or a program, or an architects' drawing (even one complete with builders). What was passed down to the developing organism involves a degree of responsiveness to conditions that surround the development process. There are single-celled organisms which become flagellated when potassium ions are more concentrated in their environment. As well as deciding whether to hang the doors inward or outward-opening, the builder's mates construct thick doors for cold climates, and ventilated doors for hot ones. The code is subject to interpretation. The cells contained the same chemicals, but chemicals don't explain form.

One classic demonstration of this feature was made as long as 100 years ago, in H. Driesch's experiments with sea-urchin embryos. An assortment of variations was shown. Removing one of the first two cells in the developing embryo gave rise, not to half a sea-urchin, but to a complete one of smaller than normal size. Similarly, if two embryos at this stage were fused, a giant, but still correctly-formed organism developed. Since that time, similar experiments have shown that a dragonfly embryo will form complete, but reduced in size, if the egg is tied around the middle. We will need to look in more detail at how form is determined.

We simply don't have enough genes

Cumulatively the above examples show several levels at which the description of the connection between "gene" and a "characteristic" misleadingly implies a close-coupled and simple chemical cause-and-effect. But perhaps the biggest nail in the coffin of close-coupling finally came from the Human Genome project – a global scientific co-operation to create a catalogue of all human genes. The assumption had been that the connection between gene and characteristic was mediated by proteins. There are about 100,000 different proteins that make up our bodies. In addition there were assumed to be perhaps 20,000 genes which orchestrate the activity of the protein-encoding genes and address the problems, such as timing, that we have just described, so science expected a minimum of 120,000 genes to be found and present in the final map.

In what he describes as a "cosmic joke" on the scale of the discovery that the earth was not flat after all, geneticists experienced a huge shock at the outcome. To quote again from Dr Lipton:-

"...contrary to their expectations of over 120,000 genes, they found that the entire human genome consists of approximately 25,000 genes. (Pennisi 2003; Pearson 2003; Goodman 2003) More than 80 percent of the presumed and required DNA does not exist! .....Now that the Human Genome Project has toppled the one-gene for one protein concept, our current theories of how life works have to be scrapped.....**There simply are not enough genes to account for the complexity of human life or of human disease**." (*our emphasis.*)

He goes on to detail how fruit flies have 15,000 genes, a nematode worm 24,000 and that rodents and humans have roughly comparable numbers of genes. In the face of this knowledge it is obvious that we have to look beyond the sequences of

DNA themselves for an explanation of how outcomes are determined.

Epigenetics and environmental influences

While it may be difficult to draw the line between control and function, builder and brick, science is beginning to rise to the challenge. In the last decade or so, the science of epigenetics has come into being, specifically to look at the phenomenon of "control above genetics". Their research has established that DNA blueprints passed down through genes are not set solid and do not dictate our destiny. Environmental influences, including nutrition, stress and emotions can modify the expression of those genes, as we have indicated.

Only half of the contents of a cell nucleus are DNA. The other half, largely ignored in the stampede toward genome decoding, is made up of regulatory proteins. These are turning out to be just as crucial to our heredity as the DNA.

In the chromosome, these proteins cover the DNA strands. Like a tattoo under a shirt-sleeve, the code under its sheath of regulatory proteins cannot be read. However, proteins can change shape under the influence of environmental signals, giving access to the gene for copying.

From half a century of assumed genetic determinism, science is moving towards an understanding that DNA does not have primacy. There is a sophisticated interplay of blueprint and environment, just as our earlier examples indicated should be expected. It is like the distinction between the hardware of your TV set and the programs which are transmitted through it. You can change what the TV set shows by switching channels. Equally, you can change the appearance of the program displayed on the screen by altering colour and contrast settings. The two are interdependent.

This relationship has been proven conclusively by an experiment with mice. A particular strain of mice carries the "agouti" gene which causes them to develop yellow coats and obesity, with a predisposition to other diseases. The scientists experimented by giving genetically identical mice different diets. One group was fed on a diet of methyl-rich food supplements such as vitamin B12, folic acid and choline. These supplements were able to attach to the gene's DNA and cause the regulatory protein to bind in such a way that the agouti gene would not activate. There were no yellow coats and no obesity. You can see a picture of the results on the American Society of Microbiology website.[14]

Just how do environmental factors influence cellular development? There are a few known mechanisms by which small changes can have impact at the level of the individual cell. We include the following as indicators of the potential for this. In one experiment, Japanese researchers used drugs to damage the insulin-producing cells of the pancreas in laboratory rats. These rats, when they bred, produced offspring in which diabetes was inclined to occur spontaneously. That is, a specific change in the organ of a parent could be seen to have an effect on the production of genetic material. (Note from earlier discussions that this does not necessarily act through the chromosomal DNA.)

In another, Andrew Maniotis experimented with the effects of mechanical force on the external cellular membrane and showed that the force could transmit to the nucleus and produce a mutation. Further work by Michael Lieber showed that other external stresses from the environment such as heat and radiation could also trigger such mutagenic and potentially adaptive response. Other experiments have shown that when plants and insects are subjected to toxic substances, they often mutate in such a way as to confer increased resistance. In another area of study, researchers at the Renssaeler Institute have shown that external chemical and mechanical factors influence the ways in which developing stem-cells differentiate and become osteogenic (that is, bone-producing).

To these examples we should also add the recognition from the work of Barbara McClintock, and later by Temin and Engels, that large proportions of DNA are indeed not directly genes (in the sense of sequences that code for proteins) but transposable elements and that these move around in response to stress on the organism, a further indication that changes can be generated non-randomly in the genetic material.

As well as influence on the development of the cell, there can be environmental influences on the proteins themselves. Dr Judyth Sassoon gives the following description:-

> "Proteins are long chains of chemical units called amino acids, strung together in the order specified by DNA and then folded up into active conformations. They are the molecular components that accomplish almost all the essential tasks in living cells. For example, proteins catch other molecules and build them into cellular

---

[14] http://discovermagazine.com/2013/may/13-grandmas-experiences-leave-epigenetic-mark-on-your-genes#.UbSQLNjmOEX has more recent discoveries of epigenetic activity. This is likely to be a burgeoning field in coming years.

structures or take them apart and extract their energy. They also carry atoms to precise locations inside or outside the cell. They are able to behave, in the metaphorical sense, as "pumps" or "motors" or form receptors that trap specific molecules.

They can even act as "antennae" that conduct electrical charge. In order to perform their particular tasks, proteins must have the correct shape and the way they are folded in space determines whether they are active or not. Most biology textbooks declare that protein folding is due almost entirely to the chemical sequence of its component amino acids, also known as the primary structure."

She quotes a standard biochemistry text that states that a protein's primary chemical structure dictates its three dimensional structure and goes on to say:-

"This is a very misleading statement because it lays all the emphasis on the protein's intrinsic chemistry and does not stress the importance of the "proper conditions". Yet every biochemist knows that proteins in different environments behave differently. …. External forces clearly play a very significant role in determining correct protein conformation and activity."

She then describes an experimental process with lysozomes from egg-white. She describes the crucial part played in this by water and the factors affecting solubility, drawing the following conclusion :-

"There is, in fact, a mass of scientific literature indicating that protein structures are dependent upon their relationship with water, but this fundamental detail is rarely stated explicitly enough. ….It is clear that the forces giving rise to protein structures in nature are, in actuality, external. Biochemists lay so much emphasis on the chemistry of the amino acid sequence because they consider the external environment to be fixed and the sequence to be the only variable between proteins. This way of thinking is totally in accord with the tendency of modern science to limit and simplify nature and completely obscures the essential relationship between biological systems and their surroundings."

The private life of the cell

In the previous chapter we took a brief look at the process by which organic life developed – from the first amino acids which gave rise to protein formation and to the first cell, surrounded by its membrane. We then saw how important the cell membrane is, not just as a bag that holds the cell together, but as an active chemical process through which receptor proteins detect the environment and effector proteins cause change.

Within this we need to recognise that a small number of proteins are doing a much greater number of jobs. Proteins are therefore being used for multiple purposes throughout the body and a linear (A causes B causes C) model of the biochemistry is inaccurate and over-simplified. This is also why pharmaceutical medicines inevitably have effects other than those which they are designed for. A causes B and D, which cause E, F, G and H, which may well feed back into A and C. And so on. Pharmaceutically the only target may have been B.

In our earlier description we referred with deliberate over-simplification to the cell's "jelly". In fact what is contained within the membrane is far from jelly-like. Be it ever so tiny, the cell contains various structures (organelles) with specific functions separated by a further membrane (the reticulum). This membrane allows us to distinguish organelles such as Lysosomes, Golgi apparatus, the ribosomes which mediate much of the protein decoding and transcription process, and the mitochondria. Fortunately we do not need to know about all of these, but it is worthwhile to discuss mitochondria, the cellular power-pack which we referred to above when mentioning the Krebs cycle.

According to a widely accepted theory put forward by Lynn Margulis, all eukaryotic cells (ones which have a nucleus) are descended from early bacteria and result from an event (or probably several events) perhaps two billion years ago where different bacteria combined their material, forming a co-operative unit. This may have come about by bacteria with stronger "electric motors" invading slower cells. Over time what perhaps began as a competitive or parasitic process evolved into a co-operative one where both parts survived together with mutual benefit. The theme of co-operation is important and will re-emerge regularly in our story. The organelles and structures described above may well have been incorporated over many millennia, or even eons as it is apparent that the additional complexity facilitated many more design possibilities, including that of multi-celled organisms.

Mitochondria contain their own complement of DNA (referred to as mtDNA) and they replicate independently of the cell chromosomal DNA. (In plants this is also true of chloroplasts). For a while it was believed that all mitochondrial DNA passes only through the egg. Richard Dawkins states ("Blind Watchmaker", P176) that "Sperms are too small to contain mitochondria so mitochondria passes exclusively down the female line. …. Incidentally, this means that we can use mitochondria to trace our ancestry strictly down the female line" More recently in "River out of Eden" he modifies this view to state "Sperms are too small to contain more than a few mitochondria … these mitochondria are cast away with the tail when the sperm head is absorbed in the egg at fertilisation." Of course he is entitled to modify his view, but this so-far unproven belief system (that the sperm's tail is discarded and no male mtDNA present afterwards) has led to an entire evolutionary hypothesis ("mitochondrial eve") which is used to justify the statement that all human life originated in Africa and that there were several subsequent bifurcations in human inheritance which can be traced through the human population. This may as a result be an over-simplification.

That particular debate while interesting, is not crucial to our argument. But we point to the mtDNA issue as an indicator that there is much more passing from generation to generation than chromosomal DNA. And potentially, even if it turns out that much, most or even all of this non-chromosomal material comes from the mother, it supports our case for looking beyond the chromosomes for some elements of inheritance. Take note also that some of it may indeed be more stable over generations than the chromosomal DNA if it has not undergone sexual splitting and recombination. As we will see, the balance of stability and opportunity for change is important in the evolutionary process.

You may recall how we noted earlier that homeopathy shows the potential for non-genetic information to be passed from one generation to the next and also the indirect influencing of the spiral direction in a snail-shell. The non-chromosomal material is one of the places where such factors might be carried through.

We are saying not only that the cell is home to other parts of the material which governs human development but that each cell has a life of its own. Beings live and replicate but they do so through their individual cells. The receptor and effector proteins (IMP's) are the means of regulating the cell's individual life. They are also a means for one cell to sense and communicate to another, though they are not the only means, as we shall see later.

We want to remind you that we are telling three parallel and interlinked stories, all mediated by "genetics". Above, we have mainly focussed on the genetic role in building the organism, from one cell to many through replication, a process which includes their differentiation into functional units such as organs. A second function of living cells is that once built, each is a "photocopier" which produces a replacement for itself within the body when needed. These aspects of first building and then maintaining stability are the way in which the organism "becomes". We must progress our narrative now towards the third function – that of "being". Now the organism is keeping itself alive and "doing stuff" which in most cases will eventually include reproducing itself. From there we can lead towards the long-term developmental process for the species as it deals with environmental change. That is, we can shift from genetics to evolution.

Just how does the cell relate to its environment? Crudely speaking, the ability to manage that relationship is critical to an organism's survival. Can it find an energy source? Can it avoid toxicity or attack? Can it find the right conditions to reproduce? We can see much of this in our own behaviour or that of other multi-celled creatures, but how do single-celled organisms do this? How do our own cells? Bear in mind that your lungs or liver only do what they do through the activity of the individual cells they are composed of.

If an organism is to locate an energy source or avoid attack it first has to sense it. In the previous chapter we briefly described the amoeba, its ability to sense food chemically and for the effector proteins to then bring about a change of shape, moving to surround its prey. A single-celled organism is taking in chemicals to consume for itself, or passing waste-product from its metabolism back out into its environment. Inside our bodies, individual cells do this too, but they are also co-operating with other cells so that not only their own internal environment is managed, but the collective environment too. For example, waste carbon dioxide, like oxygen, travels through the blood to be exchanged in the lungs. Other waste products may ultimately be excreted via the kidneys or passed into the bowel.

Thus within our bodies, cells are maintaining themselves and they are also maintaining us. But in addition to that environmental regulation they are also working together to create our "behaviour". Collectively they allow us to walk, talk, eat, and have sex. There is collective "doing stuff". The power of these big cellular collectives greatly exceeds that which the cells possess individually.

We are not used to thinking about the way in which trillions of cells achieve this collectivisation. We are accustomed to imagine that decisions are made in our brains and that muscles act to carry them out. We are led to believe that our brains work out what to do and then we do it. But if you have been following our story through the "mind all over" aspects of neuro-endocrine function and you have grasped the scale of chemical activity at the cell boundary, perhaps it is apparent to you that much more is happening besides our mental process. You may think that it is your brain which tells your lungs that it is time to breathe, but there are multiple control systems working together.

Small indications

People, scientists included, like simplicity. This is often a good thing and brings the most elegant solutions. However it sometimes leads us all into over-simplification. When we are faced with such monumental complexity and intricate detail it becomes more than the average mind can handle. We struggle to comprehend that we have tens of trillions of cells in our bodies. How do we grasp that 20-30 million of our skin cells replaced themselves in the time it took to read this sentence?

Science compensates for information overload in two ways. The first is through specialisation. The rule for specialisation is that a person knows more and more about less and less until they eventually know everything about nothing at all. Joking aside, the risk (and in our view the reality) is that many scientists cannot see the forest because they are looking at one leaf on a tree. The second way of compensating is that they resort to crude models. Even when they are as complex and brilliantly argued as a Stephen Jay Gould opus, those models are still unable to encompass in full the implications that arise from a process as subtle as a single protein change in a single cell.

As authors we face here the opposite risk to specialisation. We are generalists looking at the big picture. We know less and less about more and more until we finally know nothing at all about everything. The subtlety can also be lost in this way. The picture is painted with big brush strokes. We are attempting to cover a huge canvas without losing the resolution of a 1200 dots per inch ink-jet and we ask for you to hold these two perspectives with us simultaneously.

The case we are making is for the presence of consciousness; you cannot get more subtle than this. We are talking about influences at the finest grain of detail, and yet with powerful implications. We will meet this problem of scale again when we talk about quantum physics and deal with the deep paradoxes that it presents us with. It is not critical that you understand the detail of everything that is being presented here. We are not intentionally glossing over anything, and yet we have to feel our way towards the answer, pixel by pixel, or like building up the brushstrokes in an impressionist painting.

What we have to show is a paradoxical combination – the subtlety of very small and non-visible changes combined with the potential for them to influence larger outcomes. This could happen in several ways.

- A small change in one area could "ripple" quickly through a system and become visible.

- A single small change could be a trigger for other processes which would unfold over an extended time-frame.

- A multitude of small changes happening in the same or similar ways could accumulate and have a large impact over time.

Space restricts the depth of analysis we can give to this area, and we thoroughly recommend Dr Bruce Lipton's very entertaining full text. We are now part of the way in our journey through genetics and evolution. We have shown the need for greater recognition of environmental influence but we have not yet shown all of the ways in which this relationship with the environment is mediated and we still have only described a small portion of the processes which make life actually work. These are described in the next chapter. Beyond that in the chapter that follows we still have to examine the process of change over long time-scales and address the dogma that the mechanisms for it all arose entirely as a result of random chance and competition to survive. And we have yet to bring in the possible areas in which consciousness or spirituality may play a part. The next two chapters together will achieve this.

Review

In this chapter we have addressed the impression that science has given of a simple one-to-one relationship between genes and characteristics and shown that:-

The chemical "code" is only part of the means by which development is accomplished

The notion of "genes" is poorly defined, and that there are too few of them to manage the processes for which they are claimed to be responsible.

The outcomes are complex and most often cannot be related to a single gene – indeed, most of the mechanisms have yet to be understood.

There are complex relationships with the environment and intricate controls of sequence and timing that are still not known in detail and sometimes clearly not managed by the DNA code.

The current idea of what "genes" are and what they do is an oversimplification and the way they are typically presented should not be accepted as a full picture.

We have indicated the great complexity of a genetic code, both in relation to its capability to achieve several different functions and its capacity to extend that control over long developmental time-scales. This provides a platform for greater understanding of the way in which genes are influenced by the environment, and for the flexibility that would required for any of that environmental influence to be derived from a layer of information in the realm of consciousness.

We have touched upon the complexity of the cell and the part played in reproduction by other parts than the cell nucleus. We have also given indications of the relationship between activity in individual cells and the overall workings of a multi-celled creature. These aspects of cell function provide the beginnings of a deepened understanding of how coherence may be brought about.

11.   The rhythm of life

The hostess at a dinner party introduced the distinguished-looking gentleman at the table. "I would like you to meet Professor Feinstein. He is a world expert on crocodiles." "My dear lady," responds the Professor, "I fear you exaggerate my knowledge. I am only an expert on the crocodile's eyelids."

Anon

In science, a generalization means a principle that has been found to hold true in every special case. Mind is the weightless and uniquely human faculty that surveys the ever larger inventory of special-case experiences stored in the brain bank and from time to time discovers one of the rare scientifically generalizable principles running consistently through all the relevant experience set.

The thoughts that discover these principles are weightless and tentative and may also be eternal. Mind's relentless reviewing of the comprehensive brain bank's storage of all our special-case experiences tends both to progressive enlargement and definitive refinement of the catalogue of generalized principles that interaccommodatively govern all transactions of Universe.

Specialization tends to shut off the wide-band tuning searches and thus to preclude further discovery of the all-powerful generalized principles. Only a comprehensive switch from the narrowing specialization and toward an ever more inclusive and refining comprehension by all humanity-regarding all the factors governing omnicontinuing life aboard our spaceship Earth - can bring about reorientation from the self-extinction-bound human trending, and do so within the critical time remaining before we have passed the point of chemical process irretrievability.

Buckminster Fuller   *Synergetics (Introduction – edited)*

"I hold to the presupposition that our loss of the sense of aesthetic unity was, quite simply, an epistemological mistake. I believe the mistake may be more serious than all the minor insanities that characterised those older epistemologies which agreed upon the fundamental unity."

Gregory Bateson   Mind and Nature

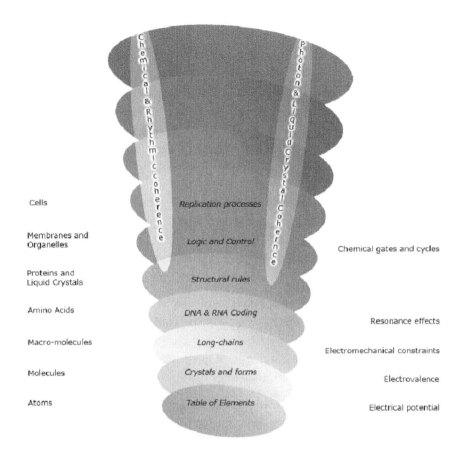

Cells — Replication processes

Membranes and Organelles — Logic and Control — Chemical gates and cycles

Proteins and Liquid Crystals — Structural rules

Amino Acids — DNA & RNA Coding — Resonance effects

Macro-molecules — Long-chains — Electromechanical constraints

Molecules — Crystals and forms — Electrovalence

Atoms — Table of Elements — Electrical potential

Chemical & Rhythmic coherence

Photon & Liquid Crystal Coherence

Theme

In this chapter we carry forward our journey into uncertainties and build on the image of cells managing their individual integrity and interacting with their environments. We can now examine further the way in which this process of interaction is then extended into a systematic, coherent and co-ordinated relationship. This co-ordination encompasses billions or trillions of cells, all working together with a mix of differentiated function (lungs, kidneys, blood-cells muscles etc.) that nevertheless function as a unified whole. We will see that this requires several layers of connection, viewing the body as a biological computer and as a system exhibiting the qualities of laser-light focus through organically derived coherence processes. These are added to the body rhythms synchronised by our beating heart and the decision-making capacities of our brains.

Quantum Biology and Coherence

The ability for a human body composed of 50+ trillion cells to function as one single entity is another daily miracle. Don't forget that this functionality runs all the way from the basics of heartbeat, breathing and digestion to the sophistications of peak athletic co-ordination, consummate musicianship, poetry and intellectual achievement. We continue here to ask fundamental questions about life, many of which are built in to the related words we are employing – organ, organic, organism, organelle (within the cell) and above all - organisation. Where does the organisation come from, and how does it work?

When it comes to scale, you can't get smaller than the quantum level. We were tempted to place the chapter on quantum physics earlier in our story in order that some terms used here would have a deeper context. Instead we will attempt to indicate the flavour of the quantum world, and expand on that later. Quanta are the stuff of the universe which are even smaller than atoms, and which make up just about everything. Put another way, different atoms are composed of different combinations of quanta. We are talking of electrons, protons, neutrons and many others. Their behaviour is weird, in some ways unpredictable and still not entirely understood.

When matter was originally created, the forces in the core of stars caused increasing densities of matter to form over billions of years, from small atoms with a small nucleus and one or two electrons like Hydrogen and Helium to bigger ones like Iron with 26 electrons or Gold with 79. This has its own place in the spiral of evolution, potentially its own spiral with a layer to represent each of the electron shells, but all this is abbreviated to the single "table of elements" band in our overview diagram.

Underlying this quantum unpredictability, and beneath their role in coalescing into stable material forms there are properties which cause the behaviour of one quantum to be related to another, and to bring about forms of "coherence". To have the feeling for this we would ask you to recall a film you may have seen showing large shoals of fish which respond to the presence of a predator with a seemingly instantaneous collective shift of direction. (Some bird species, when in flocks, behave in a similar way that might also be familiar to you.) That is, coherence provides the possibility for large numbers of potentially free events to be co-ordinated in such a way as to act together. This coherence is a somewhat paradoxical state which maximises local freedom at the same time as facilitating global cohesion.

Jumping in with a more fulsome description, here is an extract from an article by Dr Mae-Wan Ho. This description is at the molecular rather than the quantum level, but illustrates the concept, as well as showing that coherence is significant at many levels of existence. (A fuller text can be viewed on www.i-sis.org.uk and complete references downloaded by becoming a member.)

> "The macromolecules, associated with lots of water, are in a dynamic liquid crystalline state, where all the molecules are macroscopically aligned to form a continuum that links up through the whole body, permeating through the connective tissues, the extracellular matrix, and into the interior of every single cell. <u>And all the molecules, including the water, are moving coherently together as a whole</u>.
>
> The liquid crystalline continuum enables every single molecule to communicate with every other. The water, constituting some 70 percent by weight of the organism, is also the most important for forming the liquid crystalline matrix, for intercommunication and for the macromolecules to function at all.

Coherence is a term that was developed in association with superconductivity and laser light rather than living systems. If you have seen a neon or fluorescent light you have experienced the way in which an atom can be stimulated by energy to emit light (photons). Normally these photons are emitted in a scattered way in all directions. In contrast, if you have seen a laser light-show you will have seen that the light from a laser stays together in a beam, does not scatter, and can travel for long distances maintaining this visibly coherent direction."

In a laser, the atoms or molecules of a crystal, such as ruby or garnet -- or of a gas, liquid, or other substance -- are excited in what is called the *laser cavity* so that more of them are at higher energy levels than are at lower energy levels. Reflective surfaces at both ends of the cavity permit energy to reflect back and forth, building up in each passage and raising the energy level.

In a ruby laser, light from a flash lamp surrounding the ruby, in what is called "optical pumping", excites the molecules in the ruby rod, and they bounce back and forth between two mirrors until something happens called a "phase shift". When this happens, the photons become coherent (vibrating in phase with each other) at which point coherent light escapes from the cavity. For a simple image of phase, think of an army marching in time and of the way that they have to break step in order not to set up vibrations that will crack bridges when they cross. That is, they deliberately shift to being out of phase. An alternative image would be Busby Berkeley choreography or the difference between successful synchro-swimming and splashing around.

Self-organising systems

Another way of describing coherence is in the capacity of a system to be self-organising. As the laser shows, self-organisation is not restricted to organic life. We are used to thinking of chemical reactions as irreversible processes. Iron rusts, but rust does not turn to iron. But there are reactions which run counter to this expectation, and one example is known as "the chemical clock".

There are certain kinds of chemical reactions which change colour. Picture then, a reaction which changes colour between two different states, in a cycle which is regular, and which also displays beautiful spiralling structures. This involves a reaction which can be reversed, and reversed again. It also requires that the whole mass of chemicals is performing this cycle in a co-ordinated way. You don't have a soupy, undifferentiated mixture, the way that you do when cream mixes into coffee. Rather, it is as if the colour of your coffee were to switch from brown to cream, and back to brown. There are billions of molecules, operating as one "unit".[15] Instead of an apparently random and disorganised state, we have a form of order - an order which contains some self-organising principle. (For the sake of accuracy we should point out that the "soupy" state does eventually occur, but the cycle continues for as long as fresh ingredients are being added to the mixture.) Another example is more purely in the realm of physics, and concerns what happens when we heat a layer of liquid that is sandwiched between two sheets of glass. One might expect that when we apply heat all the molecules of the liquid would become more energised, and show greater and more random movement. It is not to be expected then, that what happens is the appearance of a honeycomb pattern, made up of hexagonal cells of convecting liquid. That this happens, was discovered nearly one hundred years ago, but its significance in regard to dynamic, non-equilibrium states, was not seen until much later. Here again, the phenomenon of self-organisation applies only while energy continues to be supplied. Without heat, the liquid reverts to an unstructured state. (Bear in mind that there is always new energy being introduced to a live body.)

Both of these examples exhibit the same phenomenon of self-organising "communication". In the liquid honeycombs, a hexagonal cell is the width of millions of molecules, but all those millions become co-ordinated in the process. In the chemical clock (also known as the Belousov-Zhabotinsky reaction), the distances involved are even greater, but the molecules appear to "know" when to

---

[15] An example can be seen here
http://aulascienze.scuola.zanichelli.it/fisicamente/2009/12/13/entropia-4-la-creazione-di-ordine/

turn red, and when to turn blue. They work together in an apparently instantaneous series of changes, or phase-shifts.

How can we prove coherence?

We are not aware of any experiment yet which conclusively demonstrates a similar co-ordination in the human body. Indeed, it may prove quite hard to show this in a complex living system. You might ask how an experimental design would isolate the variables concerned? It would certainly be a challenge. Consequently we must look at what might constitute the necessary evidence.

We referred earlier to the existence of cycles of activity in the body and used the example of the Krebs cycle to illustrate one with a time-period measured in minutes. In practice, living systems have a multiplicity of cycles within cycles and you can work through biology down to events of energy-exchange which, represent an electronic vibration that has a period of femto-seconds (1/ 1,000,000,000,000,000 of a second).

As a result, we have to build our picture of the mechanisms which could maintain coherence in small stages and at different levels, and once again we will skim the surface of the material. A very full and comprehensive technical view of both the physics and the biology that we are summarising here is presented in Dr Mae-Wan Ho's book "The Rainbow and the Worm", which we would recommend particularly to those who really want a much deeper picture of the science.

What then are the elements that would demonstrate coherence? What would indicate that there is a co-ordination and organisation taking place across our trillions of cells that would be akin to the almost instantaneous change of the molecules in a B-Z reaction? What mechanisms would we expect to see?

By way of answer, and as Dr. Ho suggests, the components we are looking for might include:

1. Maintenance of order over long range and at high speed

2. Rapid and efficient energy transduction

3. High sensitivity to external cues and triggers

4. Symmetrical coupling of energy transfer

5. Populations functioning together without "noise" (a term communications engineers use to denote the energy fluctuations which are not part of the

intended signal – like radios before FM.)

Let's look at these briefly, one by one.

1. Order We used the image earlier of the shoal of fish changing direction. When required (particularly in emergency) organisms can mobilise great amounts of energy in an instant, and nerve communication to muscles is incredibly fast. Yet it appears that the muscle responds in advance of the nerve signals for enhanced co-ordination being received. This suggests that there is a system of communication which sends emergency messages simultaneously to all organs, including those not directly connected with the nerve network. It acts at a speed which appears to rule out conventional nerve mechanisms.

There are indications that there are electro-dynamic signals involved. It has been shown that Daphnia emit light and that the rate of emission is related to their distance from each other. Such a mechanism would enable collective behaviour such as the fish shoals described and could also take place cell-to-cell in a multi-celled organism. Fritz Popp, a quantum physicist turned biophysicist and a pioneer in the investigation of photon communication, has shown direct communication between separated containers of luminescent bacteria, which synchronise their light flashing when there is no light barrier between them.

There are numerous examples of phase-locked oscillations within organisms, one such being the way insect wing-beats are governed. Similarly patterns have been detected in brain-wave activity which shows rapid coherent changes across large areas. The pacemaker cells of the heart and the insulin-secretion cells of the pancreas likewise show synchronised electrical activity.

2. Energy transduction Muscle activity in mammals is highly efficient. Outside of living organisms, chemical reactions lose 70% of the energy available (or supplied), which would cause mammals to overheat very rapidly. Muscle contraction is triggered by an instant of electrical discharge at the point where the nerve meets the muscle-cell membrane; within a millisecond calcium ions are released to trigger contraction of the entire cell. In a typical muscle contraction, all the cells, often numbered in billions, are executing such contractions together and the chemical energy, which is stored in a molecule called Adenosine TriPhosphate (ATP) is converted into mechanical energy. Thousands of billions (Dr. Ho uses the figure of 10^{19}) of molecules are utilised rapidly and co-ordinated over distances ranging from the microscopic to the length of a long muscle (e.g. calf or thigh). This energy is supplied at close to 100% efficiency

3. Sensitivity to cues The eye is highly sensitive, and in some species can detect a single quantum of light falling on the retina. The signal that this triggers in the nerve contains perhaps a million times more energy through a molecular cascade. The muscular activity described above mobilises vastly more energy than the nerve impulse. Thus a minuscule trigger generates a major use of energy.

Another aspect of sensitivity would be the ability of the cell, or groups of cells, to respond to very weak electromagnetic signals. Think of the way in which a radio, when tuned to the right frequency, can pick up a radio station and deliver the co-ordinated information that it contains. Irena Cosic[16] has shown that groups of proteins which share the same function also share a periodicity in electronic potential and exhibit a form of common recognition of an electromagnetic frequency. That is, there are fluctuations which are co-ordinated in response to electromagnetic signals. Such an effect would create a crucial relationship between communication of information and organisation of energy and function.

4. Coupled symmetry The process in the body that creates ATP (above) from ADP (which has one phosphate group less) is reversible. We just described the release of energy. But when energy is supplied (e.g. from food) the ADP adds a phosphate group and becomes ATP. So there is a repeatable cycle of energy store and release. This is just one example from many where the body has chemical cycles which fulfil the criterion of symmetrical coupling.

5. Noiseless communication The examples such as pacemaker cells in the heart given under point 1 show noiseless functioning, but this can also be demonstrated at the molecular level where high-speed ultra-sensitive instruments indicate that the contractions in fine muscle activity (such as the beating of cilia in mussels) show synchronised quantal behaviour with little or no fluctuation.

Thus each of the individual elements which would be required to produce coherence are shown to exist. What remains is to show that these are combining – that synergistically they produce an overall functional effect that is more than the sum of the individual components. One illustration of synergy might be the way that two gases, hydrogen and oxygen, when combined produce properties of stability and chemical behaviour (such as holding other substances in solution) that neither can do on their own. Another would be the way that several soft metals like Iron, Chromium and Nickel, when combined in the correct proportions, can produce a steel that has three times the tensile strength of any of them individually. We believe strongly that equivalent synergies are present in the body.

---

16 Cosic, I (1994) Macromolecular Bioactivity.

The Liquid Crystal bio-computer

We appreciate that we are piling one level of detail upon another, and adopting multiple modes of description for the observations of bodily activity. While we are sorry if this is potentially confusing, we are only describing what is actually there. This is how nature works, building layer upon layer. All of life exists in nested hierarchies, as our spiral layered diagram illustrates. It is complex, it is multi-layered and it does involve many different processes taking place at small and large scales of activity from the quantum to the cell to the organism. Please allow us therefore to introduce one more layer.

It is quite common for people to draw analogies between a brain and a computer. It is more of a surprise for someone to present the whole body as resembling one. This is not as far-fetched as it might sound, and the reason for that lies in "liquid crystals". You will have come across these in a laptop display, and some flat screen TV's and monitors, but they have properties beyond display technologies.

Bruce Lipton tells the story very entertainingly of the moment over 20 years ago when he had insight into the role of liquid crystals in cells. A crystal is a structure where molecules are arranged in regular and repeated patterns. We are accustomed to diamonds or table salt as crystals which are solid, but fluid molecules can also adopt regular and repeated patterns. Even though they are flowing they retain their organisation. They can alter their shape and yet maintain integrity, and this is just what the phospholipid molecules that make up cell membranes do.

We described earlier the receptor and effector proteins that may be conducted across the membrane. Bruce Lipton perceived that the fact that the membrane conducts some things and not others made it a semiconductor. He further perceived that effector proteins formed gates and channels in the membrane and arrived at the description "The membrane is a liquid crystal semiconductor with gates and channels." He then recognised this as just the description that was used in describing the microprocessor chip inside a computer.

He also describes how 12 years later, in 1997 B. A Cornell[17] and colleagues isolated a cell membrane and placed a piece of gold foil under it. They then flooded the space between with an electrolyte solution. When stimulated by an electrical signal, the membrane's receptors opened up and allowed the electrolyte solution across the membrane. Through the foil, the electrical signal could be picked up and displayed on a screen as a digital readout. This device demonstrated

---

17 "Nature 387. 1997

that the membrane can function like a chip. As a result there is a potential for levels of interrelationship between cells that would be the equivalent of multi-processor computing. Nature did it first!

Dr Mae-Wan Ho follows a parallel train of thought in her recognition of a potential relationship between the liquid crystal continuum and consciousness. She suggests that it is not appropriate to locate consciousness as a property of the brain, but that it has to be seen as throughout the entire liquid crystalline continuum of the body. Further than this, she describes the importance of connective tissues in bringing this about.

The connective tissues include the extra-cellular matrix that surrounds all cells, along with skin, bones, cartilage, tendons, veins, arteries, air-passages and more. These tissues are also liquid crystalline in nature and are ideal for mediating the rapid intercommunication that would be required for the efficient energy transduction that enables coherent behaviour across the whole organism. This is facilitated by the properties of collagen, a protein that provides an ordered network that can orient itself in response to electrical signals and in addition binds water molecules in such a way as to support rapid jump-conduction of protons.

Jump-conduction of protons is faster than electrical conduction in the nerves. As a result, the connective tissue of the body provides a superior intercommunication system to that the nervous system offers by itself. The collagens also supply an element of structural stability which enables tissues to retain memory of previous events. This adds a further element to the bio-computer, that it has processing (the yes/no of the membrane gate), rapid connectivity, a form of input (senses), output (change and behaviour) and now, memory.

Phase-shifts and pumped systems in the body

When discussing lasers and coherence, we referred to the process of "pumping" which is applied to the photon emissions within the laser crystal and the eventual jump, or phase-shift, into coherence.

Herbert Frohlich, a solid-state physicist, has suggested that the dense packing of dielectric molecules in living organisms could lead to the condensed-matter conditions where metabolic pumping could result in a build-up to collective modes of vibration. Taken together with the rapid energy transfer features of the connective tissue and also with the jump-conduction of protons, there is potential for the conditions that facilitate phase-shift and coherence to occur.

We will return to this area in our discussion of physics, but for now would merely note that other researchers have supported this conclusion. Additionally, there is work which demonstrates the greater degree of coherence present in healthy people compared with cardiac patients, and also showing the coherence associated with varied forms of meditation.

We said that this chapter would concentrate on the processes which enable life as we know it, especially the kind of life which large multi-celled organisms exhibit. We also said that we would indicate how the facts we see would create a space in which consciousness could become an active part of the picture.

What is beginning to reveal itself, is a combination of necessary factors which would enable this to occur. We are seeing:-

sensitivity to external conditions

the possibility that such sensitivity includes detection of subtle electro-magnetic or other low-energy changes

the capability for such low-energy changes to affect low-level processes in the cell

the existence of a mechanism or mechanisms that would propagate a low-level change throughout the organism

In the previous chapter we painted the picture of flexibility in the mechanisms which manage the building of cells and of multi-celled creatures. In this chapter we have begun to view those cells firstly as individual units with their own mechanisms for maintaining their integrity and balance, and then as a set of coordinated components which communicate with each other and work together. There are several aspects to this co-ordination – electrical, electromagnetic, bio-computer, liquid crystal continuum and laser-like phase coherence. Many of these are not understood in detail, and the balances and interplay even less clear. When they eventually emerge, these details will be fascinating but are not essential at this stage. The overall picture is sufficient to indicate that there are several ways in which individual organisms are sensitive to very small changes of energy and can co-ordinate their responses to those changes. Such mechanisms are also sufficient to provide physiological mechanisms which are capable of explaining many features of life that we are describing – intuition, and energy-healing for example. The only reason that we do not have such explanations is that the research is yet to be done (a good idea, we suggest!)

This is merely a beginning, but it is a gateway to a scientific picture which can explain some of the relationships that were evidenced in Section 1. We will explore these connections more in the context of physics. In the meantime, having dealt here with Uncertainties 2 and 3 we should take a look at evolution and the fourth uncertainty that we set out to address in this section. We are returning to the theme of fundamentalist Darwinism in order to complete this discussion. We will present the evidence against a completely random universe, and the flaws in a purely competitive view of evolution. Into this we will weave a potential for consciousness to enter the relationship between the long-term development of life, species differentiation and the emergence of ecological balance.

Review

In this chapter we set out to show how individual cell function could extend into a systematic, coherent and co-ordinated relationship that encompasses billions or trillions of cells, all working together with a mix of differentiated function (lungs, kidneys, blood-cells, muscles etc.) that nevertheless function as a unified whole.

We looked at the phenomenon of coherence to understand what might demonstrate our body's ability to demonstrate laser-like organisation and found several features that indicate this. We found evidence of systems in the body to be self-organising and saw mechanisms which would contribute to this, jump conduction of protons, the liquid crystal membrane "biocomputer" and metabolic pumping of the sort required to bring about phase-shift.

In all of these we see the level of sensitivity to small energy triggers which would enable us to detect very small changes in the non-visible world and for them to affect our function, in the way required for examples of energy healing such as homeopathy.

12.   Not staggering, but dancing

To be properly expressed, a thing must proceed from within, moved by its form

Meister Eckhart

"…[The] place of the embryonic formative process is a field (in the usage of physicists) the boundaries of which, in general, do not coincide with those of the embryo but surpass them. Embryogenesis, in other words, comes to pass inside the fields. … Thus what is given to us as a living system would consist of the visible embryo (or egg, respectively) and a field."

Gurwitsch, A.G.   *The Theory of the Biological Field.*

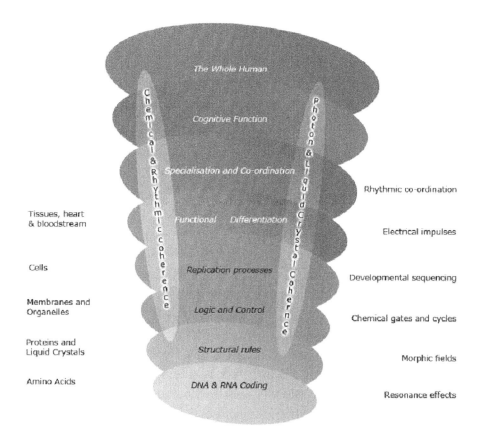

Theme

In this chapter we reach the conclusion of our journey through biology in which we have been exploring the themes of who we are and how we came to be. We have examined how organisms are "built", and how there is more to this than DNA / genes. We have seen that once an organism is built, cells replicate and interact in the long-term maintenance of the living body and we have seen how trillions of cells can be co-ordinating in a coherent, functioning whole. We now complete the circle of life, with the story of procreation and change over time. We return to the theme of how multiple species on the Earth have come into being, how and why their differences arose, and what is the driving force for increasing complexity of form and behaviour.

Here we encounter the crucial issue of whether the absolute purity of proposed randomness as championed by the Darwinian fundamentalists is justified. We look for whether there is even a tiny thread of "guidance" present and if it is, at how that might happen. Lastly we give indications of what this means for the spiritual context, and show that while the biblical creation metaphor and other stories that describe an external creator-god may be unsustainable, there is a core truth which maintains the rationality of perceiving a creative spirit, and a higher consciousness in the universe with which all life is connected.

Just how special are we?

In a book called "Life's Grandeur" or "Full House", depending where it was published, Stephen Jay Gould dissects our natural tendency to view evolution as naturally leading towards increasing complexity of form, behaviour or larger size. In face of the "obvious" truth that evolution began with bacteria and single-celled organisms and now displays the richness which includes whales, elephants and us – wonderful humanity – he provides analysis, backed up by statistical evidence, that we too are merely an outcome of the mathematics of variation. His contrasting image is that of a drunkard's walk along a street, with a wall one side and a gutter on the other. Random staggering in a space constrained on one side by a limitation (i.e. you can't get smaller than one cell) and open on the other (size and complexity have no simple upper limit) will eventually see the drunkard fall in the gutter. This is an argument for humility. We are not the perfected pinnacle of God's plan. We are just one example of a fallen inebriate – even if we are looking at the stars.

Humility is good for us. It does us no harm to reflect that our time of imagined superiority on the earth is a mere eye-blink in comparison to the 150+ millions of years for which dinosaurs were believed dominant. We should also recognize that this generic reference to dinosaurs is misleading too. No particular dinosaur is known to have been around for so long. Species came and went, usually in less than 10 million years. For example the records for the most famous of all, T. Rex, are based on only 20 finds, with three complete skulls in total, and span just four million years or so. What we think of as "humans" can change too.

Similarly, in face of the apparent success of multi-celled forms, Gould produces evidence that there is a greater biomass of bacteria than there is of surface flora and fauna. Much of this bacterial life lives within the porosity of rocks and in underground water. In addition, bacteria colonise a wide range of environments, including many which would kill animals, such as the deep-ocean fissures in the earth's crust where they live at temperatures above 450°Fahrenheit and pressures 265 times that which we are used to. This is why it is possible that bacterial life could be found on Mars, and Giuseppe Galletta of Padova University has demonstrated this in the laboratory. So who's the big success? Bacteria have been around for over 4 billion years and the more complex single-celled organisms (eukaryotes) had a 2.75 billion year head-start on multi-celled creatures. It is hardly surprising if a great deal was achieved in this vast time-span, especially when bacterial generations succeed each other in hours rather than years. That's a lot of opportunity to try things out.

Although the evidence indicates that humanity has been physically evolved (e.g. upright posture and brain size) for more than a million years, our putative "more than animal" status is quite hard to quantify or demonstrate. While some evidence of aboriginal cultures may go back as far as 40-50,000 years and be shown for instance in quite remarkable cave paintings, the vast majority of all we are accustomed to think of as human culture and civilisation arises since the end of the last ice-age – about 12,000 years.

Only within this brief time-span has mankind made the journey from hunting cave-dweller or nomadic herd-follower into settled village-dweller with agriculture and animal husbandry. The growth has been rapid and accelerating, from stone tools through pottery to metal-working. Even a technology as fundamental as the wheel (with extensions such as mills and lathes) arose half-way through this settled period. Only recently have we become numerous. At the end of the ice-age there were perhaps a few million of "us".

So where does this leave our theme of spirituality? However hard we might look we will not find fossil evidence for experience of soul or spirit. We might try to assign a spiritual agenda to cave-paintings, but it cannot be more than conjecture. Cultural evidence is limited to the time since settlement and anyway does not constitute scientific proof of anything. Clearly if we are to place spirituality into the realm of genetics and evolution it will not be easy. Nevertheless, we will once again show that there are gaps which undermine the scientific dogma and demand further investigation.

It should not come as a surprise if our answers reflect and repeat the recognition contained in our chapters on plant spirits and shamanism that there is a form of consciousness throughout the living world. There is pride and wishful-thinking that leads us into the falsehoods which Gould points out that derive from the human desire to see our species as a peak achievement. However this same pride also manifests as hubris (often masquerading as humility) which elevates the intellect above all other forms of knowing. It causes us to ignore other evidence. In our opening chapter we referred to the mistake of placing the rational mind on a pedestal. Here we must look to the other tools in our toolbox.

Gould's arguments are so strong, so powerful in their statistics and logical inference as to be totally convincing. Indeed, we have to concede his case that there is no statistical evidence showing any tendency towards complexity of form or function, none for the inbuilt trend towards larger size and none for increased behavioural complexity. Yet absence of proof is not the same as proof of absence; we are certain, and will prove from other angles, that evolution does have a "guidance system". So we must and will also show that the guidance system would by its nature have exactly the statistical characteristics that Gould presents. Unfortunately Gould's conclusion is the price we pay for looking assiduously in the wrong place.

Some readers may hope that the line we are about to pursue will be supportive of the recently popular "Intelligent Design" approach to evolution. We are reluctant to adopt such a label. For some who use it, ID is simply a more sophisticated version of the biblical creation story. For others ID may not imply a "God" but it carries some vaguer implication that evolution has been guided by an intelligence which sits outside of the living process. We cannot simply align ourselves with either of these since for us nothing is outside. What we see is built in to life; it is more subtle and in our eyes more wonderful than any of those views of why we are who we are.

It would be helpful to retain the use of the word intelligence, so please understand that for us, intelligence is a quality that is inherent in the living world. This will be one of many areas where we will step outside of a tradition of polarised thinking – science / spirituality, genes / environment, matter / energy, mind / body and in this case, God the external creator as opposed to random chance. As Robert Pirsig argues so powerfully in his sadly neglected classic "Zen and the Art of Motorcycle Maintenance", quality comes first of all, and ahead of the dichotomies of dialectical thought.

Elisabet Sahtouris has pointed toward the simple contra-distinction between science, which starts with matter to generate consciousness, and spirituality, which starts with consciousness to generate matter. The two cannot be separated – at least not in us or the world we live in. Each is inherent in the other. She also points to the parallel distinction between Eastern and Western forms of science. For Western science whatever you can touch is real and for Eastern Science whatever you cannot touch is real (and the touchable is the world of illusion). Here too (using Bede Griffiths' phrase), we see a need for "The Marriage of East and West". We are asking you to see both as equally real and equally unreal. This will require that you step outside your cultural conditioning – not in making that cultural view wrong, but in expanding and adding to it. It is to be hoped that you are by now getting good at this type of exercise.

We would also like to introduce an idea here which we will explore in greater depth in our remaining chapters. The universe we are presenting is one which is self-creating and self-creative. We would borrow from physics the idea that the universe as we know it started from a "big bang", from a time when nothing existed (in any form which we would recognise). Everything has come about from that starting point and we would wish you to think afresh of the big bang as a creative act, one in which consciousness was present, an act of primal spirit in self-expression, of the first matter expressing its being. We would like you to see every part of creation from then on as an extension of that auto-creative and auto-actualising process in which all that is, experiences itself.

When we come to the living world, and to the development of life and the evolution of planetary ecology inhabited by multiple species, then we are describing what Nobel prize-winning biologist George Wald called the presence of "creative mind throughout biology". It is in this sense that we attribute "intelligence" to the process and **only** in this sense that we would see a process of "design". However we see this as an increase in the importance and power of consciousness, not a diminution.

The alternative to entropy and competition

A creative view such as this flies in the face of two fundamental principles which have come to us through science and which now inhabit Western culture. The view of Newtonian Physics is that the universe is in a state of gradual run-down. In a process known as "entropy", all the energy from the big bang will go from its extreme of intense compression toward another extreme of dispersion and dissipation. Heat cannot pass from a cooler to a hotter body, so ultimately all matter must eventually end at a uniform temperature. Everything that we experience as life is a merely a blip in this miserable decay. Supposedly.

The biological model is equally grim and gloomy. Alongside the randomness of change is the over-simplified reproductive "survival of the fittest", which arose in Darwin's mind as an expression of the endless competitive struggle against scarcity which was postulated by Thomas Malthus. Malthus predicted that population growth eventually leads to the point where there are too many people for our resources.

> "The power of population is indefinitely greater than the power in the earth to produce subsistence for man. Population, when unchecked, increases in a geometrical ratio."

This is quite explicit in Darwin's work, as illustrated by the following quotes:-

> "Nothing is easier than to admit the truth of the universal struggle for life, or more difficult than to constantly bear that conclusion in mind. Yet unless it be thoroughly engrained in the mind I am convinced that the whole economy of nature, with every fact on distribution, rarity, abundance, extinction and variation, will be dimly seen or quite misunderstood."

from the introduction to "On the origin of species" – and later

> "It is the doctrine of Malthus applied with manifold force to the whole animal and vegetable kingdoms".

We referred above to Gould's presentation of the success of bacteria. Earlier we cited Lynn Margulis' widely-accepted view of the development of eukaryotic cells from prokaryotes and suggested a process in which the eukaryotes became beneficiaries of a co-operation or symbiosis between previously separate micro-organisms. Arguably, the further development of multi-celled creatures extended that process of cellular co-operation to a very advanced level. How does it feel to think of yourself as trillions of bacteria all stacked co-operatively together? We recognise that the notion is slightly whimsical but it is also not without a grain of truth; that we are bacteria's greatest creation. Not only are we single cells in massive co-operation and a long-term host that provides a stable living environment for decades at a time, we have even been perfected to the level where we will purposely manufacture more of their number and sell them by the billion in pots for consumption. Mmm Danone!

It is worth noting in addition, that every bacteria can exchange DNA directly with any other (Margulis again) and that effectively they can be regarded not as species, but as genome-shifting strains. In her more recent work she suggests that evolution includes the trade of entire genomes in more complex organisms, particularly the metamorphosing insects. As mentioned before, the genetic material also includes transposable elements which can move within the genome. It is also known that the protein sequences can edit and repair themselves. There is no evidence we know of yet that material is exchanged (other than destructively) with bacteria and viruses that inhabit human bodies but it is entirely possible.

Mae-Wan Ho ("Rainbow and the Worm") deals extensively with the physics of an anti-entropic view of life and we have no space for detailed exposition here. Hugely simplified, we will just say that the second law of thermodynamics, from which the idea of run-down derives, deals with isolated or closed systems and deals at the level of heat energy. It does not deal adequately either with systems which store energy and it does not include systems which retain information about themselves and which can choose to change their environments. Simply put, Newton's second law is about thermodynamics and not about living systems. Nor does the Law address the question of what life indicates for the boundaries of closed systems or what a conscious or connected universe does to the definition of "isolated". Suffice it to say that we do not accept that the second law provides any useful guidance for our planet to live by. We would prefer to concentrate on the alternative.

What is more apposite here is the biological view that Dr. Elisabet Sahtouris presents ("Earthdance: Living Systems in Evolution"). She identifies cycles of evolution which can be demonstrated in different types of ecosystems. These have strong implications for our view of a competitive living world and for the notion of limitation and scarcity.

In her view, immature ecosystems (Type I) are characterised by organisms which are themselves not mature and are competing to find space. In more mature ecosystems (Type III) the later descendants of these organisms have evolved in a way which feeds their competitors and which makes them collaborators. The process of life creates a format of built-in negotiation leading to greater stability. Her view transcends the traditional Darwinian view of random mutation and natural selection among individuals. It also takes us beyond the extension of natural selection to the species operating as a whole, as well as the inverse Dawkins view that the competition is between genes, via the species which carry them (as indicated by his book title, "The Selfish Gene").

Sahtouris' view is that the survival of the whole ecosystem depends on the balance of tensions at all these levels. The ecosystem as a whole survives when the balancing process itself works. There is a tension between the impulse for the

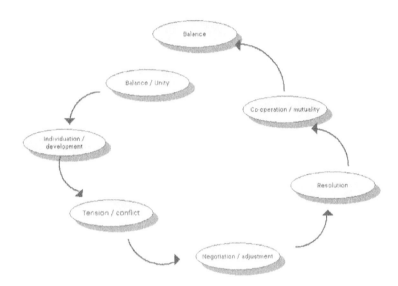

individual or the species to gain its maximum benefit, and the need for such development not to undermine the success of the whole. The individual or species cannot ultimately be successful at the expense of the whole. For this reason, a plague bacillus which regularly kills its hosts is less well adapted than the bacterium which gives you a sore throat year after year; the probiotic that we sell in pots is arguably the best-adapted of all.

We will revisit this view later, as it is a mirror of other theories which are valuable at the societal level and which describe the evolution of cultures. This balance between competition and co-operation is also critical to the functioning of human economic systems, as discussed in Jon's book "Future Money". The diagram here which depicts the cycle Sahtouris describes should therefore be seen as an example of a more generic process, present throughout life. We will stay with the biological evolution focus for now although we have added to her labels for some of the stages to widen their context.

Perhaps you can visualise two or three cycles super-imposed and viewed from above, forming a helix (cylindrical spiral) – inhabiting the spiral of development that we are depicting or overlaying our 'target' diagram, propelling its growth from centre to periphery. We have attempted to show this below and a colour version is on the website[18]. Perhaps you can also visualise that in the area of individuation and development there might be many ways in which this would happen – the development of new versions of an organism alongside the old, the arrival of an additional organism from elsewhere which adds to the ecosystem complexity, or the development of more advanced characteristics to which other species in the system have to adjust.

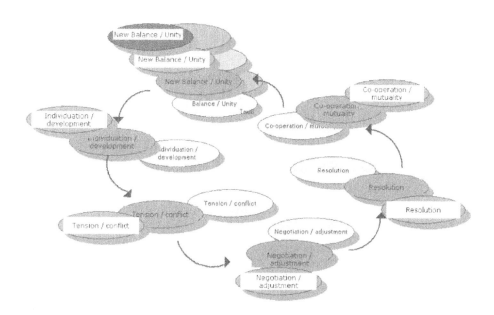

Added to this, there is a need to imagine that the changes in such a cycle may be happening at any of the levels which we encountered when describing the

---

18   www.scienceofpossibility.net/resources

biological complexity. For example an organism might become more toxic or less digestible by developing a new piece of chemistry that requires adaptation on the part of another species which feeds on it or from it. At another level it might develop a capability such that it is faster, or has better sensing mechanisms. Either instance places new adaptive pressures on any species that is higher in the food chain. Its rhythms and cycles might change, for instance breeding slower or faster. There are species of cicada (locust) which lay dormant for 13 or 17 years at a time and emerge at the end of the cycle to breed and then burrow again as larvae. It is suggested these prime number cycles reduce the likelihood that another species can evolve a corresponding shorter cycle and that this increases the likelihood of the cicada's flourishing. A 15-year cycle might have overlapped with 3 or 5-year ones or a 16-year cycle with 4 and 8. So change may be generated at any of those levels, chemical, behavioural or reproductive and may have its impact on other organisms at any of those levels.[19]

Fundamentalist Darwinians might argue that such features are properties which emerge as by-products of a competitive system. This cannot be simply disproved by external observation and this gives their case its appearance of strength. Like Gould's statistics it is true, but is it the whole truth? When deeper relationships or the existence of consciousness are not part of the picture, the fundamentalist view can appear sufficient. Our view is that there is abundant cause to see eco-systems as co-creative cycles, not linear phenomena.

Within this co-creation, the system is likely in our view to explore and exploit any direction and colonise any niche available to it. For this reason, it is to be expected that Stephen Jay Gould's statistics would not reveal a compulsive trend toward complexity. However, this is not the same as saying that the process is purely random – merely that it could not be distinguished statistically from it. It is not the same as saying that complexity is simply a manifestation of "the drunkards walk" –

[19] **Link to Sahtouris Spiral on website**

staggering between the minimal wall of single-celled simplicity and the open road of complexity. It is equally the case that an inherently auto-creative and self-exploring consciousness would behave in the way that Gould observes – not staggering, but dancing in free-form both close to and away from the wall – inhabiting increasing amounts of the open territory.

Readers will make their own judgements whether fundamentalist Darwinism makes a sufficiently convincing case once all the information is included, and when the full picture of the inheritance process is truly observed. For instance, if genomes repair mutations and bring new "genes" (or expressions) into service, what governs the editing process. How do the "editor" genes know which mutations to keep?

Equally, returning to the underlying claim that the evolutionary drive consists entirely of competition for scarce resources, how does Malthus' doctrine stand up when tested against a planet which started from zero in terms of life and food, and which has steadily increased in abundance of forms and complexity of expression? As with Newton's law, there is an underlying view which treats Earth as a closed energy system. As well as not recognising the developmental process of life itself, this apparently ignores the huge input of energy that we receive each day from the Sun. For there to be scarcity in the face of such daily abundance requires either huge failure of vision or massive incompetence. Malthusian thinking gives us a lame excuse for competitive attitudes that we really don't need. In case it is not obvious we should point out that there has been massive population growth – beyond anything that he imagined – since his time. Nature has shown how it can take such energy and create abundant life from nothing. We need simply and gratefully to join in. This means learning to co-operate.

Similarly the doctrine of "survival of the fittest", which is a very crude expression even of Darwin's view, has been responsible for an entire culture of disrespect for nature. We are all familiar with the ecological planetary outcomes of this and it has been used to justify all kinds of abuse from Nazi genocide to corporate greed. It results in a business model which runs counter to the interests of humanity as a whole and leads to a corruption of social values. Malthusian thinking has undermined the core wisdom of Darwin's thesis and left it with a significant weakness. Darwin's theory as interpreted by fundamentalists cannot encompass evolution at the level of entire ecosystems that display intelligent dynamic harmony.

Clearly it is our contention that there is a component within this intelligence that is related to the presence of consciousness. Equally clearly we cannot prove such a thing purely by observation. We need to be participants in the natural world, and it is here that we must remember the views represented in our "talking to trees" chapter – views which were part of wide human understanding in the centuries before science chose to limit what is "knowable". The world has need of a restored balance in this arena and of Goethe's ways of seeing, and pursuing the scientific endeavour.

The bigger picture of coherence

In the previous chapter we detailed the kinds of characteristic that are necessary for coherence to be demonstrated in a large organism such as the human body. But internal coherence also relates to these other questions about development and change at the level of the species. Ultimately it must bring us back to our core theme of the pervading presence of consciousness or information in the universe. We are still pursuing our claim that subtle change can be amplified and reflected in a larger way. The presence of coherent energy-states, and their ability to resonate, radio-like, to the appropriately tuned signal is one of the key elements in this picture.

We referred previously to the difficult task which has faced biologists when they attempt to understand how a single cell differentiates into the many different organs and structures of a complex animal. We indicated earlier how Driesch's experiments on sea-urchins, as well as other's work, showed that an organism can still develop, even after the initial cell is interfered with.

At the start of development, all parts of the embryo have wide-potential to become any structure in the adult organism and there is no indication at that stage as to which part will become what. A piece of embryo removed and transplanted before cell differentiation has occurred will develop in harmony with its new surroundings. However, after differentiation has begun a transplanted cell will continue on its selected course, regardless of where it is grafted. It has also been shown that even if interrupted in its development by freezing or preservation in oxygen-free conditions, an embryo carries some kind of "memory" of which structures its cells are committed to develop into.

Science has searched in vain for chemicals that might be signals for differentiation (morphogens). What seems instead to be the case is that the determined state possesses properties which more greatly resemble a response to the presence of a "field". What seems to be happening, as first suggested by Joseph Needham in 1936, and more recently borne out in experiments by Tatafurno and Trainor, is that there is a vector field present which causes major polarities of electrical activity and major body axes to be oriented according to fields which run first anterior-posterior (that is, mouth to anus along the central tube of ingestion and elimination) and then subsequently dorsal-ventral (that is, between back and front). The third dimension, side-to-side is, as your mirror tells you, typically symmetrical in many respects in most multi-celled creatures.

However, such indications leave much detail yet to be found. In addition, none of the above gives a clear indication regarding the development of behaviour. There is room to consider other possible influences. One such has been proposed and is our next topic.

Development of form and behaviour : Morphogenetic fields

A strong challenge to the purity of the view that everything is in the genes, or in the physics and biology, is that which is made by Rupert Sheldrake, firstly in "A New Science of Life", and then in "The Presence of the Past", a hypothesis which he calls "Formative Causation". This theory offers one example of the kind of mechanism that is required for consciousness to direct the process we are discussing. All the detail that we have so far examined still fails to supply us with a real understanding of how an oak-tree is oak-shaped or a liver-cell not a toenail-cell. We have shown that there is no one-for-one correspondence shown between bits of DNA and most observable functions. As Sheldrake points out, all plants contain the same chemicals and the chemicals don't explain form.

There is no correspondence between the physical shape of DNA, or the form of proteins, and the shape of your nose. The genes are carried identically in your liver and your toenail cells. As Sheldrake points out and as we have indicated, the explanation of a purely genetic program for these factors becomes very weak. One formulation of the conventional answer would ask us to be satisfied by such statements as "complex spatio-temporal patterns of physico-chemical activity not yet fully understood", as explanations for quite crucial elements of the process. It would be easier and more open to say "I'm sorry, I haven't a clue."

Biological science developed in opposition to notions such as vitalism, the perception that there was a "current of life", and to un-provable assertions regarding the mind of God. Yet it needed some kind of explanation for the development of form. We have seen that genes may determine many differences in form, but that does not mean for sure that they determine the whole structure, nor that they are the only determinant of the form. We discussed the limitations of purely genetic explanations.

There is some explanation emerging in science that explains increasing size in organisms. There is a particular set of genes known as the Hox cluster that are responsible for staking out the spatial domains of the body plan along the longitudinal axis. This cluster is repeated in increasing numbers in larger animals. (J. Garcia-Fernàndez). No doubt more will be discovered to account for a number of features.

Nevertheless, it is clear that a patterning influence is at work and that DNA is not showing itself adequate to the task. Something makes sure that both your liver and ours are on the right hand side of our bodies. This is generally thought of as some sort of "positional" information which tells cells where they are, and enables them to manufacture the proteins that are correct for different organs and functions. We referred above to the postulation of chemical "morphogens", graded concentrations of chemical guides. None have so far been found, which we believe tells a story in itself but even if they could be found this would still leave the basic question unanswered. What determines the gradient and location of chemical concentration in the first place?

If you have ever seen sped-up video of organism development you might recall that the early cell mass does not stay still and develop in a static, linear way. There are stages where the whole "blob" will rotate and fold in on itself, like turning a split football inside-out. This creates a challenge for all attempts at explanation, such as how do the genes tell a whole mass of cells to fold, but is a particular challenge for gradient models.

It is also obviously difficult to establish how such measurements would be made by a developing cell. Where would a chemical gradient start, or end? Relative to what starting point does an abdominal cell detect that it is spleen, liver, lung or rib, when all these are so close. Furthermore there is a duodenum winding past these, and many other organs. There is added difficulty regarding the number of gradient chemicals required, when there are many different structures involved, and the difficulty of preserving gradients in a way for example, that allows ribs to be separate. Even with electrical fields to support such chemicals, the embryo is not an obviously promising environment for precise positional information and the presence of large amounts of fluid not supportive of positional stability. Once again, all of these are good reasons to doubt strongly whether the DNA and cell chemistry alone can explain what is happening here.

An alternative conventional explanation involves the suggestion that there is some form of counting mechanism, by which cells know how many replications they have performed. According to the tally, they select what type of cell they should now form into. We do not know how cells would count, unless of course they are conscious of one other, but that would support our approach, not theirs. It is unlikely though as this theory is hard to reconcile with experimental evidence. It does not fit well with the results of Driesch's egg and larva-tying experiments described earlier, and does not explain at all how undifferentiated tissue moved from one area of the organism to another will develop the characteristics appropriate to its new location, as for example other muscle tissue does when attached to the heart.

As a result of the problems inherent in making the chemical, positional or counting models of development work, and the lack of hard evidence for them, it is appropriate to take a look at the most credible alternative. This is known as the "morphogenetic field", and dates from the 1920's. Physics at that time would have been providing many reasons to think of fields. Electromagnetic fields had been shown in many guises, and are known to be integral to many aspects of physical form. Fields are a helpful means of conceptualising actions which take place at a distance, and with no visible connection.

You cannot see the field which causes a magnet to point north-south It is only the magnet (or other detectors) which are evidence that a magnetic field exists. Science is comfortable with magnetic fields, because it has discovered a wealth of detail regarding the relationship between this field, and electrical activity. It is comfortable because it has generated deep and wide-ranging theories according to which we can successfully predict results.

Morphogenetic fields are no less plausible than magnetic ones, though you would think otherwise if you read the objections of biologists. They are no less visible, and no more so. And if you accept that the detector mechanism could be an in-built intelligence in the developing organism itself, then we have potentially as reliable a detector as the sliver of iron in a compass. This is not by any means enough of a reason to believe that the theory holds true, since the argument is circular. Much more is required. But it is not in itself implausible, any more than magnetic fields are, and does not warrant ridicule. If you had been told that birds can navigate 5000 mile journeys on the basis of invisible lines of force round the globe, prior to the discovery of magnetism, you might have had difficulty with that notion too. We don't yet fully understand how birds detect those fields or the in-built intelligence that navigates by them, only that they do so. But at this point in time morphogenetic fields have still not received anything like the level of research attention that have been applied to genetic mechanisms, and are hard to provide further evidence for. So what evidence, what indicators can we find?

One indication that there may be such a field comes from Kirlian photography. The Kirlian energy field is not normally visible, but can be made to display on a photographic film, when electromagnetic energy is applied to the subject. When humans are photographed in this way, an energy field is shown to surround the body, in the manner of an aura.

Kirlian photographs of leaves show the "aura" of a complete leaf, and then the similar shape which persists after a part of the leaf has been removed. There are plenty of photos of this on the web. It has been shown that similar effects occur with other organisms where a part of the body has been removed or damaged. This is not by any means proof of a morphogenetic field, but it is a strong indication that some kind of energy field is present around living systems and that this field persists even (as with these leaves which are detached from their plant) when the organism is not fully alive.

There is a large gap between knowing that there is a field, and showing that this field somehow conditions the development of form. One feature of growth that we are not used to in humans, which occurs in certain animal species as well as in plants, is the ability to regrow limbs, as a starfish or salamander can. A nematode worm will grow a replacement body - almost from any part. It will grow head and tail from a middle-section, and it can grow a right half from a left half (or vice versa), if sliced lengthways. Of course, any gardener knows that plants will grow from cuttings, and one of the common frustrations of British life is that the smallest piece of dandelion root or ground elder that you fail to pull out, will re-grow from deep beneath your lawn. This is not the same as growth from seeds. Whatever "program" is involved, is clearly capable of taking account of what already exists. A true "program", as is attributed to a seed or an egg, starts from a known place. Re-growth starts from anywhere, and yet is responsive to some kind of condition regarding what is, and what is not, already present.

Research shows that such re-growth can occur in humans as when a child's finger-tip was regenerated; it appears that this can take place if the wound is prevented from healing over. We conjecture that this prevents conflict between the protective impulses behind rapid healing and the more open but less safe programs governing re-growth. One way of making this difference is known to be the presence of an electric current. But this still would not explain how the cells know what forms to take up, which once again cannot be done by counting.

In the absence of research resources being applied to this field of study, there is not much more to say about it. It is as much as we can do, to put it forward as an equally credible alternative to the orthodox assumptions.

Formative Causation

As stated, morphogenetic fields are not a new theory. The quotation from Alexander Gurwitsch used at the head of this chapter is from the book describing his theory which is now 100 years old. Gurwitsch was also one of those first to detect and investigate the bio-photons which Fritz Popp later demonstrated and which we discussed as one potential synchronizing mechanism. For Dr. Sheldrake though, beneath the old theory of morphogenetic fields lies a deeper explanation which he calls the hypothesis of formative causation. This hypothesis begins from the assumption that morphogenetic fields are as physically real as gravitational or electromagnetic fields. Each kind of cell, tissue or organism has its own kind of field. The field is taken to supply the "memory" of form which would be required to guide the organisation of developing organisms over time.

What Sheldrake sees as being new in his hypothesis, is that the structure of the fields is not taken to be determined by mathematical formulae (which inadequately explain evolution), nor from some simplistic transcending idea in the "Mind of God". The field results from the form of preceding organism, and is akin to a collective memory, one which changes and develops with the contribution of each member of the species. He postulates that the way this would work would depend on a kind of "morphic resonance". Resonance effects occur when the vibrational frequency of one form of energy is picked up by another object, and it too takes on some of the same vibration. When a lorry passes your window, and that window-pane hums the note of the lorry-engine, that is resonance.

Resonance, as a descriptor of action at a distance such as is necessary here has the benefit of supplying a mechanism by which the "sound" of a form might be transmitted. However, there is no obvious equivalent in this context to the air molecules which transmit sound vibrations from one place to another. We would need to find a medium of transmission. Sheldrake states that morphic resonance is different from other resonances, in that it is required to transmit only information, and not energy. This could be true, and it implies that information has no energy value. This asks a deeper question regarding the nature of information and organisation and whether ideas have energy, which we will avoid for now but which you might choose to think about. If information has no energy / is not energy, what is it? So we will look further at the plausibility of morphic resonance while accepting for now that we do have not shown what the mechanism might be.

In general, resonance effects occur on the basis of similarity. That is, the size of the window-pane is such that, like a guitar string of the right length, its natural frequency corresponds to one emitted by the lorry-engine. A tuning fork is a piece of metal shaped in such a way as to have a very precise resonant frequency. Resonance occurs on the basis of a rhythmic activity (the movement back and forwards of a piston in an engine, or a pane of glass in the air). All substances are in some form of vibratory movement, from the electrons in an atom to the cycles of cell division, to the motion of planets. Thus all are in theory susceptible to resonance.

This means of course that the morphic field does not operate on its own. Otherwise a mustard seed in the child's mustard-and-cress tray would just as likely grow up as cress. It is a requirement of this process that the morphogenetic field should act on the ready-made operating structure of the developing system, which is already organised. The image of this is that a cress seed is already "tuned" to receive and respond to "cress vibrations", and mustard correspondingly to its own kind. The postulate of morphic resonance does not deny the chemical activity of genetic material, but works with it, providing an external guide to the development process.

As Sheldrake describes it, all past members of a species influence these fields and do so with a cumulative effect, which increases as the total number of members of the species grows. Under their collective influence, the morphogenetic fields in a new organism are not sharply defined, but consist of a composite of previous similar forms. He describes morphogenetic fields as "probability structures", in which the influences of common past types combine to increase the chances that such types will occur again."

Available space limits the depth to which we can pursue this subject. But to gain a full flavour of the nature of morphic resonance, and to understand a little more of the justification for the theory, we need to step briefly outside of biology and into chemistry.

Rupert Sheldrake asserts that morphic fields operate at all levels of matter. Two areas in which there are reasons to think that they might have application concern the formation of crystals, and the folding of protein molecules. With protein molecules, we are dealing with very long chains of amino-acids. These chains spontaneously fold up into typical three-dimensional structures. Scientists in the laboratory can cause them to unfold, but unless damaged, they will refold into their original conformations.

The surprising thing about this process is that there are a very large number of ways in which molecules of this size might fold. The folding process takes time, and there are so many combinations that to try all of them might take several years - probably large numbers of years. The protein cannot test out which ones require the least energy to maintain the structure in existence and are thus the most energetically stable. The modelling that has been done for this process indicates that there are many different structures that would all fulfill this condition. No theory adequately explains how it is that the protein consistently adopts the correct configuration. It has been suggested that some routes may require less kinetic energy to adopt, but this has not been proven, largely because the calculations required are unbelievably complex. Accordingly, Sheldrake suggests that the hypothesis of formative causation would be appropriate to supply the "memory" which the molecule uses in making its folding "choices".

The other example in chemistry concerns crystallisation. Crystals are regular structures of chemical compounds less complex than proteins. In a similar manner though, it is not possible to predict from first principles what would be the "correct" crystal lattice for a particular molecule to arrange itself into. Nor is there a way of testing whether the selected structure is the most stable. In general, molecules simply will not adopt any other than their normal choice. In that respect crystallisation is a similar phenomenon to protein-folding.

The interesting feature in relation to morphic fields is the creation of new crystal forms, from new substances. Laboratories the world over are engaged in synthesising new kinds of substances, thousands of new molecules each year. So there cannot be any field in prior existence for this new substance. It is consistent with the theory then, that the first attempts to cause new substances to crystallise typically take a long time - sometimes weeks or months. It would also be expected that once this has taken place, the existence of a morphic field would cause subsequent attempts at the same process to happen more quickly. This is the case, and it has been observed many times, without any other sensible explanation.

Morphic resonance and behaviour

One other area where there are phenomena that are not mediated purely by genes and therefore hard to explain without some recourse to something like morphic resonance, is that of memory and learned behaviours. Although it is possible that Dr Sheldrake underestimates what the brain is capable of, particularly in the light of recent work on neural networks, he quotes research that is not easy to account for.

How would a behavioural habit be passed on genetically? It is one thing to show that a behaviour could arise over generations by evolution, and be selectively beneficial, and from that to conclude that it is passed on in the genes (even if we do not know how). But what of a habit that was trained in the laboratory, and found in the next generation? The famous Prof. Pavlov, pioneer of the conditioned reflex, trained white mice to run to a feeding place when an electric bell was rung. In his results, the first generation of mice took 300 trials to learn. The second generation took 100, the third 30, and the fourth a mere 10 trials. This experiment was not repeated successfully by him - which could be because the morphic field was already established, and would affect any future group of mice.

Another relevant set of experiments began in Harvard. William McDougall, a fierce opponent of behaviourism and mechanistic views of nature, trained white laboratory rats in a water maze where they had to associate the correct choice from two exits with the presence of a light. The first generation of rats learned the maze after 165 errors, on average. Each generation learned more quickly, until after 30 generations only an average of 20 errors occurred. This was shown not to be in any way due to selection for intelligence. Even when McDougall selected the less successful rats to breed from, there was still a progressive improvement from generation to generation. McDougall himself was interested in phenomena of human "group mind" but also regarded his results as indications of Lamarckian inheritance.

This experiment was repeated in Edinburgh and Australia. In both cases the rats learned very quickly, and appeared to be starting where McDougall's rats left off. In the Australian tests, they also used rats that were not descendants of previously tested generations, and even then the behaviour was learned faster. This removed any potential, not only for the inheritance through genes (which in the absence of a spontaneous mutation is anyway not compatible with normal scientific descriptions of genetic change) but also for a Lamarckian inheritance of acquired characteristics.

The last example of this that we offer was in the work of R.C. Tryon in California. Tryon was attempting to establish "bright" and "dull" strains of rats, and was selectively breeding to this purpose. He achieved his aims in the sense that he did show that "bright" parents had "bright" children. But he also found that in both strains, the maze-learning results improved from generation to generation.

It is close to impossible to account for results like these without recourse to some form of transmission that is not based in genes and physiology. There are alternatives that would not involve morphic fields - for example that the experimenters were influencing the rats psychically. But it is not clear how this would progress from one laboratory to the next, nor is it apparent why the experimenters (who knew the right answer from the beginning) would increase their influence over time. The only way this would happen, is if the rats were becoming psychically "attuned" to their masters. This is a feasible mechanism, but in terms of normal science equally unacceptable, and desperately weak.

There is a lot more that could be said about morphic resonance and formative causation, and it is worth reading Dr. Sheldrake's full accounts, which go into areas such as language, human learning, and Jung's collective unconscious. He also sees connections with David Bohm's implicate order, which we are not presenting in this book, but recommend as an interesting area for anyone who wants to delve deeper. We are not convinced that the balance between genetics and formative causation lies where Rupert Sheldrake sees it. We think it likely that morphic fields are subtler, and have less effect than he claims, and that more is occurring at the biochemical level as is indicated in Dr Mae-Wan Ho's work. However, for them to exist, and have any effect at all, is a major challenge to scientific orthodoxy. It is probable that something more is involved than is accepted by orthodox science and that "morphic resonance" is somewhere close to the target.

The hypothesis of formative causation is a further example to be added to others in this book, showing that there are forms of energy and connection which must be taken into account in the proper development of scientific models of the world. These forms of connection share the requirement for "resonance" and for communication processes that are not recognised in scientific orthodoxy. More than that, morphic resonance is entirely consistent with the notion that consciousness is an active guiding force, and it provides precisely the kind of mechanism that would be required in the unseen realms. It fits well with all that we will go on to say about the nature of information in the structure of the physical universe.

The existence of such processes as intuition, resonance, healing triggered by pure energy, not to say those experiences narrated in the context of shamanism, come together with the small-scale biological features which maintain coherence and which are responsive to very small changes of energy. The scientific arguments for genetic determinism and randomness in evolution have been used to reject any possibility of an external creator. We have already hinted that we regard the biblical explanation as a metaphor. It seems unreasonable to us that such beliefs are regarded as primitive or merely irrational. To unsophisticated and pre-scientific cultures they may simply be the best formulations that have been available to encompass relationships that are beyond the intellectual frameworks they are exposed to, as well as being experienced in subtle and internal ways. Bear in mind that the descriptions of many aboriginal cultures and their world-views have been given to us through translations made by Western academics. They are likely to have been filtered through a world-view and language that does not encompass the original expressions. This is one reason we picked Malidoma Somé to speak on behalf of that reality. Such metaphors may deserve review in the light of modern understandings, but those who adhere to them do not deserve to be regarded as moronic.

At the opening of this section we listed five "certainties" that we were seeking to challenge. The last of these is that by showing an external creator-god as unnecessary to explain our existence in scientific terms, we should therefore write off all spiritual experience as irrational, or as a side-effect of an incompletely evolved brain. Millions of people have experiences that correspond to our view of a scientific reality. It is very important to recognise that making a creator-god non-essential does not by any means destroy all of the potential for "spiritual" consciousness to have meaningful involvement in a scientific reality. We have demonstrated where this can fit within biology, genetics and evolution. Our next task is to fit it in the non-living context of physics, and forms the third section of this book.

Review

We have reached the close of our section on biology. Having covered a vast and complex area, we would like to attempt a summary of the key elements.

The accepted view in science has become a widespread doctrine which presents Darwin's theory of evolution together with recent theories of genetics in a narrow and misleading way. We are told, or at least encouraged to believe that:-

Some chemical DNA sequences determine how organisms develop and grow

There is a clear and simple relationship between sections of genetic code and the development of characteristics

The environment in which development occurs is insignificant

Cells are simple blobs of jelly which host the DNA sequences – created by those sequences in order to pass them on

Living systems derive their complex co-ordination through the action of the brain and nervous system

Any appearance of consciousness, intelligence or co-operation is an accidental by-product of random activity in self-replicating chemicals

Changes between successive generations arise from random fluctuations in that process of replication

Any increase in complexity, growth in size or development of capability is a random stagger into a space of available possibilities

This all leads to an accidental evolutionary process in which the forces of competition for scarce resources determine which species will survive.

We have gone to some lengths to summarise the evidence for an alternative view in which:-

Creation can be seen as the out-working of a primary creative consciousness inherent in matter

The choice of which species survive is driven by an exploration by that consciousness of an abundant field of possibility

That exploration is a fundamentally co-operative engagement between the multiple forms of consciousness which inhabit the various species, leading to a spiraling dance into increased complexity and capability

The creative exploration led through self-replicating protoplasm and simple prokaryotic bacteria to more complex co-operative eukaryotes

Single cells are vastly more rich and sophisticated in behaviour and ability than blobs of jelly, are capable of sustaining life without their replicating nucleus and that they are in fact complex living organisms in their own right

When the limits to single-celled capability were reached, nature's exploration continued first with co-operation between cells such as slime moulds and then towards the development of multi-cellular plants and animals of increasing size and complexity

That both the single celled and multi-celled organisms are characterized by processes which drive coherence of function and multiple levels including:-

- vibrational quantum-level coherence and non-local connection

- sensitivity to electromagnetic fields and resonance effects

- photon communication

- bio-computer capability through the liquid crystal nature of cell membranes

- environmental influences on protein folding and gene expression

- rhythmic cycles of function at every level

- chemical messenger systems

- electrical nervous systems

Chromosome DNA is not the only part of the cell which affects development

The relationship between any "code" and the characteristics which develop is hard to determine

There are environmental influences on development and "editing" processes which intervene

The mechanics of timing and control are as yet poorly understood

That there are fields of information (parts of the creative consciousness) which maintain information about the forms that may develop – from crystals to organisms

All of this biological richness combines with perspectives from our section on consciousness, intuition and inner knowing. The presence in the world of nature spirits, devas, kontomblés and plant spirits is a demonstration of the ways in which human beings have experienced and continue to experience direct connection with the realm of creative consciousness. The informational connectivity demonstrated by Cleve Backster's plants and bacteria, and the intuitive knowledge of one human being by another are further demonstrations of the universality of this consciousness and of its ability to know itself in all parts from the largest to the smallest.

The cumulative effect of these facts is to show that all levels of living systems are working together in a harmonic and coherent process that consciously created life. This auto-creative or self-creative activity starts from the most simple level of replication. Like a computer in its boot-up process, the first instruction drives the second and third, until millions of instructions are loading and executing in an elaborate operating system. Life-consciousness has developed its own program, trialling exploring and intuiting its way into material existence, developing and embedding its self-knowledge in material form, increasing over time in abundance, complexity and physical capability.

The entire spectrum of this intelligence and creativity is inherent in every one of us. Each of us can connect with the World-Wide-Web of creative consciousness if we choose to do so. Each of us can experience ourselves more and more as a part of that whole. In our next section we will further explore the physics which underlies all of this. We will explore some of the implications for the way we think and for the philosophical views which have embedded themselves in our culture. We will also look at what we understand as "reality". We will see just how science got itself bent out of shape and find some very powerful messages for the ways in which we view ourselves and for what we are as individuals and as collectives in society.

Section 3 : Cosmology

"The day science begins to study non-physical phenomena, it will make more progress in one decade than in all the previous centuries of its existence."

Nikola Tesla: The Forgotten Genius

"What pattern connects the crab to the lobster and the orchid to the primrose and all the four of them to me? And me to you?"

Gregory Bateson. Mind and Nature

The Big Questions

Why does anything exist? How do we come to be here? These questions are the biggest that humans ask.

No-one can offer an ultimate answer to the question "why"? We do think we can offer a richer and more satisfying picture of all the rest.

Here are the questions that we would like to frame. We believe they will lead to a very full and nourishing answer. Having acknowledged that everything in the universe is energy

1. what is it that causes that energy to form itself into matter?

2. what governs the forms that matter then takes?

3. what does this tell us about who we are and the kind of cosmos we inhabit?

4. what are we able to be, think and do as a result?

What if much of what you have been told about the cosmos, about energy and about the reality of the quantum world, is seriously misleading? We intend to show how this is so, and to offer something fresh in its place.

How would it be if the shape of the material world was driven by patterns that we can see and recognise? How would it be if those patterns were built in to the very nature of our Universe, formed by information that the Universe knows about itself, held and sustained by patterns of self-knowledge that are the outcome of a long process of self-exploration, self-discovery, self-creation

on the part of All That Is?

In our section on evolution we described a co-creative process in which all living forms are in a dance with one another and with their environment. Every change sends its ripples through the ecosystem. New patterns form; new life conditions, ecological niches and life-forms become possible. That which works sustains and becomes long-term. Cells have a track record over several billions of years. That which does not work does not last. There are creative experiments which are never birthed, or which die very young.

Our view of the cosmos is that the co-creative dance started a long time before organic life. Universe itself is an also evolutionary process, a creative act. In that process, matter forms from energy. Some results vanish while others show stability. Those with stability become templates for others and then form the building blocks of a material creation. Patterns are formed, and those patterns are part of the self-knowledge that the Universe has of itself.

It is a fundamental law of physics that while matter can be created and destroyed, energy cannot be destroyed. All that can change is how it shows up. It can be heat or light or x-rays, or it can become matter. We do not and cannot know where energy came from. Its existence is our starting point.

It is not the whole of our starting point, however. All that we have said in this book and all that we will describe in this chapter leads to recognition that consciousness is everywhere. That prompts an obvious and crucial question; what do we mean by consciousness? The answer is elusive, but the description we choose here is that consciousness is a mixture of information, memory and communication. This description works for individual personal consciousness as well as for the whole.

The **information** is a description. It is a statement about something. Some of it is accessible to human awareness. Some of it may be a property of the universe itself and not knowable by us. This texture will emerge as we progress. **Memory** is the ability to record and retain that description, that information, through time. This takes place both personally in brain-patterns and in the whole in ways to be discussed. **Communication** is the ability to connect one item of information with another, or to propagate the information and memory through space.

The story we are about to tell shows that this consciousness, this assembly of information, recorded and propagated, is as real and fundamental to the Universe as energy is. It is a truism to say that "everything is energy", but this is not enough to explain the nature of the world. There may be people whose response to our use of consciousness in this way is a demand to "prove it". We cannot "prove" the existence of consciousness; we can provide evidence and we will use the assumption of its existence to show how it delivers helpful answers to the big questions.

Others may seek to know where consciousness is located and how it is stored. We will not answer this question either, except to say that we believe that it is everywhere and all-pervading. Beyond that it would be like asking the air where the weather is, or asking the electron where the energy is. Consciousness is the awareness of information and memory and it is located both within us and in the universe as a whole. We will explore location more in Section 4.

Descartes said "I think, therefore I am" in response to a theoretical doubt about the nature of personal existence. We do not doubt that the universe exists. We see little point in this book if we did. Instead we are taking its existence and ours as given, and showing what kind of "thinking", what kind of consciousness on its part would produce the outcomes that we see. We find this description very satisfying both because it answers a lot of the big questions and also makes sense of a great deal that to date has seemed difficult to explain at all. Having the description then makes it possible for us to engage actively with consciousness. It is a route to empowered choice and action.

Pattern is fundamental to our perspective. Gregory Bateson's statement "The pattern that connects, is the pattern of patterns" deserves constant repetition. Every aspect of creation reveals this patterning process. Patterns which work become templates and building blocks for bigger patterns. Patterns are central to the way in which a piece of information is recorded and propagated. Patterns are the mnemonics of creation.

One of the problems with quantum physics is that it does not supply patterns easily. It is too full of uncertainties. Despite its power and its areas of extraordinary insight it takes us away from what most of us would wish to know. Its mind-boggling nature has made understanding harder to find. Part of our task in the coming chapters is to say how and why this happened. How did science go so far into uncertainty that it can't understand how we can even be?

We cannot tell that story without referring back to some history. There are concepts to be explained that are still unfamiliar to many people. There are misunderstandings that have arisen in physics which are presented as truth. Some are leading-edge speculations. Others are mistakes and journalistic oversimplifications. We need to sift the useful from the unhelpful. As previously we need to unravel the truths that "just ain't so."

We also need to introduce some old and some new ideas, and some updated versions of very old ideas which have been neglected. We have to make a connection between what happens at the largest scale of the Universe and what takes place at the smallest scale where an energy wave turns into the tiniest particle of matter. It's a lot to encompass but we will endeavour to weave these multiple strands into an attractive and convincing tapestry. This is our cosmology.

13.   The hunting of the quark

If your Snark be a Snark,that is right:
Fetch it home by all means
 - you may serve it with greens,
and it's handy for striking a light...

But oh, beamish nephew, beware of the day,
If your Snark be a Boojum! For then
You will softly and suddenly vanish away,
And never be met with again.

"The Hunting of the Snark" – Lewis Carroll

All truths are easy to understand once they are discovered: the point is to discover them.

Galileo Galilei

The Challenge of Scale

Just how big is the cosmos, and how small are the energy waves and the fundamental particles that they may turn into?

Humans sit midway along this range, so we can look at what's bigger and what's smaller than us.

The largest scale we know is 3 followed by 28 zeros centimetres in size. This is the size of the known universe and might be spoken of as a million, trillion, trillion metres though technically it is called an octillion.

The smallest scale is that attributed to energy strings in string theory, whose order of magnitude is 1/ 1,000...000 (33 zeros in all) – which is a decillionth of a centimetre. This is written as 10^{-33} where the minus denotes that we are dividing by a number with that number of zeros after the 1. Each added zero denotes that we have divided by another 10. For comparison a virus, which we typically think of as very small, is only 1/10,000 cm (10^{-4}). Such small scales as 10^{-33} are not directly measurable (what could you measure them with?) and are derived mathematically.

The answers that physics seeks are required to hold true at both ends of this scale. Physicists must explain the nature of matter, how it is formed from energy, what causes it to generate, absorb or radiate heat, what it does in relation to dynamic, electrical, magnetic and gravitational forces. They also attempt to explain where it all began, how stars and galaxies are formed, and how it all might come to an end. Their challenge is to do this with one all-encompassing theory. That challenge is at the heart of where Physics finds itself just now.

When you consider that we are looking at things that are more than an octillion times bigger and smaller than we are, you realize that what we can measure is very dependent on the kind of apparatus we use. You can't bounce a laser off the "edges" of the universe to determine its width and the particles or strings at the smallest scale are bundles of energy that cannot be measured at all. They are only inferred to be present by tracking the movement of larger entities in very special apparatus. Even these larger entities are not all that large; for example protons are less than one trillionth of a centimetre in diameter.

We must also say something about "Universe" since this frames the boundary for our largest scale. We have accepted the convention of saying "the universe" and will continue to do so, but not without warning you that there is a potential flaw. It gives an impression that "the" universe is a thing in itself rather than the designation for the sum of all things.

Even if you think of it as "a" universe it still misleads us if we see universe as a thing. That could lead us to ask what is outside its outside, or if there might be others. Universe is everything that is, so it cannot have an outside or another. If there was something not included then Universe would not be Universe! What kind of thing has no outside! When we say Universe, you need to hear "All that is".

As Buckminster Fuller observes, to each of us Universe must be all that isn't me, plus me. It therefore has twoness and otherness inside, but literally the word means "one turn" or "one side" and by extension "(turning) towards oneness". That also makes Universe metaphysical; it is conceptual, abstract and philosophical. It is the essence of one-ness. Since it is the sum of all that is, it must also contain everything else which is metaphysical. As a result we need to note that there are boundaries to physics and see how it quickly and unavoidably becomes metaphysical. This is relevant for three reasons. Once again we encounter the myth of objectivity. Here too we will see that there is a pretence that science is only dealing with demonstrable material reality when in fact it is full of mathematical symbolic representations, artefacts created by methodology and by interpretations. The second reason it that the boundaries which science would like to set around material reality simply do not exist. They cannot exist because all that is interesting about how energy becomes matter is taking place at those blurred and fluctuating edges. Thirdly this book is concerned with spirituality and consciousness and these are metaphysical notions. We will show that metaphysics is more important than physics because that is where life is.

Physics, Metaphysics, and the God Particle

At the time of writing, many scientists are excited about certain events in the mountains of Switzerland. Around $4 Billion has been spent there in order to build the Large Hadron Collider. A primary reason for this construction was that it would enable physicists to investigate some very small particles, and one in particular known as the Higgs boson.

The standard model of quantum theory predicts for this particle it to have a mean life time of about 1.6×10^{-22} Seconds. That is 1/10,000,000,000,000,000,000,000th of a second. The amount of energy it contains (viewed as mass) is estimated (mid-point) as 1.26 Billion-electron-Volts / $9 * 10^{20}$ (approx velocity of light, squared). So it is very small and doesn't last for long at all.

So why would society spend so much to prove that this particle exists? The Higgs boson can be produced inside this huge apparatus which fires particles together in order to produce collisions at high energy. A single Higgs boson is produced approximately once in every 10 billion collisions. The recent proof of the Higgs boson's existence took around 300 Trillion collisions and a vast network of computers to analyse the results. You do get a lot of bangs per buck.

Apparently many brilliant people are persuaded that the Higgs boson is important. It is central to what Physics believes it needs to know if it is to reconcile the large and small scales. It also has acquired the nickname "The God Particle" which is very unhelpful, inaccurate and not at all what Sir Peter Higgs had in mind, but the fact that this name has stuck tells us something about the territory these investigations occupy.

In the coming chapters we will hear statements from Einstein and others about God. In theory, physics is about the material world; that is what people believe "physical" means. Perhaps that is how physics once was but now this notion is more of a mythology. In Newton's time physics concerned itself with planets and with scales of energy that we could detect directly (heat, light, magnetism, gravitational pull). Ever since Einstein came up with a theory that told us energy can be turned into matter and ever since Max Planck showed that waves of energy don't always behave like waves, the conversation has been about a world that is barely material at all, taking place at the lower extremes of scale just described.

Physics can define a physical universe. It has proved that energy must be preserved and cannot be created or destroyed but only changed from one form to another. That means there must be a finite amount of energy. This includes matter which is energy that has taken form. In this sense a physical Universe is finite. All the energy that exists now, wherever it came from, was present at the dawn of creation.

The prefix "meta" means something like behind, or above, or beyond, so metaphysics must by definition be outside of physics. Whether dealing with either the quantum world of the very small or with the cosmic scale of galaxies and stars it is not possible to avoid questions which are beyond matter. The attempt by others to deny or avoid this truth causes some of the difficulties with the science.

A metaphysical Universe would have to include all that is not material, such as experiences. Physics plainly does not do this. As we have seen, science does not engage with experiences; physics least of all. It would also have to include that which is conceptual, such as time and space, which it does, but without acknowledging that they are metaphysical.

So can you define the contents of a metaphysical universe? Clearly not. The metaphysical Universe can be encapsulated only at each infinitesimal snapshot point in time when it contains everything that already is, but of course it instantly moves on. Like the particle whose speed and position cannot be determined at the same time, and like Einstein's relativity picture of the cosmos, the metaphysical universe is restless and not to be pinned down.

Whereas the physical universe is finite because energy cannot be created or destroyed, the metaphysical universe is potentially infinite because something new is being created in every moment, an idea, an experience, a new moment of time or definition of dimensional space. Nature is not a thing, nature is creation. As Fuller states ("Synergetics" 321.02) "You cannot get out of Universe. Universe is not a system. Universe is not a shape. Universe is a scenario. You are always in Universe. You can only get out of systems."

Philosophers have sought Truth that would exist independent of sensation and be somehow unchanging. Yet Truth is dynamic because creation is dynamic, and it is our creation, our scenario, our sensation and our experience. There is a great deal of knowledge but very few genuinely objective Truths. It is the nature of being human that everything is sensation, experience and information. We will see that the same applies to the universe.

Physicists have skirted the dilemma of metaphysics by making their conversations mathematical, a sleight of hand which helps their work to appear scientific, rational and objective. Yet mathematics is entirely in the realm of ideas; it involves principles and transformations and relationships. Mathematics cannot be physical. It is a representation; it is about the physical and therefore, metaphysical. Mathematicians like to give the impression that mathematics is pure and absolute, that the answers it gives are Truths. They are only mathematical Truths and before very long we will discover that it is not possible for them even to be entirely right about those because there are limits to mathematical Truth. A mathematician proved it!

Are you boggled yet? We are asking that you get your mind around that which is abstract and not aligned with a material view of "all that is". The downside might be that this makes you uncomfortable, that it throws you into the unknowable. The upside, we suggest, is that it also propels us into infinite possibility. There are no boundaries to creation and no limits to scenario. The truth shall set you free, because the truth is that life is creation, in physics as much as anywhere. Freedom may not always feel comfortable but it does expand the range of choices. We find it valuable to know that, and even exciting.

You might think, or might still be affected by the general conditioning that we are all subjected to, that the physical is more important than the metaphysical. As we said in an earlier chapter "For Western science whatever you can touch is real and for Eastern Science whatever you cannot touch is real (and the touchable is the world of illusion)." Buckminster Fuller would go even further. He asserts

> "Metaphysical has been science's designation for all weightless phenomena such as thought. But science has made no experimental finding of any phenomena that can be described as a solid, or as continuous, or as a straight surface plane, or as a straight line, or as infinite anything. We are now synergetically forced to conclude that all phenomena are metaphysical; wherefore, as many have long suspected – like it or not – life is but a dream."

So let us dream into possibility.

Elusive Concepts: Framing the conversation

Little by little we are working our way towards a picture of the material world. As we do so we would like to continue setting a framework so that we understand just which aspects of that material world must be encompassed by the description we arrive at.

Physicists, mathematicians, laymen, we all face the same problem when we deal with the boundaries of material reality. We suspect that the mathematician, playful logician and philosophically curious Reverend Charles Dodgson, the real person behind the pen-name Lewis Carroll, would have loved the world that we encounter in this section.

We also suspect that one interpretation of his hunt for the snark is that mathematical truths are elusive. If the snark turns out to be a boojum, it is the hunter who vanishes away, rather like the career of a mathematician who pursues the wrong theory. There is a lot at stake.

This section will take us all to the very edges of creation. There, in the deep mystery of what happens in the instant that energy becomes matter one is confronted by the smallest and briefest slices of existence, attempting to form a picture of creation that will also remain true on a cosmic and eternal scale. We are all quite literally lost in time and space. Yet if we want to reach for the fundamentals of where our world comes from, this is where we must explore.

All mathematicians will tell you is that there has to be a kind of simplicity in the answer, a certain elegance of the kind that was present when Einstein distilled $E=MC^2$. They understand that nature cannot be clumsy because that would make it difficult to be sustainable or scalable. The medieval earth-centred model of the solar system had planets behaving in some very strange ways – looping trajectories that lacked some elementary reason. The calculations looked clumsy because they were based on a flawed model. When all planets were seen to orbit the sun those strange trajectories were replaced by elegant ellipses. Quantum physics seems to lack elegance and to be ripe for a Copernicus or a Galileo. What if the quark is a boojum?

How do we present the weird, complex, mysterious and paradoxical world of quantum mechanics and black holes without triggering your belief that you can't possibly understand such things? And why should you even bother trying? There will be very many who feel that they can manage without such knowledge and if you are one such you will of course be right. Whatever problems you think you have in your life, it is probable that you don't look for your answers to come from physics.

And yet, what if everything that you have believe to be real has rested on supposed Truths about the nature of creation which, if you could see beyond them, might transform into a whole new world? What if you were to discover that not only is there more to the world than you have ever known, but more to human potential, including your own, than you have ever imagined? Would that make it worth exploring?

It is said of Prof. Stephen Hawking's book "A Brief History of Time" that it was bought by many, but read by few. He himself makes the comment that his publishers told him that for every equation he put in, his sales would halve. Millions wanted to get close to what this brilliant man had to say, but few imagine themselves capable of doing so. It's a mass math-phobia. Our education systems have so much to answer for, and generations of teachers have contrived to convince us that the failing is ours. You don't need to be good at math to understand the world; what you need is to imagine more than you ever have before.

While we will talk about quantum physics (which you won't need to understand) and black holes (which are beyond our understanding) we will also present something that you will probably have not encountered before, something simpler, more elegant, more powerful, more empowering, a description which answers the question of what governs the forms that matter takes.

Among the many paradoxes of this section, we will address your logical mind at the same time as hoping to persuade you that your linear, rational thinking will not tell you what you need to know. Both hemispheres of the brain are required and getting your right-brain to supply as much information as your left will be a start. Even more than that, we invite you to allow yourself to engage Big Mind, the mind of the All. We ask you to seek that third dimension, to release your hold on the conditioned materialist perspective in order to touch something new, something more. It should not shock you at this stage for us to suggest this. After all, we have been showing that it is normal to be psychic (yes even you) that the intuitive world has provable existence, that your biological form is made in such a way that it is a transmitter / receiver into a field of information in exploring physical structure and the energy-matter realm. Why would we not invite you to be part of that?

This is an onion-peeling process. We make no apology for occasionally going around in circles, or returning repeatedly to different aspects. Everything that humankind has engaged with through millennia of philosophy, religious exploration, scientific endeavour and mystical experience lives here. All of our repeated themes are present, who we are, what made us, where we come from, where we might go and what is possible for us here and now.

Let's look at the areas that we will need to touch on:-

- Being and becoming: the tension between stability and creative possibility

- Knowable and unknowable: recognising the limits to our minds, methods and techniques

- Energy and matter: the very fabric of Universe

- Time and Space: again visiting our sensory and conceptual limitations

- Randomness and purpose: re-visiting familiar territory to ask whether it is all accidental or is there some element of "design"?

- Multi-dimensional views of reality: how many different points of view can you hold?

- Is Universe expanding? If so, what is it expanding into, and what would be its limits?

- The small and the large, the finite, the evanescent and the infinite

- The significance of paradox: It's OK that we can't get our minds into certainty so let's get over it!

- The nature of the creative process: Paradox and uncertainty are what give us possibility and choice.

- The mind of God: Are we ready to reframe how we view this?

- Sacred geometry: Some grounded principles that underlie big and small patterns in creation. A fresh way to connect the big and the small.

- Synergetics and syntegrity: The patterns that balance the material world in its dynamic tension between explosive expansion and constraining stability

- Elegance and simplexity: Finding something beyond quantum uncertainty, the power of a conscious creative impulse

The greatest minds that the world has known have wrestled with these questions without resolving them. Einstein, Newton and Galileo are all here along with dozens of others who are less known, but scarcely less brilliant. Some of the greatest technological achievements have been employed in the attempt to provide more information to work with or to test theories against observation. From the Fermi laboratory, the Mt. Wilson observatory and Jodrell Bank to the CERN Large Hadron Collider and the Hubble Space telescope, humanity has leveraged everything that it knows in order to find out more. But know from the beginning that every one of those individuals, supported by the most advanced of technology remains confounded by aspects of the mystery. If you feel confounded too then you are in good company. We all need to get comfortable with not knowing, and with having our minds boggled. As they used to say about Northern Ireland, if you're not confused you don't understand the problem. If we can increase the clarity just a little then we will have done well.

From here our journey is in four parts.

1. We will look at the large scale and in particular at what Relativity has been able to tell us

2. We will delve into the very small and look at what the Quantum model of the world has to say

3. We will look at what has been needed to bring these two views into alignment, at why that has not happened and at some fundamentals of living in paradox

4. We will describe an alternative and radical view that makes sense of the patterns, resolves some of the incompatible features and reconnects physics with the realm of consciousness that is our theme.

In the end, it's all quite simple. And if it's not simple, then it isn't the end.

Bigger and Bigger: Einstein's journey into the expanding universe

As the twentieth century dawned, previous views in physics were breaking down. Michael Faraday and James Clerk Maxwell presented electromagnetic activity as waves but as the century drew to a close, some of the experimental data was showing that light and heat did not behave in the continuous way that waves would indicate. Instead, some forms of radiation were subject to step-like characteristics that were at odds with classical physics. Energy was only being emitted as discrete units, described by Max Planck as "quanta". Physics needed to replace Newtonian mechanics with a new theory that would reconcile the laws of classical mechanics with the laws of the electromagnetic field.

Einstein later said of this

> "(it became) ..clear to me as long ago as 1900, i.e., shortly after Planck's trailblazing work, that neither mechanics nor electrodynamics could (except in limiting cases) claim exact validity. Gradually I despaired of the possibility of discovering the true laws by means of constructive efforts based on known facts. The longer and the more desperately I tried, the more I came to the conviction that only the discovery of a universal formal principle could lead us to assured results... How, then, could such a universal principle be found?"

A few years later, Einstein introduced his special theory of relativity. Like all other physicists he was searching for laws to describe how the world works. More than most he was seeking answers that would lead to the heart of matter.

Einstein's expression "true laws" point to his deep belief in a world which would have such things. His genius is that in the search for such laws he was willing to imagine a world quite different from our normal way of thinking. This divergence from what we typically think is the reason most people experience physics as hard to understand. Mathematics is not the problem, imagination is. You have to change how you think about reality.

Here's an example. You probably think that time doesn't change, that it is constant and always runs at the same speed. Sir Isaac Newton would have said the same so you are in good company. But it isn't that simple.

In 1971 two physicists, J.C. Hafele and R.E. Keating took atomic clocks aboard commercial airliners, and went around the world, one from East to West and once from West to East. On completion they observed that there was a discrepancy between the times measured by the travelling clocks and the times measured by similar clocks that stayed home at the U.S. Naval Observatory in Washington. The East-going clock lost time, ending up off by minus 59 nanoseconds, while the West-going one gained 273ns. This demonstrated what relativity had predicted decades earlier, namely that time is affected by speed of motion. It isn't constant.

In the Special Theory of Relativity, Einstein showed that time and length are not as absolute as our everyday experience would suggest; with high-speed motion moving clocks run slower, and moving objects are shorter. They don't just look shorter; they are <u>actually shortened</u>. In discovering laws that would not change –

universal principles – Einstein created rules that were no longer right by normal human perceptions and propelled us towards the boundaries of reality. Special Theory, and later the General Theory worked really well, the former providing us with the fundamentals of a relationship between energy and matter and the latter presenting gravity as a distortion in space-time.

The General Theory challenges our senses too because humans see space as an emptiness, an absence, and time does not seem like a thing. It is hard to conceive that space-time is something that can be "bent", but experiments prove that gravity does exactly that. Nor is it easy to understand that as a result gravity also affects light. When light from the stars passes by the sun, the sun's gravitational field bends space-time. Light travels in space-time; if that is bent, so too is the light passing through it. Shadows tell our eyes that light only travels in straight lines. If there is enough gravity, light is measurably curved.

From before the time of Newton's legendary musings on the falling apple, gravity had been a mystery. Newton was able to name the effect and describe how it worked in such a way as to make new sense of planetary motion. We are all used to the idea that there is attraction between the Sun and the Earth that prevents the Earth's speed from causing it to fly off through the rest of space. But why do objects have this property? What is it about matter that causes it to exert a gravitational force? This is the big unknown, the key problem in physics.

One of gravity's mysteries is that it is effective over large distances. If all material objects cause gravitational attraction then the universe should be shrinking. Everything should be getting closer together until all matter fuses into one huge lump. Is this what is happening, only taking place so slowly that we cannot see it? If not, why not?

The answer appears to be that the universe is not shrinking. That leaves three other possibilities. The first is that it is stable, a steady-state condition that neither expands nor contracts. That would call for an expansive force equal to the contraction of gravitational pull.

The second is that the universe is expanding, perhaps under the influence of an initial explosive force that is bigger than gravity. There are three sub-possibilities within expansion – continuous creation, oscillatory (will eventually contract again) and the popular "big bang".

The third possibility is that the question itself is flawed, that it assumes something that turns out to be untrue.

At the time of his General Theory, Einstein seems to have had a preference for a stable, steady-state universe. It caused some discomfort to him that his own theory indicated that the universe was not necessarily static and that it either had to be expanding or contracting. Astronomers were looking and as far as they could tell at the time it was doing neither. Looked at through relativity theory it seemed that such stability was like balancing a pencil on its point – not impossible but highly unlikely. So was relativity wrong?

Einstein's response was that this meant there must be some mysterious force propping up the universe, a sort of antigravity that pushed outward just hard enough to balance the gravity that was trying to pull it inward. He didn't like this idea because it spoiled the mathematical elegance of general relativity equations. An extra force would tarnish their beauty. But since the universe was apparently not matching his theory it was the theory that must change and Einstein being an honest scientist introduced "the cosmological constant" into the theory.

A decade or so later, the great astronomer Edwin Hubble was using the Mt. Wilson observatory, the world's most powerful telescope, to make very detailed measurements of the galaxies he could see beyond the Milky Way. He discovered that they weren't stationary at all. The galaxies were in motion, each moving apart from the other, like dots on the surface of a balloon being inflated. It was this discovery which ultimately led to the Big Bang theory, since reversing such expansion as if looking backwards in time suggested that the cosmos had once been tiny, with all matter packed tightly together.

Einstein was relieved to find that the cosmological constant wasn't needed and that his beautiful equations had been right to begin with. In 1931 he journeyed to Mt. Wilson to thank Hubble for saving relativity from the cosmological constant, whose invention Einstein later denounced as "the greatest blunder of my life."

Shouldn't it be simple to determine whether the universe expands or not? Why has this question continued to occupy astrophysics for most of the twentieth century and beyond?

It is hard for scientists based in one tiny location to measure what is happening over the vastness of the cosmos and there must be an element of doubt as to whether what they see fully represents the whole. We can suggest several reasons why this is so.

- Such measurements as are made use equipment and concepts that reflect what we think we know so far and are interpreted in the light of those theories.

- Even when we have confidence in what we are measuring, the reasons for the results we see may be complex; they may or may not be within the current bounds of knowledge.

- There is more unknown than known, such as what might be happening inside portions of the universe like black holes that are not visible to us.

- We continue to face the issue of a material-based perspective, where whatever is unable to be measured is not deemed suitable for inclusion in any physics theory. Under this sits the recurring deeper question discussed above of whether the universe is physical, or meta-physical.

- Science is more likely to find what it is actively looking for than that which it is not looking for or even actively seeking to avoid, thus creating inbuilt bias

- In the quantum context that we will discuss later it is possible that our observations are influencing the results. This point is less relevant to cosmological measurements but may well be reflected in particle collider interpretations.

In practice, as with the story of evolution, huge amounts of brilliant speculation are piled on fragmentary data with the danger of over-selling what is truly proven.

So, working with all of these vulnerabilities, what has physics achieved since 1931 and what is the current view of an expanding universe? Certainly knowledge has increased throughout the last century. When Einstein formulated relativity it was not even known that there were millions of galaxies. That fact was not discovered until 1924 – also by Edwin Hubble. Indeed, the Hooker 100-inch telescope that Hubble used was not built until 1917.

Science now believes that the answer to the question "is the universe expanding" is "yes". But it is more than that. The world now has telescopes yet more powerful than those available to Hubble. Using these, astronomers at two different locations competed to measure the light from supernovae at varying distances from us. In 1998 the results were published and they showed that not only was the universe expanding, the rate of expansion was increasing[20].

---

20   In 2005 that discovery of the accelerating universe earned the Lawrence Berkeley Lab's Saul

It appeared from these observations that Einstein had been right when he introduced the cosmological constant and upset the elegance of his relativity equations. But it would seem the size of the cosmological constant must have been wrong. If the expansion of the universe is accelerating, that would require a greater amount of anti-gravity than if it was expanding at a constant rate. What was Physics to do? Where would it get additional anti-gravity from? To see how it approached this problem we need to follow the story from the other end of the scale.

Perlmutter half of that year's Nobel prize. His competitors — Brian Schmidt of the Australian National University and Adam Riess of the Space Telescope Science Institute, split the other half.

14.   Curiouser and Curiouser

"If it was so, it might be; and if it were so, it would be: but as it isn't, it ain't. That's logic."

Lewis Carroll   Alice in Wonderland

Absurdity is the only reality

Frank Zappa

In this chapter we look at creation from its smallest elements and encounter the difficulties that Physics finds when it attempts to understand the event that take place at the boundary between energy and matter. We see that in the attempt to reconcile the almost impossible, Physics took pathways which then became traps to its thinking. In order to come up with something new, we first have to unthink some very powerful theoretical perceptions.

We have written the word "paradox" at many points in this book. In these chapters we will deal not only with deep paradox, we will enter step by step into a world-view which is also bizarre, strange and often only describable in mathematical terms. We are not mathematicians, still less quantum physicists and it is likely that neither are you. We start with the paradoxical and mysterious. We try to communicate abstract concepts in inadequate images and words. Then we will try to say why those images are leading us into failure even to understand what might be understood, while intentionally leaving you with mysteries rather than certainties. We remind you that this is the key to possibility; it is easy but it is not simple and maybe not comfortable. Forgive us, for we know what we do.

It is most important that you do not understand this chapter.

Our goal at this point is to set out with as much clarity as we can manage, just why this is. If it helps at all, we will be in good company. Richard Feynman, who as much as anyone developed the science of quantum physics, said of it:

> "It is my task to convince you not to turn away because you don't understand it.
> You see, my physics students don't understand it either. That is because I don't
> understand it. Nobody does."

Feynman was a Nobel prize winner and probably the most capable of all physicists of providing a clear explanation. His short book "QED, the strange theory of light and matter" from which the above quote is taken, comes as close to an explanation as any you will read, and manages to be witty and entertaining as well. And yet it is not understandable, because it is fundamentally paradoxical.[21]

Quantum Electrodynamics (QED) has been the most complete explanation to date of how the universe works, how the matter that makes it up is formed, how it behaves. It can explain everything in physics except gravity and believes that it may be getting closer to that since scientists at the CERN large hadron collider have proof that the Higgs boson exists. Its mathematics are precise in predicting the outcome of experiments on the smallest known elements of matter and the measurements of the universe. It is more precise than Newton or even Einstein's relativity in describing the motion of the planets. To give a feeling of this accuracy, Feynman says

> "If you were to measure the distance from Los Angeles to New York to this accuracy, it would be exact to the thickness of a human hair." But he goes on to say "The theory of quantum electrodynamics describes Nature as absurd from the point of view of common sense. So I hope you can accept Nature as She is – absurd."

We are not going to attempt to explain quantum physics in great detail. We are also not going to accept it in its entirety. Our task is to show where the brilliant accuracy that Feynman describes begins to shade into something less convincing, as he too acknowledged. What we first need you to grasp is what the quantum view tells you about the world, and how tricky common sense perspectives can be. Common sense and science both rely on what you can observe and measure. The paradoxes in QED tell you that there are aspects to the world that will never be directly observable and measurable. We hope that by the time you finish this section, you will see that as a cause for celebration.

Paradoxes in general also tell you about the limitations of the human mind, of the five senses and of living in a world of time. A reading of the physics literature would also tell you that the cleverest people on the planet (in this arena at least) don't agree about what QED means, or how the gaps can be filled. They can't resolve the paradoxes either, though we will hear that they are still trying. We will also hear why it may well be impossible.

At the core of the problem is the interchange between energy and matter. In simplistic terms, when energy is being energy, it moves through the universe as a

---

21 If you are so inclined, as a special treat you can experience Feynman's lectures direct on www.vega.org.uk

wave, like light or radio signals. When it becomes matter it can be recognised as particles like photons (particles of light) and protons (parts of the nucleus of atoms). Particles have mass and so one would think that they could be described as having a location at any point in time.

The problem for physics has been that while that may be true, you can't measure it. Put simply, the apparatus that you would have to use would affect the particle. This recognition led to the formulation in 1926 by Werner Heisenberg of his now notorious "uncertainty principle", which has since been shown experimentally to be true.[22]

This principle says that the more accurately you try to measure the speed of a particle, the less accurately you can measure its position, and vice versa. Heisenberg's formulation showed that the uncertainty in the position of the particle, times the uncertainty in its velocity, times the mass of the particle, has a lower limit, which is known as Planck's constant. This says in mathematical terms that there is no way of measuring that provides certainty, and you can't get closer than that limiting point. It does not matter how one tries to measure the position or velocity of the particle, or what type of particle it is. Heisenberg's uncertainty principle is an inescapable fundamental of our ability to know our world.

Mechanistic and deterministic views die with this observation and physics is obliged to work with what's left – our limited capacity to measure the universe. The theory that arises from this, put forward in the 1920's by Heisenberg, together with Erwin Schrödinger and Paul Dirac, is quantum mechanics. Quantum mechanics gets around the inability to provide exact predictions by forecasting a number of possible results for an observation, and saying how probable each outcome is. In so doing, quantum mechanics gives the universe an element of randomness.

It might seem that it is not important to us as individuals if unpredictability is present at the level of particles humans cannot see or imagine. In fact it is critical. It will have been clear throughout this book that we are seeking to undermine the mechanistic perceptions of our world and point out the spaces for something else besides material-based thinking. In biology we spent several pages unravelling the fixed package of genetic predetermination in order to provide an alternative to the purely random view of change. What we are doing here is similar. The physical universe is also not determined. However, while there is an element of randomness

---

22 For a graphic illustration of the indeterminacy problem, you can look on "YouTube" for Dr Quantum's video of the two-slit experiment

to the quantum mechanical world its randomness is affected by "something besides".

Once again we need to go a little deeper into the detail to understand the flexibility and the space for "something besides". In describing quanta as either energy or matter we skirted QED's view that they are essentially both at the same time. You could equally well say that they are neither one nor the other. Alongside the Uncertainty principle, quantum theory has a "Complementarity principle" which states that the two ways of describing, as a wave or a particle, complement each other and a whole picture only emerges from the two together.

Quantum physics confounds common sense by being more complex than we are capable of imagining. Common sense and our physical senses simply don't allow for anything to be two different things at the same time. It is likewise difficult to put into everyday language, since our perceptions and language are based in our experience as physical beings. In consequence the means of description ultimately has to be mathematical which is to say, symbolic. Even when the symbols match measurable reality, they are not the reality itself. They are the map, not the territory.

In quantum terms, subatomic entities are neither fully particles nor fully waves, but rather a confusing mixture known as a "wave packet" in which we can achieve only fuzzy measurements of both. This is why quantum theory came to find it necessary to predict the probabilities of outcome, depending on the experimental situation.

Being and Becoming

Nowhere more than here do we encounter the blurred edge between physics and metaphysics. The perceived requirement for a probability-based approach to creation led to an intense debate between two different views of quantum theory. How does anything ever come into being? What stops objects simply fading in and out, or keeps us whole?

One view led by Einstein saw this uncertain state of affairs as a weakness in quantum theory – that it was only a problem because of our inability to study nature without disturbing her, i.e. nature doesn't have a problem; we do. In his view, the discovery of quantum theory has not made the world suddenly unstable – a view which he famously encapsulated with the statement "God does not play dice with the universe" –a metaphysical observation of the highest order.

Nils Bohr told Einstein that he should "stop telling God what to do". He and

Heisenberg maintained that quantum theory was complete and that the instability is fundamental, even at the expense of an adequate understanding of how things become fixed and real. The developmental line followed subsequently by David Bohm argues that some states are more stable than others and that the transition involved

> "resembles the idea of evolution in biology, which states that all kinds of species can appear as a result of mutations, but only certain species can survive indefinitely, namely those satisfying certain requirements for survival..."

When introducing this section we drew some parallels between biology and physics in the balance of exploratory creativity with stability that is required in order for life as we know it to exist. How does it happen that something, be it particle, star or living creature, comes into being and survives, yet its survival not fixed for all time but remains capable of further change? We are asserting that these laws of nature would be in some way reflective of each other. We have more to resolve if we are to establish that adequately.

In the crucial "two-slit" experiment which was set up to prove whether light was behaving as a wave or a particle it proved impossible to say either, because the answer you get depends completely on when and where and how it is measured. This is how indeterminacy was discovered. It is only when it is measured that it becomes describable as one thing or the other (wave or particle). Until then the particle carries, in theory, all possible states in "superposition". It is the act of measurement (or "observation") which "collapses" the probability into a definite state.

This leads us directly into two areas of paradox. First is the phenomenon known variously as "non-locality" or "quantum entanglement", or the EPR paradox. This key unlocks a wide door to what is possible.

EPR represents the initials of the three physicists, Einstein, Podolsky and Rosen who were attempting to prove that a definite structure of reality has to exist, in line with Einstein's views given above. They proposed a thought experiment which considered what happens when two particles (for example an electron and a positron) are prepared for the experiment as a pair with complementary properties (e.g. positive or negative "spin"). These properties are not precisely known at the time of preparation, but the fact that they are complementary means that whatever one of them has, the other will be the opposite. You don't need to know what "spin" means for an electron, just that the two particles have to correspond.

In the thought experiment, the particles are seen to leave the point of preparation in opposite directions. At some point, when the properties of one are measured, this collapses its probability so that its spin is defined. By implication, the paired particle also becomes instantaneously defined in its properties as having the opposite spin, no matter how far apart they are.

According to relativity theory, nothing can travel faster than light. The information therefore cannot pass from one particle to the other in the instant required. There is no form of preparation which can make the outcome predetermined. As a result, Erwin Schrödinger (whose cat we will meet shortly) suggested that the particles need to be regarded as "entangled" – connected in a single coherent state. What could that mean? How could they be separate and connected at the same time?

This result might seem meaningless, trivial or even an irrelevant mind-game were it not for two facts. Firstly, David Bohm eventually found a way to turn this thought experiment into a real one and proved that the outcome is as predicted. We will discuss this further in the next chapter. Secondly – while the EPR proposal involves particles that have been artificially prepared together, it has subsequently been shown that any pair of particles, once they have interacted, remains entangled after they have separated. They become one quantum system, connected even though separated by distance. Quantum entanglement is a feature of reality.

This feature is referred to as non-locality, which means that there are connections at the quantum level which potentially span galaxies and the whole of universe, and which we cannot see or detect. There is a form of connection that allows information about one element of Universe to be known to another, regardless of distance. Thus conventional physics theory acknowledges mechanisms by which one particle can "know" something about another, irrespective of time and space – the same sort of "impossible" connectedness that were present in our stories of intuitives. In introducing our cosmology we indicated our view that universe is a self-knowing scenario. Even though it does not know where information about entanglements might be "held" physics is supporting the possibility for this view to be correct. Consciousness is entangled: or perhaps entanglement **is** consciousness.

That damned cat!

And now for Schrödinger's imaginary cat, almost as famous as Lewis Carroll's Cheshire Cat and whose existence is almost as inconstant. This parable will help us explore the nature of observation and "collapse". It is another thought experiment, one in which an imaginary moggy is shut inside a box. The box is imagined as one we cannot see into, hear from or otherwise detect anything regarding what has happened inside. Inside the box with the cat is a mechanism which, according to the completely random behaviour of a radioactive substance, might trigger the release of a poisonous gas, thus killing the cat. (In another version, there are two triggers such that there is a 50% probability of releasing food or poison.) Thankfully, unlike the EPR experiment, this has remained in the realms of thought only.

The question posed concerns the quantum state of the cat. According to the conventional interpretation, the cat remains in its state of superposition of possibilities for as long as the box is unopened. That is, it is neither dead nor alive, but in the "both / neither" quantum probability state. The probability wave collapses only when we observe the cat. Zen Buddhist readers will notice a certain similarity with the Zen question regarding whether a cat has a "Buddha nature".

There are many interpretations of this conundrum. One is to suggest that since we don't know and can't know, science has nothing to say about it and thus un-asks the question. This would be similar to the master's response to the zen question, which is neither yes nor no but "Mu" – a term which simultaneously translates as "both, neither and not applicable". Another is the core quantum theorist's insistence that the cat's existence is mathematically indeterminate, which offends our common sense since we know (in any normal sense of that word) that its fate was determined before we opened the box, even if we cannot prove it. A third imagines the generation of parallel universes in some of which the cat lives and in others, dies. Under this third theory all of us also branch multiply at every decision point, and so there are billions of Jon and billions of you the reader, in billions of universes. This makes for some good film plots, but has little else to recommend it.

Idealism and the role of consciousness

A more interesting notion is that it is our knowledge, our human idea about the cat, our consciousness that collapses its wave function. This has been resisted by many because it allows consciousness to interfere in the realm of materialist realism, but has achieved acceptance by some, including John Wheeler and Eugene Wigner, who have proposed that human consciousness is the missing link between the quantum particle world and everyday reality. It might initially seem that this is the road we are following as it ties in quite well with the views we have expressed so far. For us though, it cannot be a satisfactory solution – at least not in the way others are expressing it. Our disagreements would include the following.

1.   Even if you are not taking a purely materialist view, it defies common sense to suggest that if dead, the cat was not determined to be so until the instant we looked. "Reality" knew it was already dead even if we didn't. To suggest that the timing is dependent on us assigns an unwarranted importance to human knowledge and is true only as an artificial mathematical abstraction.

2.   The leap from determining the state of a particle by measurement to determining by distant observation the state of being for a complex life-form such as a cat that contains trillions of quanta is a misleading over-simplification. The quantum-level event was the either-or of the radioactive particle action. That has been determined and has been "observed" (i.e. measured) by the poison-release device. The deliberate omission of an external read-out from this device prevents human observers from sharing the experimental observation. The cat's response to poison is something quite different and takes place in a much more complex realm.

3.   The view that it is only <u>human</u> consciousness which can supply the missing link of observation that collapses the probability wave is clearly contrary to the evidence that we give from the plant world, and to state the obvious, it is likely that the cat itself had some feline equivalent of an "Oh bugger" experience before it ceased breathing. No human would have anything to add to that determining observation.

4.   The view we have been constructing is that consciousness is an inherent property of matter – that it has been present from the moment of the big bang and is creatively engaged in self-construction. That being so, even if something else changed the cat's state inside the box – by definition the realm of consciousness contained the knowledge of what it had done. This is a little like saying that "God knew", but the knowing is "within" rather than somehow held by an external entity. If you choose to ascribe some kind of Divinity to this reality, then that Divinity is inherent in all things.

It is only fair to Shrödinger and to Einstein too, to note that Shrödinger himself prefaced his introduction of this thought experiment with the sentence "One can even set up quite ridiculous cases." Ridiculous or not, the relationship between human reality and quantum indeterminacy has caused debate for decades and the cat is frequently discussed. We believe that our view will resolve the disquiet which gripped Einstein; there is no dice-rolling involved in our cosmology. Further, it overcomes the difficulty in resolving the questions of how anything initially comes into "being", of how the world acquires the kind of stability which we experience of it and of how we ourselves as experiencers are in a position to do so.

Quantum uncertainty opens up a scientific space. You can push almost anything that you wish through the gap that it creates and many people have done so. Almost any theory can be "justified" in this way, regardless of whether that theory has any substance or independent body of evidence to support it. Unfortunately many of the people who make such statements understand little of either quantum theory or science. Rather than adding to the credibility of alternative viewpoints the result is often to make them appear ridiculous and to give fodder to the sceptical extremists. If we are to do better than this, we have to go deeper; we must delve below the surface of quantum reality.

The key to material existence

In closing our previous chapter we indicated that the quantum view of gravity would be the next step in our story. Beneath the uncertainty sits the quantum physical exploration of what makes up the world of the very small. What is it that led to a theory that requires notions like the Higgs boson and from there to $4Bn particle accelerators in order to prove its existence? Quantum gravity lies at the end of this thread.

Max Planck's name has already appeared a few times in our narrative and it is about time that we explained why. His work introduced a significant shift in how the world was seen and laid the foundations for all that Quantum theory has become.

Newtonian physics was concerned with objects and with the energy of objects in motion, as well as with other forms that energy can take such as heat and light. By the 19th Century there was considerable focus on electricity and magnetism too. The idea that there is a spectrum of electromagnetic energy began with Herschel's discovery in 1800 that infra-red light conveyed heat energy. The theory developed over the next hundred years through Faraday, Maxwell and others to include a continuum of differing wavelengths and frequencies for different forms of electromagnetic energy. Regardless of what has come since this theoretical perspective works and continues to be used extensively.

Waves are continuous. They propagate through space; indeed you cannot conceive a wave that doesn't. Think of this continuity as like running your finger up a violin string. There are no gaps in the spectrum of wavelengths. Yet by the end of the century some results were showing up that did not fit this model. It seemed that light and heat radiation would also behave discretely; there were gaps in the continuum.

As referred to earlier this led Max Planck to a theory that electromagnetic energy could only be a multiple of an elementary unit – a constant which has become known as Planck's constant. Through a temperamental conservatism he was at first reluctant to depart from Maxwell's formulations and it was Einstein who eventually persuaded him that light was in fact acting as particles, an idea that Louis de Broglie (1924) later extended to encompass all types of sub-atomic particles in a wave-particle duality.

Since wave-particle duality is at the centre of the matter-energy shift, the way that we conceive of the transition that takes energy and turns it into sustained form holds the key to material existence. To get to the quantum view of this we need to divide the world into finer and finer pieces until we reach the boundary of material existence.

The first step is to look at what atoms are. For centuries these were the smallest conceivable amount of matter, and each substance in the world viewed as different. Prior to quantum theory, J.J. Thomson (1897) had already identified the electron as the first particle which is smaller than an atom, and during the early twentieth century other sub-atomic particles proposed by Ernest Rutherford followed, like the proton (1920) and the neutron (1932). These particles provided a model for the construction of matter in which all the different chemical elements have common sub-components. The difference between iron and sulphur consists of how many neutrons, protons and electrons (collectively termed "hadrons") they are made of and the whole table of elements can be constructed in this way.

That seems relatively straightforward and even elegant but unfortunately the subdivision of matter did not stop there. Step 2 in subdivision is that in order to discover what hadrons themselves were made of scientists figured out that it would be a good idea to fire them at each other until they broke. In the early 1960's the first atom smashers were built and at the 2-mile long Stanford Linear Accelerator this led to the discovery of quarks by Murray Gell-Mann and George Zweig who had first proposed their existence in 1964.

There are two forms of quark that are stable and many that are not. Neutrons and Proton are formed from different combinations of the stable, or "first generation" ones called "up" and "down" quarks. But after this, things get difficult. Many other kinds of quark are not stable and tend to decay into stable forms. Only first generation quarks occur in nature and to create 2nd generation and above takes much more energy. Arguably any type of quark after the first generation should be regarded as a "virtual" particle, a kind of artificially created Step 3 of sub-division.

The current interpretation is that the hadrons which form atoms do not interact directly, but rather manifest "virtual" particles that mediate the actual interactions. Virtual particles are so named because they don't stick around for long. They blink in and out of existence and are very hard to detect. They are a momentary vibration in the matter-energy dynamic interplay. They represent or even embody the shifting sands of quantum uncertainty. We believe that it is legitimate, even crucial, to ask "how real are they?"

What is happening in this arena of huge energy and tiny time? Let's first frame the dimensions, using just one of these virtual particles – the Higgs Boson – as an example. As we described earlier "finding" this particle was a major reason for spending close to $4 Billion to build the Large Hadron Collider. The Higgs boson can be produced by creating particle collisions inside this apparatus. Once in every 10 billion collisions or so, a Higgs boson will be produced and will exists for less than one septillionth of a second.

Bear with us please, even if there is a strong temptation to glaze over or to think What the Bleep....! There is method in our madness and we are coming close to the question of gravity. We are describing the world that scientists are experimenting with – things that may exist and which they choose to define as particles (real or virtual) even though they maintain their mass for much less than nanoseconds. In the standard model for particle physics the Higgs boson is just one of 17 identified fundamental particles, all of similar dimensions. There are 6 quarks, 6 leptons, plus five "bosons" - the photon, gluon, Higgs, W, and Z particles. Various particles have complex and exotic properties of spin, charge, charm and "colour" and in some cases combine to form other classes of particle.

Bosons are named after the Indian physicist and mathematician Satyendra Nath Bose whose name re-occurs later in our reference to Bose-Einstein condensates. Bosons are essentially force and radiation carriers. The photon is the particle that carries light when that is not being treated as a wave. Gluons mediate strong nuclear interactions. W and Z bosons mediate weak interactions. The Higgs particle is proposed as the mediator for gravity. Collectively bosons carry energy away from interactions or they carry energy to particles. In a sense they are descriptors of interactions between "types" of matter / energy as fleeting and virtual as a thought. This is the world that quantum physics has cooked up for us – one which treats a relationship as a particle. It is the ultimate absurdity in the attempt to make material and physical that which is metaphysical, a ridiculous solution to a problem that materialist thinking has created for itself by its refusal to accept the intangible, its attempt to deny that relationships are fundamental.

Connecting quantum and cosmic scales

Many brilliant people have tried to bring relativity and quantum models together and many more are still trying, with gravity at the heart of the mystery. Of course everything in this section has a large element of mystery, but it is possible to find some resemblance to our daily reality in the notion that "something" has come into existence when energy forms itself into something that has a boundary around it, which maintains mass and which can be thought of as a particle rather than a wave.

Effects over distance are somewhat less easy to explain. Scientists don't know why a magnet generates a field. At this level of reality everything is being described in terms of patterns of relationships. That is all there is. In fact the world this book describes is composed entirely of patterns of relationships. Quantum theory can tell us in some detail what happens in terms of a relationship between particles. In theory, relationships between elements of matter have been distilled to just four fundamental forces, the electromagnetic, the weak and strong interactions (sometimes called "nuclear forces") and gravity. Physicists can describe in terms of the quantum conventions which of their virtual "particles" mediate the relationships between the first three forces. But they can't explain gravity. How can it be that a vanishingly small and brief-lived entity like the Higgs boson would mediate the pull between our Sun and our planet?

All theories are operating at the boundaries between physical matter and raw energy and are attempting to describe what produces the relationship between them. In our view the term "virtual" particles is profoundly unhelpful here. We suggest that it would be much clearer if only the elements which are capable of sustaining their existence over time – photons, atoms, protons and electrons - would be thought of as particles. Anything that does not sustain over time is energy. And anything that mediates the relationships between particles is neither matter nor raw energy. What quantum theory calls "virtual" particles is a form of information, something that describes the forces that energy is creating, something that affects the pattern of relationships.

We remind you that the strong interaction is what keeps the protons and neutrons in the nucleus of an atom from separating, even when the proton electrical charges would repel, just as two magnetic north poles would. And while Electromagnetism can operate over quite large distances – like the magnetic field around the earth –

and can be detected outside laboratory environments, only gravity is observed to operate at interplanetary or galactic distances and that, to put it technically, is a bitch of a problem. Physics has got itself stuck on that and we need to look at how and where it has done that.

Why Physics can't get it together

To tell this story, we will take a brief journey through some crucial moments at which decisions were taken and how those decisions then became embedded in all that followed.

With all theories there is an ever-present risk that something which does not belong will attach to it, like the absolutist view of randomness in neo-Darwinism, and become a constriction to future theories. We will also see that things can be excluded for convenience and to simplify the conversation. These exclusions may then be forgotten. The choices of viewpoint that result affect how the cosmos is perceived, possibly for centuries.

Choice 1. Newton's frame of reference was limited. Let's start with motion and with Isaac Newton. In an earlier chapter we played with the experience of being in a car and not being sure whether another car is moving or we are, until the background is seen. This tells us that motion can only be assessed relative to something else - other bodies, observers, or a set of space-time coordinates. Physics calls these "frames of reference". If the coordinates are chosen badly, the laws of motion may be more complex than necessary.

Newton's first law of motion states that an object will remain at rest or continue in uniform motion until acted upon by a force. Newton's laws apply to isolated objects, or apply within the limits of an isolated frame of reference. That is, they apply to objects and forces when one excludes such facts as

a) the objects are on or near the Earth, and the Earth is moving through space

b) the speeds concerned are not close to the speed of light and

c) we are dealing with linear motion and not spinning-in-place or subject to coriolis forces (we will define these later).

These exclusions work just fine for planetary mechanics, but not when dealing with particle-level energies.

Choice 2. Waves are not presented in 3D. This might also be described as a habit or an accident. If you have ever seen a mathematical drawing of a wave, it is likely to look something like this:

This is convenient when drawing waves on paper, which is how a lot of mathematics is done. When physicists describe electromagnetic waves, they go a little further, showing the electrical and magnetic components separately as follows:

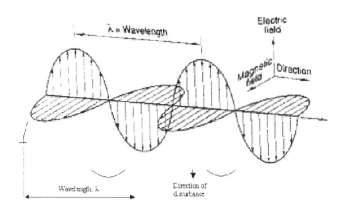

We will see that this choice, while it assists explanation, can cause some false or limited images to arise.

Choice 3. The strong nuclear force causes difficulties. We have described how an atomic nucleus is formed from protons and neutrons. Orbited at a distance by rings of electrons these varying numbers of protons and electrons determine how the table of chemical elements is built up, from Hydrogen with its single proton and single electron to uranium 238 with 146 neutrons, 92 protons, and 92 electrons.

Protons are positively charged particles and should repel each other quite strongly, which would make atomic nuclei impossible to sustain. Neutrons have no charge, but that would not prevent the proton repulsion. This caused physicists to assume the existence of a force of attraction and to include it in their theoretical model. They call that force the strong nuclear force. In due course you will see where this assumption fits in and the difficulties it causes.

Choice 4. Inventing particles We have seen that from the time of Max Planck a purely wave-based model of the universe has not been possible. He broke the link with the classical physics of Newton, Faraday and Maxwell and brought us into the quantum world. If energy as matter remained continuous rather than discrete you could not have an atomic table of elements with differing properties and chemistry would be impossible.

Planck described the fundamental relationship between the frequency of an electromagnetic wave and the energy of the photon when it is viewed as a particle. His work defined the process at the boundary between energy and matter and Planck's constant sets the limit for the smallest amounts of energy that can be viewed as matter. That is, it lies at the extreme end of our small scale. These inconceivably small-scale predictions correspond to what can be measured. Thus Planck's theory seems to fit how light and heat radiation behave and quantum theory begins in a grounding of physical reality. However it didn't stop there.

Choice 4 is about what physics has done since with quantum electrodynamics. As noted above, one hundred years after Planck the standard model of QED presents its theory of the Universe in terms of seventeen different subatomic particles, some of which are "virtual". We repeat our question as to whether it is legitimate to treat an evanescent burst of energy in a particular range as a particle. We suggest that this isn't working out as well as people hoped and that since these energies mediate relationships between the more stable parts of the material world, the energy patterns concerned would more legitimately be treated as information.

Choice 5. Convenient but false limitations to mathematics of infinity. As noted we are approaching infinity at both ends of the physical scale. The large end is approximately 10^{61} times bigger than the small, a magnitude so huge it can seem infinitely large to us. So we need to be aware of what infinity is or might be, and how it is treated. This leads us to the heart of choice number 5. It is by now well-known (though easily forgotten) that what we see as matter consists of 99.999% space. This is because of the distance between the electron "shell" and the nucleus of the atom. The rest of the atom appears to have nothing in it and so is called a vacuum. We think of a vacuum as empty and it is indeed empty of matter but it is **not** empty of energy.

Physicists performing calculations with this vacuum found it to have an infinite energy density. It was not possible to calculate the energy at the subatomic level of the atom using this infinite density. The standard textbook on "Gravitation" by Mesner, Thorne and Wheeler states

> "present-day quantum field theory gets rid by a renormalisation process of an energy density that would otherwise be infinite if not removed by this renormalisation".

What does this mean?

Infinity is awkward for mathematicians as it tends to make equations unworkable. "Renormalisation" is a procedure that replaces infinity with a limiting value taken to be an effective approximation so that mathematics can continue. This can work acceptably when the infinity is at the small scale where it is perhaps reasonable to ignore small things. It does not work at the large scale where potentially vast amounts of energy are involved. So it does not work for gravitation.

Several early formulators of QED and other quantum field theories expressed dissatisfaction with this state of affairs. It seemed illegitimate to do something tantamount to subtracting infinities from infinities to get finite answers. Freeman Dyson, Paul Dirac and Richard Feynman all voiced reservations. Dirac said

> "Sensible mathematics involves neglecting a quantity when it is small - not neglecting it just because it is infinitely great and you do not want it!"

Feynman called it a "shell game" and a "dippy process", saying

> "Having to resort to such hocus pocus has prevented us from proving that the theory of quantum electrodynamics is mathematically consistent. .. I suspect that renormalization is not mathematically legitimate!"[23]

However, the machine of academic theory found no alternative, so it has rolled on regardless since their deaths.

The implications for this choice in terms of how much energy there might be "somewhere in space" and occupying what is often presented as a vacuum (containing nothing at all) are considerable. Of course there would have been an alternative. That would have been to go back to basics and look afresh at the path that theory was taking. Physics didn't do this so instead, the path led to:-

Choice 6.   The Dawn of Creation We have referred regularly to the "Big Bang" during this book so far, because it is the best known and most accepted term for the beginning of creation. Conventionally it describes a time, or a time before time, when everything that the Universe is – all its energy, all that eventually became matter - was compressed into an unfathomably tiny amount of what we call space. We suspect that there is no-one who can really imagine this, even those who are expert enough to model it in mathematical terms.

---

23   Feynman, Richard P. ; QED, The Strange Theory of Light and Matter, Penguin 1990, p. 128

We don't know how the Universe was created and don't know for sure that we can do better than the big bang. Any such description is inevitably conceptual – yet we believe that it is essential to challenge the conventional description and to put forward something different than you will find in all the physics textbooks, because a familiar pattern re-appears here. The conventional presentation has nothing to do with Consciousness, so it is unsatisfactory in relation to our story. As previously the appearance is of strong consensus presented as a high probability of its being "the answer" despite the gaps and possible alternatives. The conventional presentation would again seem to offer a random and accidental reality where our view presents something which, while not calling for a "creator", expresses some sense of underlying reason, some equivalent of logos.

You may be familiar with the child's excuse that "a big boy did it and then he ran away" (the grown-up equivalent being "it was the previous administration"). Likewise, prior to Planck and Einstein theories of creation were more or less limited to the various versions of the familiar idea that "God did it". The Big Bang is no less mysterious than "God" because we cannot fathom the "before". What if there is no "before"? This may be hard for us to imagine but it is no different than knowing that there is no "outside" to universe.

Humans will always be called upon to accept that some things are not intellectually knowable. There is a boundary here between what the logical, analytical mind can grasp, and what our access to the "big mind", the mind of the All will convey to us. This is the territory of the mystics and the meditators and it would appear from their reports that one can experience being directly connected – being a part of the All – but that this does not provide any rational answers to how creation happened. My colleague Ian McDonald refers to this as "trying to "eff" the ineffable". The 1960's philosopher guru Alan Watts used the vivid expression "like trying to bite your own teeth".

Choice 7. When formulating relativity theory, Einstein adopted the viewpoint that nothing in the Universe can travel faster than light does. While this works very well in many applications, in quantum entanglement we have already encountered the possibility of faster-than-light transmission of information and connectedness at a very fundamental level. Current presentations do not deal with this.

Taking all of these together, physics has got itself stuck, unable to break out of the constrictions imposed by these choices in order to resolve the question of how gravity operates. We need to look for the reset button.

And so to gravity

Relativity describes how the existence of Mass warps space-time, causing gravitational pull so that planets orbit stars and light bends around both. The act of being material, of having large amounts of energy bound up in form, distorts the very fabric of the cosmos. Remember, to create just 1Kg of matter (a typical bag of flour or sugar) requires almost 90 quadrillion joules of energy—the same amount as is used by the entire world in roughly an hour and a quarter (2008 figures). Matter looks to us to be static but energy doesn't stop being "energetic" even when it is in form. The one thing in Universe which cannot be destroyed is also never settled. It is in a continuous state of vibration – like the palpable nervous tension emanating from a hyperactive child who is trying really hard to sit still.

Thinking of it this way it may be easier to imagine that there will be non-visible effects from that vibration which can be felt across distance. But how do we translate the minute quantum-scale model of what matter-energy is into a specific description of precisely how that vibrating reality causes the outward effects that we call a gravitational field?

At the larger scale we know that this field affects the whole solar system. Conventional science also believes that it ultimately affects the entire cosmos – that in the absence of something else to counteract it, gravity would eventually pull all matter back into a single pile, like a cosmic-scale suction-cleaner. At the same time the evidence seems to say that in place of that contraction, universe is expanding at an accelerating rate. Maybe something is being missed.

Viewed from the perspective of the big bang model, something caused an explosion the force of which propelled everything in the universe outwards. But a big explosion is a one-off event. It is like pushing a marble across a wooden floor. Obeying Newton's law that a body will continue in uniform motion unless acted upon by another force, it would continue to expand at a constant rate. Acted upon by gravitational pull it would gradually slow down, eventually stopping and reversing that expansion. For a long time that is what physics thought was happening.

Since 1998, and the observation of accelerating expansion, physics has needed something extra to make the big bang model work. Because Newton's law is not breakable, there has to be something else happening – a force which continues to supply energy to the expansion, like putting your foot on the throttle of your car to shift it from cruise into overtake mode. What could cause that force?

Some while back we asked the question "what was physics to do?" to supply the anti-gravity force that would make sense of an accelerating expansion. Here is the answer. Their response was that there must be something that we cannot see or measure which provides the universe with its expansive thrust. Accordingly they have taken up the notion of "Dark Energy" and "Dark Matter". Understand that the darkness is not some attribution of evil or hostility. They might just as well have said "invisible" or "undetectable", but that might have been too much of a giveaway.

Now you or we might think that coming up with the notion of something we cannot see as the cause of a major phenomenon in the universe is rather like inventing a boogie man. The big boy did it and then hid. The positivist approach to science does rather discourage choices that rest in things that cannot be measured so physics makes the convenient assumption that one day we will find out how to detect the undetectable. You might also think that they would have taken this situation as providing strong cause to go back to fundamentals and see if there might be a better model for the universe. However there are a lot of research dollars at stake, a lot of peer pressure and some very large contracts for the construction of facilities like the CERN LHC. There is an awful lot of intellectual ballast against it and isn't an easy shift to make. As with the "renormalization" choice 5 referred to earlier the physics juggernaut rolls on.

In fairness it also does so because despite such weaknesses, it retains a kind of consistency. Often there are new elements to the theory which accurately match previous observations and regularly there are new observations which fit predictions of new theoretical ideas. We are not saying that physicists are anything other than intelligent and nor are we saying that they are intentionally dishonest, but a trap is present here. The more elements that are added to a theory, and the more pieces of apparatus are constructed in line with those elements, the more difficult it becomes to roll back the perspective and the more difficult it becomes to distinguish genuine events and phenomena from events and results which are artefacts of the method. The risk is that the entire framework, while meaningful to the practitioners, drifts ever further from a meaningful depiction of the Universe.

In this chapter we have engaged with the paradoxical and mysterious world-view that emerges from quantum uncertainty, and the evanescent nature of the boundary between energy and matter. We have looked at the difficulties that the quantum view presents in determining what is real, and at the mysteries of wave-particle duality that arise at this boundary, a place where potentially both are simultaneously correct descriptions   This led to an examination of the approximations, assumptions and constrictions that have brought Physics to its current difficulty in reconciling the quantum and cosmic scales of existence, and making sense of gravity.

15.   From paralysing paradox to creative possibility

Deep in the human unconscious is a pervasive need for a logical universe that makes sense. But the real universe is always one step beyond logic.

Frank Herbert

"...A single group of atoms existing only in one copy produces orderly events, marvelously tuned in with each other and with the environment according to most subtle laws... we are here obviously faced with events whose regular and lawful unfolding is guided by a 'mechanism' entirely different from the 'probability mechanism' of physics."

E. Schrödinger. What is Life? 1944

Theme

This chapter brings several strands together. It deals with the advanced physics and mathematics which take us beyond conventional views of reality to reveal the creative flexibility in the universe which is available when we accept its uncertainty and paradox. We find that theoretical physics can support our theme that information and consciousness connect and define the world, and discuss the ways in which this might work. We explore the boundary between physics and biology and examine the processes which facilitate the connections between the information realm and our cognitive experience and which enable small thoughts to have the potential to offer larger consequences.

Untangling non-locality and resolving paradoxes

In Chapter 14 we introduced the EPR experiment which proved that two particles could share information, either by faster-than-light communication or by some mysterious "non-locality" or "entanglement". How should we understand "non-locality"? What does this kind of connectedness tell us about our world and what does it mean in practice?

The "faster-than-light" communication between the complementary or "entangled" particles is measurably real. In the version of the experiment conducted in 1997 by Nicolas Gisin[24], particles ten kilometres apart appeared to be in communication at **20,000 times** the speed of light. In effect, this is as near to instantaneous as our

---

24 Salart; Baas; Branciard; Gisin; Zbinden (2008). "Testing spooky action at a distance". Nature 454

accuracy of measurement can determine. This speed would mean that communication across our globe takes one-millionth of a second or less.

It is a fundamental tenet of relativity theory that nothing can travel faster than light. This limit effectively defines the nature of time within the theory. It was Einstein's leap of imagination to picture the universe when viewed from a particle travelling at that speed which led him to derive the theory, and with speed of light as the constant, it has to be time (or 4-dimensional space-time as a whole) that bends and becomes relative.

How are these two things, a defined limit and experimental proof that it is being broken, both true at the same time? The first suggestion we would make is the one which typically is required to resolve paradoxes and we will use the famous Zeno's paradox as an example.

This story pits the famously swift Achilles against that poor plodder, the tortoise. The problem is set out like a series of cartoon-frames. In the true and politically correct spirit of positive discrimination, the tortoise is given a start. Now imagine a series of strip cartoon "frames". In the first, the tortoise begins 20 metres in front of Achilles. In the first part of the race, Achilles dashes to the place where the tortoise began. In that time, the tortoise has advanced a little (frame 2). In the next view, Achilles again reaches the tortoise's second mark, only to find that the creature has plodded a little ahead (frame 3). This process is repeated, and Achilles edges closer and closer to the carapace but never, even in an infinite number of such steps, catches up. Each frame would have to be drawn bigger, to allow a scale which even makes visible, the ever-decreasing distance between the contestants, since by the last frame the distance between them is microscopic.

Whether or not you can describe what the precise flaw is in this story, you know that something is wrong with it. It does not match the reality we all live in. You may be mathematically confused, but when a call of nature comes you will still make it to the bathroom in time, even if required to overtake a tortoise. What Zeno has done is to take one of the dimensions in the world, and to cause it gradually to vanish. The time dimension in the first frame might occupy several seconds. In the second it would be a fraction of a second, and that fraction would get smaller in each subsequent frame. Not only would you have to draw each frame larger in order to even see the distance between the competitors, but meanwhile the time dimension has become vanishingly small.

The paradox is resolved when we change perspective and re-introduce the dimension of time according to our usual view. If each frame is of the same duration, say one second, and Achilles is travelling at 10 metres per second then Achilles is just behind the tortoise in frame 3 and well past it in the fourth.

Solution 1. An information dimension

So if paradoxes resolve when we change perspective, what new perspective would we need to resolve the faster-than-light communication that is implied by non-locality? One possibility is that the information which we are detecting in our entangled particles exists in a dimension (or dimensions) of its own. Rauscher and Targ have put forward an extension to relativity theory along these lines. Let's try to explain what that would look like.

If we again use the example of the "spin" property then the two particles are in an indeterminate state prior to measurement as far as the experimenter is concerned. At the point where the measurement is made of one particle, and the experimenter knows the outcome, the other particle is also instantaneously in a determinate state.

This means that the two particles are in a constant state of knowledge about the other. Each possesses the necessary information about their combined / shared state, irrespective of their position in space.

If we were to accept as true the limitation imposed by the speed of light, then we are by definition saying that the information does not actually travel in four-dimensional space-time. It has to be present in a different dimension which contains the others so that there is no distance to travel.

The existence of other dimensions is not unfamiliar to physics. Indeed, when we were discussing J.W. Dunne's experiences with precognitive dreams we heard that he used relativity theory to explain the events in just such a way. Several physicists have proposed the added dimensions that would be required.[25] We are not equipped to explain these, still less to judge their respective merits. So their significance is that theoretical physics is entirely allowing of the possibility. This has led cosmologist Andre Linde to suggest that consciousness, like space-time, has its own intrinsic degrees of freedom, and that neglecting these will lead to a description of the universe that is fundamentally incomplete. We are certainly agreeing with Linde's view that existing descriptions lack something crucial. What is new in our context is the possibility that the new dimension is the container for

---

25   For example Alex Green with a five-dimensional manifold, Elisabeth Rauscher with an eight-dimensional Minkowski space.

consciousness at the quantum level.

We have referred before to the view taken by physics that there are more than four dimensions. Stephen Hawking addresses their existence by asking why we don't notice the extra dimensions, why we can't see beyond three spatial and one time dimension. He suggests that the others curve up into an unimagineably small space, less than one trillionth of an inch. That might be so but since it is not provable we find that somewhat convenient and in the end rather unsatisfying. Nevertheless, the value of an additional dimension of this kind is that it takes us outside of the constraints of space or time as we know it and so potentially frees us from the restrictions of relativity theory.

Solution 2. Zero-point energy and the Akashic field

A second and highly popular potential solution to the paradox of faster-than light communication is found in a phenomenon called zero-point energy, and leads to other descriptions involving a "field" of connectedness. Several authors, including Ervin Laszlo, Gregg Braden and Lynne McTaggart have explored this area. In particular László's 2004 book, *Science and the Akashic Field: An Integral Theory of Everything* posits a field of information as the substance of the cosmos. The term Akashic field echoes the thinking of Rudolf Steiner ("Akasha Chronicle") a century earlier with the associated idea of cosmic memory[26]. The viewpoint of this book is similar.

We are approaching the leading edge of physics in this discussion. Although the zero-point energy field has been part of physics for almost a century, it is still far from well-understood. To describe it, we will go back again to the dawn of creation.

Physicists believe that the time of the big bang all four forces in the standard model were a single "superforce". The physics of this has been demonstrated to be possible in an experiment during the 1980's at the CERN collider. However, the attempt by physicists to find a theory which describes this reality meets a major block because the fundamental particles which supply the force of connection would have to have no mass for the theory to work. You can think of that as being like something which exists physically but doesn't weigh anything, something

---

26 . Cosmic Memory: Prehistory of Earth and Man, Rudolf Steiner Archive 1904

which has barely extended beyond pure energy.

Attempts to unravel this problem are continuing and more details can be found by those with the appetite by reference to the www.calphysics.org website and others. The accepted theory since 1964 based on work by Higgs and Haisch suggests that the properties of some particles, rather than having gravitational or inertial mass as an intrinsic property, only acquire mass as a function of their interaction with their environment. In "Scientific American" author M.J.G. Veldman gave the following image:-

> "The way particles are thought to acquire mass in their interactions with the Higgs field is somewhat analogous to the way pieces of blotting paper absorb ink. In such an analogy the pieces of paper represent individual particles and the ink represents energy, or mass. Just as pieces of paper of different sizes and thicknesses soak up varying amounts of ink, different particles 'soak up' varying amounts of energy or mass. The observed mass of a particle depends on the particle's 'energy absorbing' ability, and on the strength of the Higgs field in space."

This theory places the properties of matter and energy outside of the particles themselves and make them, in effect, environmentally determined. This has huge implications. You may recognise the echoes here of the interactions at the cell membrane which led to Bruce Lipton's pithy comment "It's the environment, stupid!"

Such an explanation would mean that there is information which was previously regarded as a property of the particle, but which is in fact present in an information field in its environment. Such a field could potentially be the medium of connection between the particles in the EPR experiment and once again physics comes close to our proposed need for a field of consciousness without quite getting there. Earlier we indicated our doubts concerning the way that "virtual" particles like the Higgs Boson are proposed to mediate a gravitational field.

The existence of zero-point energy had been inferred by Einstein and others before the coming of quantum electrodynamics, but was explored more explicitly by Dirac and others. Quantum physics predicts that in the case where all other energy is removed from a system (that is, when the temperature is reduced to absolute zero because there is no motion any longer in the atoms) there is nevertheless a minimal residual energy at the quantum level because the quanta will continue to vibrate. Because of the vast number of quanta and the many modes in which vibration is possible, there is a huge quantity of zero-point energy in the universe. Even when all matter is cold, the universe is humming with energy. We know this would have to be true, because energy cannot be destroyed, but it is as if all of the heat which was contained in matter has now become part of the ground-being of Universe.

Zero-point energy is present throughout the universe at the lowest level of known existence. Thus many see it as a prime candidate to provide a field that could hold the type of extrinsic information about the universe that is required here. This has been taken up by non-physicists, and extended to support the belief that zero-point energy is the container for "consciousness" and has been present since the big bang (or day 1) as an inherent feature of the universe.

The entanglement of quanta within this field – the known connectedness of specific information that exists between one quantum and another after they have interacted - provides the means by which the universe has "evolved" from instability into partially stable physical existence. Why doesn't quantum uncertainty cause everything in the world to dissolve? Why is there matter in the universe that we can treat as "real" – ourselves included? The answer would be that it is because the zero-point field has increased steadily in the extent of its entanglement. It is as if we were to say that each particle knows more and more about what it is, and retains that relationship with the whole. The energy which allows matter to "become" and then to "be" is held together by quantum entanglement. This provides the stability of what already is, but exists alongside the flexibility of all that is not yet held – the uncertainties which allow something new to be created.

Information about information

All of the above leaves something of a question-mark over the nature of information itself. It does not appear to be physical; physics accepts that it does not have mass and there is no reason to think that it should. It is metaphysical. It is still not at all clear what the relationship is between the humming energy in the zero-point field and its capability to encode information. This is the problem that physics is struggling with when it postulates virtual particles. As we have just discussed information does not "travel" in any of the senses that other waves or particles are known to.

To be sure, there are other aspects of physics that are hard to describe and we discussed earlier the mysterious aspects of gravity and electromagnetism. We know of gravity because of what it does. We can describe its action and model the effects that a mass which exerts gravitational force will have on the space-time which surrounds it. We call that a gravitational field, because we know the effects of gravity at different distances from the mass which exerts the force. Magnetism is similar. Like gravity, we know it by what it does.

In a similar way, we need to understand and know information by what it does. Gravity and magnetism exert a force and that force can be measured so we know what the effects of that force will be on other matter. Information does not exert a force. It cannot be measured. When we discussed the quantum cat, we discussed the fact of its life or death. We did not discuss and could not discuss in physical terms, the effect that knowledge of that cat's demise would have upon an adoring owner. It is not measurable. Its force in physical terms is zero, but the effects are no smaller or less real for that.

Nor does information have a defined field in the sense that gravity or magnetism do. If the cat's owner is on the other side of the world from the cat when he or she learns of its death, the impact is not diminished by the distance. Its impact is detected and has its effect over potentially infinite amounts of space. That impact might even be physical such as an extreme case where the owner commits suicide or murders the experimenter. Thus one possible interpretation is that Universe is a field of information, omnipresent and accessible anywhere, with non-measurable effects which contribute new information. Universe is scenario as Bucky said and the scenario expands.

Information is not covered by Einstein's universe. We might say that the limits that relativity has defined for speed of travel do not apply to an entity that has no mass. Physics as a whole has nothing to work with if something exerts no force, is not a particle or a wave or even a probability superposition. The information may be real, but doing mathematics with something that you cannot locate, let alone measure, presents a challenge.[27] That's a traditional British-style understatement. We mean it's not possible.

Within this understanding of information lives the potential for a Star-trek style transporter and for teleportation. In theory, if all the information could be taken or transmitted from one location to another, it could be used to recreate the originating material pattern in that new location. The matter itself does not have to move. Instead the information is separated or extracted from the originating material form. That form returns its energy into the quantum field. New quanta are assembled at the destination. Could that be possible? How much entanglement would need to

---

27 In making this statement we exclude some limited and specific arenas like formal logic and like number theory which is referred to in the next few pages.

be deleted and restored?

At the quantum level, the mystery of entanglement is an informational event. The two particles are aware, have information about, have knowledge of each other. The information is as real and meaningful as any other part of the world being described. Arguably it is more real, since it is the information about each quantum which tells us what it is. We should also say that we may be misleading you in describing particles as entangled in pairs. It may be that the field of information has to encompass multiple entanglements for any given particle. This creates an argument in favour of the information being held externally rather than by / with the particle.

Even more, if the view that the inertial or gravitational mass of a fundamental particle are blotted up from the zero-point energy environment then the very formation of matter begins with the information. The implication is that certain fundamentals of the material world are created from that field, faintly echoing the biblical phrase "In the beginning was the Word".

We have been talking consistently here about information, but it may be apparent to you that we could as easily have been using the word "consciousness". What we are presenting here is the view that consciousness is awareness of the sum total of a body of information. Your personal consciousness is your awareness of the information that you hold about yourself and your world. The consciousness that we are suggesting was inherent at the dawn of creation is the sum total of all information about every particle and wave that came into existence. Consciousness has expanded as the universe has expanded and as the quantity of information has grown and developed itself. In our terms, creative consciousness is in a state of continuous self-determination. That auto-creation is as real at the level of particles and waves as we saw it to be in the development of the biological world.

Limits to knowledge

The universe we are depicting is as far from a determinist one as you can get. What we are discovering is that nothing is determined until consciousness creates it. However, that does not mean that everything is totally unpredictable, since everything builds upon what has gone before.

Even without the auto-creative impulse of the universe, it was never truly predictable. We have shown the difficulty determinism has in practice with a paradoxical and quantum-indeterminate world. But some people continue to believe that a determinist boat can float, in the expectation that a Theory of

Everything will eventually unite all theories underneath the quantum paradoxical world. This belief continues in spite of the work Kurt Gödel published earlier this century, which should have been seen to sink the boat irretrievably.

Gödel's specialised in a branch of mathematics called "number theory". This concerns our understanding of the properties that numbers have, and what rules there may be about types of numbers (like prime numbers, even numbers, odd numbers, squares, cubes and so on). The aim, as in the wider scientific areas, was to find and formulate universal laws, ones which would apply in all cases.

In effect, number theory is a specialised language in itself, with symbols that represent numbers. It is a way of saying things in a symbolic way about the properties of numbers. All numbers themselves are symbols. The symbol for 2 of anything can be added to another symbol for 2 of anything, and we know that the answer is 4 of anything, and we are familiar with using number-symbols for that kind of purpose. But you can also have a symbol that represents an even number, and a law presented in terms of those symbols that says, for example "if I add an even number to an even number, I always get an even-number result".

At this point we need to introduce another famous Greek paradox. Epimenides' paradox consisted of the simple statement "All Cretans are liars". There's nothing paradoxical about this unless you know that Epimenides came from Crete. If his statement is true, then he is not a liar. If he is not a liar then his statement is untrue. These conclusions are incompatible, hence the paradox.

The leap that Gödel then made, after recognising that symbols could be used in number theory, was to see whether the symbolic language of number-theory, instead of being used to establish statements about numbers (i.e. rules in number theory), could be used to establish rules about number theory. It turned out that this would be possible if numbers were used to represent statements, and what this meant in a bigger way, was that there is a way in which rules can be established that are about the process of establishing rules.

So far this is not startling. Showing that there can be rules about rules was not a big deal. But Gödel's next question was to ask whether it was possible to prove the statement "This statement of number theory does not have any proof". When he succeeded in proving it he created a paradox similar to Epimenides', because the proof contradicted the content of the statement that it proved. Unlike the Cretan statement it has much wider ramifications. It's a paradox with huge implications for what we are capable of knowing or proving.

The logical process that Gödel used can be extended and generalised so as to show that **all** formal systems would include undecidable propositions. The power of Gödel's theorem is in showing that it will **never** be possible to create a system of rules about the universe without it having the possibility of such internal contradictions and paradoxes. In this one shot, the notion of a determinist universe was blown apart. It makes it plain that the universe is not knowable - not just in the sense that we cannot encompass all knowledge ourselves nor even in the sense of quantum indeterminacy, but in the sense that **no possibility of total and absolute knowledge exists**. It simply cannot be formulated. Any set of rules would be just like the human world, full of paradox and contradiction, undetermined and undeterminable. Heisenberg had shown that we cannot measure everything. Gödel shows us that even if we could, we could not formulate a complete system of rules about it.

Gödel was one of a series of mathematicians who tackled some of the hardest concepts in the field, investigating the nature of infinity, certainty and the limits to human knowledge. Georg Kantor was driven to insanity by his attempt to encapsulate infinity – something which lesser minds might immediately see as a certain route to madness. Ludwig Boltzmann paved the way for the 20th Century investigation into a world of probabilities, proving clearly that there were limits to certainty, but was driven to eventual suicide by the determination of his world and his peers that such an idea must not prevail. Kurt Gödel eventually suffered mental collapse because he sought a way beyond his own theorem. He wished for mankind to transcend the logical limitations that he had proven. Like his friend Einstein, he knew from experience that some breakthroughs in mathematics were reached through intuition and not constrained by the limits of logical reasoning. Unfortunately he tried to prove the existence of this intuitive capability by use of logical reasoning and inevitably failed. As his own theorem and possibly a little more common sense would have indicated, the task was impossible. In a deep and desperate irony, he was trapped in his own paradox.

Alan Turing, a fourth in this series of doomed geniuses, took Gödel's theorem further and proved that it had practical consequences in the limitations to what computational machines might be capable of. He showed not only that there were limits to the extent of provability, but that these limits would also make it impossible to know in advance which problems were capable of solution, and which not. So science couldn't even say to itself "we can't know that", and put the problem to one side. Computers faced with such programs would never finish, but we would not know whether it was time to turn them off. Mathematicians might, like Kantor, be stuck forever trying to solve the unsolvable. While Turing's eventual suicide was almost certainly the consequence of appalling treatment by British security services who regarded his homosexuality as a security risk, he too was deeply troubled by questions about the thinking capability of machines, and regarding the true nature of his and other human minds. These deep and unsolvable concerns might have contributed to his despair.

This strand of intellectual struggle might be just a tragic curiosity, but for the way that it mirrors a wider human struggle to accept that we live in an uncertain world. For a few mathematicians, lovingly obsessed by a world of deep pattern, order and predictability it is perhaps understandable that the failure to create formulae for everything might be unhinging. For the rest of humanity the more mundane struggle throughout the last centuries has been to embrace the uncertainty of Gödel's world and to love the creativity that it gives to us. In the place of the order sought by mathematicians, one that is driven by God's word as expressed in numbers and symbols, there is instead a beautiful and harmonious co-creation driven by a glorious and divine spark of infinite possibility. We see this as a more than fair swap. That it also takes place within an intuitive engagement of the kind that Gödel and Einstein experienced, is icing on the cake.

Thus, although it continues to yield useful and practical results, the intention to create a totally objective science, one in which we step outside as observers and encapsulate systems to view and define rules for, is shown by Gödel to be based on a fantasy. Even those systems cannot be guaranteed to be totally consistent. Objective truth is a wet bar of soap, which will squeeze through our fingers every time. Humans must learn to love paradox and deep mystery; there is no alternative. We find that exciting. It opens the door to the creative and dynamic interplay that lives within multiple approaches which can make some sense of the mysteries that the last four chapters have faced us with.

It has taken us many pages to develop the picture of conventional physics in such a way as to show what remains unexplained by it, and what might be needed to provide a picture of creation that works, ties together the loose ends and gives us the patterns and patterning that we seek while allowing the place for consciousness. The next chapter fills in the space between the physics of the human form and the wider realm of consciousness. The final chapter in this section presents a theory that explains how consciousness enables energy to come into form.

Review

We have discussed the added dimension and the akashic field models that others have presented as possible explanations for faster-than-light communication and as the locations where information about the universe might be stored, separate from the matter and energy.

We have described the mathematical proof with which Kurt Gödel buried the notion of determinism, and shown limits to our understanding that other mathematicians have presented. It is evident that society as a whole has been reluctant to take the implications of these truths on board, but in its place we find that paradox brings with it a world of creative flexibility.

16.   Being and becoming, coherence and freedom

The philosopher is in love with truth, that is, not with the changing world of sensation, which is the object of opinion, but with the unchanging reality which is the object of knowledge.   .....The object of knowledge is what exists and its function to know about reality.

Plato, Republic

'Reality cannot be found except in One single source, because of the interconnection of all things with one another'

Leibniz

The properties of subatomic particles can therefore only be understood in a dynamic context; in terms of movement, interaction and transformation.

Fritjof Capra, The Tao of Physics

Theme

In this chapter we provide a link back to our earlier discussion of coherence in the biological brain systems with a discussion of the physics that would support it. We examine how this connects with information about heart coherence and the guiding function that the heart performs. We explore how this connects with chaos theory, fractals and a holographic image for our universe.

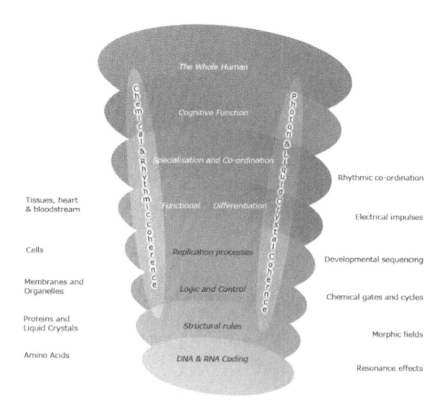

In 1990 Danah Zohar published "The Quantum Self" in which she laid out some of the ground for all that we are discussing here and anticipated much that was to be confirmed subsequently. She explored the physics which could provide the kind of mechanism by which the activity of the consciousness might influence changes which occur in matter.

We have already encountered some of this thinking when we approached the subject from the biological domain. Now we are closing a circle from the opposite direction. Dr. Zohar discusses consciousness in relation to "condensed phases", which you will recall from our biological presentation as the self-organising capability in systems which can occur naturally in materials.

Condensed phases of this kind apply to many systems, including biological ones. Professor Herbert Frohlich of Liverpool University has shown that vibrating, charged molecules in the cell walls of living tissue, which emit electromagnetic vibrations, could be caused to vibrate in unison when energy was introduced to the system. They are so much in unison that they are finally in the most ordered form of condensed phase possible - known as a "Bose-Einstein condensate". The crucial feature of a B-E condensate is that the parts of the system not only behave as a whole, but they become whole; that is, the identities of the component parts merge in such a way as to lose their individual identities entirely.

The purpose of this mechanism is not known, but biological systems have generally evolved for survival-enhancing reasons, and it is suggested that the B-E condensate might be a part of the way in which living systems create order. Other scientists have observed such effects, including bio-physicist Fritz Popp, who has discovered that living cells emit a weak "glow" which is evidence of photon radiation. Perhaps auras are not as imaginary as some people would like you to think!

Other work in recent years undertaken by Dr Stuart Hameroff and Prof. Roger Penrose has shown evidence that quantum coherence takes place in the tiny microtubule structures which provide a kind of "skeleton" within the brain's neurons. Penrose has advanced quantum theory descriptions of the ways in which this can mediate consciousness. In this, the microtubules are viewed as self-organising (orchestrated) quantum computers, providing "Quantum computation with objective reduction (OR)". The overall theory is often referred to as "Orch-OR". As Hameroff states

> "Regardless of whether or not the Orch-OR proposal turns out to be correct (and unlike most theories of consciousness it is testable), it is the type of multi-level, transdisciplinary approach needed to address the problem of consciousness."

There is no need to understand any particular theory at this point – merely to recognise that such possibilities exist scientifically and are worthy of further research. In our biological section we saw that the types of mechanism required for an interface between biology and an external information field do exist. We are now describing similar possibilities emerging from physics.

It is very clear that consciousness - at least for most humans - is a very orderly phenomenon. For example reading a book of any kind requires the ability to hold a great deal of information in relationship and context. In the case of complex books which may take extended time, even weeks to read, you are maintaining a thread which is far removed from quantum "soup". To write a book such as this one required an ordering world overview which has survived over two decades in Jon's mind, with six different drafts and continual deletions, insertions of new information, conversations with Juliana and modifications of perspective. It is nothing like the same book as at the beginning, but neither is it a different book, even though its style, order and content have changed fundamentally throughout.

This level of organisation demands a great deal of any physical processes that explain it. It is not some trivial accident - rather it is a high-order phenomenon which requires many levels of structure. The Bose-Einstein condensate, and the Orch-OR which has just been described at the cellular level may account for an ability to record states of being, or to resonate with external energies. (According to Zohar, up to ten million of our ten thousand million neurones may at any one time be in a state capable of responding to quantum-level changes.) But there is a great deal over and above this which governs the recording of information in our nervous systems, from pre-existing genetically determined structures through influences of remembered and learned patterns, to momentary modifications of emotional or mental state in response to stimulus. We are not looking for a single mechanism to explain the relationship between human consciousness and a conscious universe. We are looking for multiple layers of capability which together provide the elements in this complex interaction.

Heart Coherence

We return briefly to the scientific evidence for intuition and to recent research, particularly that from Rollin McCraty of the HeartMath institute and to the very comprehensive overview by Raymond Trevor Bradley.[28] In the tests performed by McCraty, subjects were shown randomly selected pictures, some of which were calming or neutral and others known to cause an emotional response. The subject would press a button, and the picture would be timed to display after an interval of six seconds. The results showed consistently that the brain rhythms and heart rhythms would both respond to the picture before it appeared on the screen. Perhaps most startling is that typically **the heart response occurred a full second before the brain**.

Here McCraty not only produces evidence of precognition but is able to show more of the means by which this is gathered and processed. Along with data showing the intuitive capability, experimenters were able to collect information regarding the electrophysiology of the brain and heart. This experiment shows clearly that it is not only the brain which receives informational input regarding the forthcoming stimulus. There is a Heart-mind.

---

28   World Futures Journal of General Evolution, copy on the web at http://noosphere.princeton.edu/papers/pdf/bradley.intuition.2007.pdf

In Danah Zohar's model of consciousness, she suggests that the brain has two interacting systems, the coherent Bose-Einstein states that facilitate consciousness, and the computer-like ability of neurones to store and communicate information. One benefit of this view, is that it provides a mechanism by which the "brain-waves" that can be observed on an EEG would be co-ordinated. Many people are familiar with the notion of brain rhythms - the alpha state of calm and meditation, the beta-state of activity and so on. These patterns are quite stable, in spite of the complex moment-by-moment underlying firing patterns of individual nerve-cells. The presence of waves is more easily explained by a quantum "integrating system" - an electrical field that maintains coherence between the brain as a whole, and the low-level activity of neurones.

In a different set of experiments McCraty has shown that there is a direct correlation between the patterns of heart-rate variability rhythms and the brain-wave patterns displayed by an ECG. All these indicators point towards an involvement of both heart and mind in the generation of coherence. He also provides evidence that there is a capability in trained subjects to influence a sample of placental DNA to wind tighter or to unwind in response to focussed coherent thoughts. This has been shown to happen whether the experimenters are near to the DNA or not – effectively opening up a route for both energy healing and healing at a distance.

The Silva trainings which Jon taught in the 1980's began by increasing the levels of student's alpha waves through guided meditation, and initially with assistance from sound patterns to cause brain-wave entrainment. Jose Silva was certain that psychic activity is facilitated by having the brain in an alpha-state, and the success of the technique indicates that this is another route into coherence. We should also remind you of the statements by Howard Martin (Chapter 4) regarding the way in which heart rhythms provide a synchronising pulse throughout the organism, and the correspondence this has with Mae-Wan Ho's view of multiple frequencies of coherent activity.

Holographic models

Raymond Trevor Bradley's paper presents a comprehensive attempt to formulate a theoretical base for intuition. In our view, this is equivalent to providing the mechanisms by which our inner consciousness as a whole interacts with the external world, and by extension for the process of spiritual engagement with it. He draws deeply on the work of others such as Karl Pribram, David Bohm and Edgar Mitchell in which holography provides a model which combines non-local storage of information in such a way as to allow intuitive information to propagate through the world. A full and very readable account of this model can be found in Michael Talbot's "The Holographic Universe". What follows is our very simplified overview.

Holograms are often misunderstood, and the image that most people have of them is drawn from special visual effects, such as the security foil on the back of a credit card, which displays two images, one of them in 3D. Another well-known simulation is that from the opening of "Star Wars", where a three-dimensional projection of Princess Leia appears in the room saying "Help-me Obi-wan Kenobi". Both of these are applications of holographic techniques, but neither conveys what is of special interest about holography.

To record a holographic image, a laser beam is split in two parts, each put through a lens which diffuses the beam, and one of the beams allowed to reflect from the object to be recorded. The two beams are then allowed to meet, and their waves to interact, on the holographic recording plate. The interaction of the waves resemble the interference effect that you get with two thin layers of wet silk rubbing together, known as moiré patterns. A set of visual patterns arise which are not those of either set of fabric strands. They are an effect of the interference of those patterns.

The image in a holographic film, in a similar way, stores the information in a manner which is not, to the naked eye, anything like the object being photographed. The image that reveals itself under laser light (or under the bright lights they provide for in-store displays) is three-dimensional, and clearly not located on the surface of the film as a photo would be. When you move, the position of the image relative to the surface of the hologram also moves. This image can also be projected by lasers, as with Princess Leia, in such a way that you could walk around it, viewing it from different angles, even though nothing is actually there but beams of light.

The lack of similarity between the surface of the holo-film and the original object is important because it does not store information as we traditionally would think to do ourselves. This book contains words in the order we chose to have you read them. It is linear, and provided you know the conventions of Western presentation, it is easy to start at the beginning, go to the end, and then stop. If you tear out a page, then what you have is one piece of the book. This is how we expect reality to be.

What happens in a hologram is different and non-linear. When you cut the film containing a laser holographic image, you can still see the entire image from either portion. If you cut in half, then quarter, and so on, each piece will provide you with the entire image. You do lose something with each cut, which is the fullness of the information contained. The resolution diminishes as parts are removed, like a cheap digital camera capturing fewer pixels. But the recording is not linear, and demonstrates how information can be recorded, and patterns of energy constructed that are not in a direct relation to the dimensions from which we view them. Our linear thinking and the image of books or libraries might lead us to expect that our memories would be stored at a specific location in the brain with relevant bits of information located next to each other. Not so.

In 1946 Karl Pribram went to work with one of the great neurophysiologists, Karl Lashley. Lashley had spent three decades researching memory, and in spite of Penfield's evidence, had failed to show any evidence of a linear recording. In fact all that he discovered seemed to contradict this belief.

Some particularly gruesome experiments by biologist Paul Pietsch made a real mess of linearity. Pietsch[29] had discovered that a salamander's brain can be removed in its entirety and experimented first with putting it back the wrong way round, then later with other operations in which he shuffled, sliced and eventually minced the brains. In every case when he replaced the brain, the salamander's functioning went back to normal. Other experimenters have even produced evidence that feeding with the minced brains of rats which have learned mazes, enhances the rates at which the receiving rats will themselves learn the behaviour. (We don't support these experiments, so please forgive us for quoting them so that the death of these animals is less pointless.)

Later, Pribram continued to work with the idea that memories are distributed throughout the brain - a belief which seemed to conform to the fact that human patients who had brain surgery, also never showed or reported the loss of specific memories. Similarly, accident victims were not known to forget some people but not others, or half of a novel. Even removal of the temporal lobes, where Wilder Penfield had appeared to find localised memory, did not have any effects of that kind. Pribram's conclusion was that there must be a way in which memories were distributed, but there was no understanding of how this might be accomplished by the brain.

When Pribram became aware of holography, he began to see it as a potential model for the way in which memory could operate, and contain the whole in every part. If

---

29 http://www.iub.edu/~pietsch/shufflebrain-book00.html

the brain could record in a way which was akin to holography, then the results seen by himself and Lashley would be understandable. Seeking to know how the brain would generate the waves necessary to produce an interference pattern, he observed that the electrical communications which take place in the brain do not occur singly. Neurons possess branches like trees, and an electrical message radiates out like the wave ripples in a pond. As Pribram put it

> "The hologram was there all the time in the wave-front nature of brain-cell connectivity, we simply hadn't the wit to observe it."[30]

Holography also provides a plausible explanation of how the brain stores the enormous amount of information that a human being must acquire over the course of a lifetime. Estimated by eminent mathematician John von Neumann at 280 trillion bits, it would be greater than the number of cells and neurons in the brain, but not greater than the number of potential connections between them. This also indicates that patterns, rather than individual bits of information, would be required for the storage.

Note too that the Bose-Einstein condensate discussed above, in providing a mechanism for coherence of brain-cell activity, would furnish the biological equivalent of coherent laser light that is necessary to generate the interference patterns that such a holographic process would need.

The power of the holographic model increases when we combine the understandings of the brain and memory with David Bohm's viewpoint as a quantum physicist. He drew the implication that something fundamental is emerging about the nature of the universe, and not just the brain in isolation. Bohm was deeply interested in consciousness and puzzled by the lack of interest displayed towards interconnectedness by Nils Bohr and his followers.

The work that got Bohm recognised was a special study of gases containing high densities of electrons and positive ions, known as plasmas. He found that electrons in a plasma cease to behave as if they are individuals, and start to behave as if they are part of an interconnected whole. You may recognise the resemblance with self-organising phenomena like the Belousov-Zhabotinsky reaction. Although their individual movements appeared random, large numbers of electrons were able to produce effects that were well-organised. The plasma constantly regenerated itself, and would enclose impurities inside a wall, rather as a biological organism encases foreign matter in a cyst. After he was made an assistant professor at Princeton in

---

30   (Interview, Psychology today, Feb 1979).

1947 Bohm extended his research to the study of electrons in metals. Here too, he found that the seemingly haphazard movements of individual electrons combined to produce highly organised effects. These were not entangled pairs of particles behaving as if aware of each other, but trillions of particles behaving as if part of some common process.

The classical approach to science leads along a path where the whole of a system is viewed simply as a result of the sum of all the interactions of its parts. The observations Bohm made with plasmons and the theory of quantum potential indicated that the reverse could well be true - that the whole system organises the behaviour of its parts. He likened the behaviour of electrons in a plasma to a ballet dance, rather than a crowd of unorganised people, thus suggesting that

> "such quantum wholeness of activity is closer to the organised unity of functioning of the parts of a living being than it is to the kind of unity that is obtained by putting together the parts of a machine".

As observed earlier, it was Bohm's work which led to the experimental validation of the EPR phenomenon, the paradox of particles' "knowing" of each other at faster-than-light speeds, and to the proof of quantum non-locality. He came to the conclusion that the orderliness of the universe was "hidden", just as the laser hologram hides the picture we see. He viewed order as being enfolded within the structure of the universe. This view extended as Bohm examined the hologram more deeply into the perception that the universe itself is a gigantic hologram. This universal whole is viewed as containing / embodying a vast flow of energy and events, in which there is an underlying connectedness and organisation. The details of this view were eventually published in 1980, in his book "Wholeness and the Implicate Order".

There are significant strengths to the holographic model. It provides a notion of an underlying connectedness, even if we cannot actually tell what it is. It points towards the physical mechanisms that would facilitate the brain to work in this way. It provides a framework for aspects of reality that are conflicting and paradoxical when viewed from the standpoint of other theories. It appeals to the notion many of us have that some sense of a holistic perspective, some alternative to the world-view of determinism and separation is required. It blends with more metaphysical views of the world - with the world of the Buddhist and the meditator. It is non-linear and non-cognitive in its organisation of discrete thoughts – of which more in due course.

Chaos and coherence

We have one more element to add to this picture before we return to the shape of the cosmos. This element leads again to the connection with geometry. Here too we have a parallel question in physics to that which we asked in biology. What is it that allows subtle and microscopic changes to have consequences which are visible and manifest at a macroscopic scale?

The biological answer that we described earlier was internal to the organism and utilised the capability of a small change to ripple through a field of coherence. In physics, quantum coherence in a zero-point field provides one form of equivalence on a cosmic scale, particularly if the connectivity throughout is non-local and holographic. Anything can affect anything else and the ripples are potentially unending. This could be enough of an explanation. Nevertheless we would wish to introduce an additional piece of science, which is known by the name "chaos theory".

If you have heard of chaos theory, it is likely to have been presented with an image, a typical example being that the beat of a butterfly's wing in Hawaii could bring about a hurricane in Tokyo. This is rarely explained both because it is difficult and is inclined to seem to most of us both irrelevant and ridiculous. We will try to make some sense of it here, because it could be very relevant indeed.

We have looked a little at the idea of infinite quantum possibility and at the process by which wave-function collapse occurs, and things come to "be". We looked at this alongside the idea of dynamic systems in equilibrium such as the B-Z reaction. Typically though, the world does not behave as a system in equilibrium. It changes.

The attempt to model systems which are not in equilibrium has its own branch of mathematics. Like QED it attempts to describe the realm of possibilities, but now at the molecular rather than the quantum level. In modelling the real world, there are pressures, forces, dynamics of one kind or another, which interact to make certain outcomes more likely than others. In that description the likely outcomes are ones that possible realities are "attracted" to. For a marble in a bowl, the bottom of the bowl is an attractor. If you spin the bowl enough, the marble may occupy space at the sides, and higher up. But there is a dynamic at play (gravity) which draws the marble back to the floor of the bowl. The use of the word "attractor" should not be taken to imply that there is any will involved. There is no goal. Just as the marble does not care where in the bowl it is, the fact that certain configurations of possibilities are "attractive" only tells us that certain combinations of complex dynamic influences will combine mathematically in particular ways.

Attractors can take many forms - when graphed they can look like loops and whorls and messy spider-shapes. Mathematics does not always look neat or simple. But in the world of the fractal, of complex self-organising systems, we have an extraordinary blend of mind-bending complexity with repeated simple patterns. Like a never-ending series of Russian dolls, we can go from the largest to the smallest scale, and see the same form. But looked at as a whole, the intricacy can be awe-inspiring. Nature is filled with fractals, an example being the way that a photograph of a cauliflower will look much the same regardless of whether you show 5%, 50% or 100%.

The world of fractal organisation is governed by what have become known as "strange attractors". These are still not well understood, but have a characteristic that is important to us here. The nature of the dynamics is such that although they will move into repetitive, self-replicating patterns, continuing over time to elaborate and extend, the pattern they generate will vary. Snowflakes are known to be never the same twice. The form of a snowflake depends on the initial conditions that form the first crystal. According to the precise shape of the first molecules that crystallise, the detail of the overall pattern will vary such that within the regularity of a hexagonal structure is room for almost unlimited variations.

This sensitivity to initial conditions is at the core of what is possible in the human world. The butterfly's wings image can sounds crazy and unreal, but the nature of dynamic systems that are governed by fractal development and strange attractors, is that a small change at the beginning of a process may magnify through its development such that the end-result is altered dramatically. A more familiar image of such sensitivity might be how, when listening to a public address system we may experience the point of instability at which a microscopic change to a control, or to the position of a microphone, or in the room acoustics results in an abrupt shift that produces an ear-splitting feedback howl. Just because we cannot see the way in which a butterfly might trigger such a process in planetary weather systems, does not mean it is not real.

Fractals and Geometry

The nature of a fractal is to create a small pattern and then to repeat it. The small patterns are transcended and included in a form of expansion which brings the outcome once expounded in the hermetic principle of correspondence "As above, so below". As an aside, we should note that all the seven hermetic principles are met in our presentation. We did not plan this; it simply turns out to be true. The first principle corresponds to our core message. Often and misleadingly called mentalism, it says that the physical world can be reduced to patterns of potential connection among potential concentrations of matter/energy that might or might not come into form, depending upon the introduction of consciousness.

The spiral model that we have used throughout this book is fractal in nature. Each curve is wider than the previous one. For reasons of space we don't draw a yet more ideal form here, in which each turn of the spiral would also be deeper than the one before[31].

Throughout this text we have tried to convey the dynamic sense of self-balancing the spiral image portrays. In an even more ideal presentation you would see the spiral rotating and we would show the continual adjustments between innovation and stability, between material and energetic, between being and becoming. We would see the interplay of information and consciousness in "remembering" the patterns which are sustainable, increasing their permanence.

We have seen that some aspects of Physics' theories provide the kinds of mechanism that are needed to explain what is observable. We have also seen that there are edges at which these theories look clumsy and possibly compromised; overall they now seem inelegant. It sometimes seems that every smart quote is attributed to Einstein but this one fits Physics, especially since his death – "Any intelligent fool can make things bigger, more complex, and more violent. It takes a touch of genius -- and a lot of courage -- to move in the opposite direction." We have described a number of choices that were made over centuries and which are now embedded in a model which has many impressive results, brilliant insights and stunning accuracy, but which is unable to reach completeness and fails to deliver the elegance to satisfy mathematicians or the intuitive feel that would satisfy others of us.

Somewhere within all of this complexity there must be some explanations that supply the simplexity that we are repeatedly seeking. They must still fit the material facts but also offer an explanation for the many features of the world that we listed in section 1. We are still explaining the realm of intuition, consciousness,

---

31 An image of that form is available on www.scienceofpossibility.net and also supports Chapter 19.

spiritual and shamanistic experience. We must continue to explain the world of collaborative biology, of not-totally-random evolution, of energy healings and homeopathy, of bodies which can alternate allergic personalities with non-reactive ones. And now we are adding an extra requirement, that we should also point towards a way to unify the tiniest and largest scales of physical phenomena. We have been learning new ways of seeing, as recommended by Goethe in preparation to meet Schopenhauer's challenge. "The task is not so much to see what no one yet has seen, but to think what nobody yet has thought about that which everyone sees."

Review

In this chapter we linked Physics to Biology via a presentation of connected features that make coherence possible in a way that connects human consciousness to the wider world. We viewed low-level mechanisms in the brain and discussed evidence for field effects from the heart. This led us to the emergence of order from chaos and to a holographic image of the universe.

17.   Jitterbugs, Coherence, Hamburgers and Donuts

Where there is matter, there is geometry.

Johannes Kepler

Theme

In the previous chapter we laid out the mechanisms for coherence and for fractal emergence from small building blocks towards large structures as well as chaotic responses on a large scale from minute energy impulses. Although we moved further from the paths of either relativity or quantum theory this was necessary in order to take us towards a viewpoint which will transcend and include both theories. This viewpoint will show the fractal building blocks that structure the universe all the way from the point where energy waves become particles up to the cosmological grand scale.

In what follows we will draw deeply on the work of Nassim Haramein, a physicist who works rather as Einstein initially did, on the outside of the accredited coterie of experts. Since Haramein's views are neither conventional nor supported by a University he is swimming against a very strong tide. We have described many concerns about the path physics now follows. Like us you may find that his way of looking at Universe meets some of our requirements. Haramein[32] thinks what most others have not yet thought.

Elemental form and sacred geometry

Whenever particles are depicted it is as spheres. Even though they are invisible, there is a reason for this choice. A sphere encloses the greatest volume with the least surface. We are used to raindrops; these form because of the surface tension of water that "pulls it together". We do not see precisely spherical water drops because gravity pulls them out of that natural shape, but you can see the equivalent effect of surface tension in soap bubbles. In zero gravity, this is what water would do.

Thus the sphere can be considered the most basic shape in which energy organizes

---

32   www.theresonanceproject.org

into matter. What then happens when a number of particles are placed together? You can imagine taking tennis balls and stacking them on a flat table (assume sufficient friction with the table that they don't slide). You cannot stack them as a cube – they will fall – but you can stack them as a pyramid. The simplest form of this is made with four balls, for which the closest possible stacking is a tetrahedron.

The tetrahedron is a solid that has triangular faces. As well as being the most elementary shape, the first step into two dimensions, the equal-sided triangle has structural stability. The tensions of the sides are balanced and do not deform when squeezed. The three-fold shape with its 60 degree angles is prime and indivisible. The same is not true of a square. If you were to squeeze a square, it would flex at the corners and so distort. We have placed a video showing this on the www.scienceofpossibility.net website[33]

This is the first building block in a sequence from which a progressively more complex set of forms may be built. If the uppermost ball is treated as the centre of a pack that extends in all directions, a cluster of spheres is formed that contains a further 3 balls on top and six balls around the middle – a hexagonal layer around the centre. Thus the central ball is surrounded by a pack, a shell of 12 balls.

To produce this structure for real with subatomic particles and have it maintain stability requires energy forces that hold the cluster in place, and maintain its shape. This means that there are lines of connective force between each of the balls in the shell and its neighbours, as well as between all the balls in the shell and the one in the centre. This diagram shows the patterns formed by the lines of tension, a

<hr>

33  www.scienceofpossibility.net/resources

shape known as a cubeoctahedron.

The cubeoctahedron has a special feature. The lengths of the lines (energy vectors) are the same between the centre and the vertices (corners) as they are between one vertex and the next. This symmetry causes it to be named the Vector Equilibrium (VE).

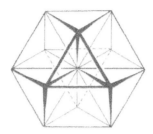

In the VE, visualise that energy radiates out from the central point, along each vector (line of force) and is countered at each vertex by the network of tensile forms on the surface. These constrain the energy and hold it in so that a dynamic equilibrium is achieved. You can think of it like an inflated balloon and in essence, the figure models a retained explosion. Since tensile and compressive forces have been equilibrated, it is quintessentially stable. However it is not static; it is energy in tension and can move. So while it is formed into matter and looks inert to us, there is no dead matter in the universe. It is always alive and humming with energy.

In the video referenced above, we show how the VE can be made to flex. The lines of force have a mix of the triangle's stability and the square's potential to deform. You can see a couple of examples of the folding that becomes possible. A larger flat triangle, a tetrahedron and an octahedron are all easily formed. More than this, as the video demonstrates, it is possible to exercise a pumping motion by which the VE transforms into an octahedron and back again. This motion is known as the jitterbug and is illustrated below.

The jitterbug concept, and much of the thinking that is described here, comes from the visionary thinker Richard Buckminster Fuller ("Bucky") whom we have referred to earlier. This section draws on his books Synergetics" and "Cosmography". What we are attempting to convey in a very short space, is the way in which energy relates to structure. This is at the heart of the pattern that shapes energy as it becomes matter, and makes it possible for those forms to acquire permanence. We are describing principles which embody a mix of rigid and flexing aspects which cause the energy to form into patterns of force. These patterns are the building blocks that shape the material universe.

As Gregory Bateson observed "The pattern that connects is the pattern of patterns". Once we have these fundamental interactions of elements the patterns of geometry are implicit. All the forms that are described in sacred geometry are in-built to the mathematics of creation. In his paper "jitterbug defined polyhedra" Robert Gray[34] lists and illustrates several complex polyhedral shapes which can be developed from those basic elements. He also illustrates how the basic jitterbug illustrated above is just one among many forms and has moving video images of the jitterbug motion which convey much more than is possible in text.

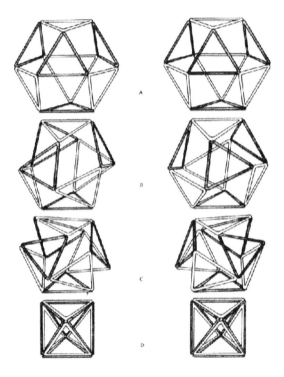

Readers may also find it interesting to explore the work of Juliet and Jiva Carter at www.thetemplate.org where some of the geometries are visible, constructed in 3D. The video of their first ceremony which can be found on Vimeo offers an opportunity to experience the influence of sacred geometry on human consciousness. The 3-dimensional experience that live ceremonies provide exceeds these videos. As well as containing energy within and in addition to transforming

34 www.rwgrayprojects.com

from one shape to another these forms can spin and when they do so a vortex of energy is created.

The source of the vortex

Vortex energy is not new; nor is it the invention of "alternative" and new-age viewpoints. As long ago as 1867 Lord Kelvin, the renowned scientist who prior to his elevation to the peerage was Professor William Thompson of Glasgow University, and who is widely viewed as the founding father of thermodynamics, formulated the theory that atoms were vortex rings. Prior to that atoms were thought of like little billiard balls, but the elasticity of their interactions could not be explained until the vortex theory was proposed. It was strongly supported by other eminent physicists, including Maxwell and remained strong until the upheaval caused by Planck and Einstein. So let us play spin doctor and try and describe where vortex energy comes from.

Let's start from fundamentals. If you try to define space, the first thing that you have is a point. A point has no dimension. It is a concept of a location, like your address or your mobile phone number, but does not have width or depth. If you are thinking of a pencil-point on a page then you have already progressed from conceptual to physical.

When you add a second point you create the possibility to join the two. That gives you a line. The line has dimension, it indicates a direction in space between the two points. This is still not a pencil-line – the conceptual line has length but no width. It is one-dimensional; a vector, a direction, an opportunity for motion, but it is not a "thing".

If you add a third point to the space, you can join all three and that gives you a triangle. A triangle is the least number of lines or points required to enclose an area of space. While the lines may have no width, conceptually they form a boundary. One way of thinking of this triangle as two-dimensional is that it can define an inside and an outside. Since none of these steps are physical, all of this is metaphysics.

We are accustomed in the human world to square and rectangular constructs that create cubes and cuboids using four lines and 90 degree angles. If you work with stone, it is difficult to do otherwise – you start with stonehenge. It took centuries before humans produced an arch.

The universe didn't start in squares. In the stacking of four particles shown above, when you connect the centre of a particle to its two neighbours you get a triangle with three equal sides and equal 60 degree angles. Any lines of force – attraction or repulsion are in this shape. We have seen above that these shapes can stabilize in polyhedrons when there is dynamic equilibrium. But what if that equilibrium is not present?

When we draw triangles we make lines on paper. These lines appear to join up, but this is misleading. Imagine that you were making the triangle from thin rods, or straws or toothpicks. You will see that the two lines can never occupy the same space. It doesn't matter how infinitesimally thin you make your rod, two things cannot pass through one point in space. This corresponds to the earlier statement that a point does not have width and it means that a triangle – even if you form it from a continuous piece of material like a paperclip, is always open at one end. Even energy, if it attempted to occupy the same point in space, would cause interference between the waves.

This has two interesting results. The first is that you can show that two open triangles are capable of interlocking in such a way that a tetrahedron is produced. Fuller notates this as 1 + 1 = 4. That is to say, out of two triangles we produce four, as per the illustration below. We have opened the initial triangles wide in order to make this transition visible.

The second result is that you can look upon the open triangle as a foundation. What happens if you put another triangle on top, and another? The rotation of energy around the triangle gains an additional vertical dimension. From this we get the initial spiralling movement of the vortex – around and up.

All of the phenomena we have been describing – stacking of spheres, the vector equilibrium, the tetrahedron and the spiral have their roots in triangles. The complementary rotations of the two triangles that form the tetrahedron – one clockwise and the other counter-clockwise – balance each other. They are opposite directions of turn, or spin and they create a stable pattern such as is required to form matter. Where energy is in a free rotational state and where clockwise is added to clockwise (or anti- to anti-) we get spin energy. The harnessed explosiveness of the jitterbug forms matter. The unfettered turn of the vortex produces dynamic results in the way that matter behaves. The universe is made of spin.

The core geometry of the universe operates in spirals and in the frequencies that are created by the jitterbug energies and spins of the various polyhedrons. For completeness and to assist those who think of sacred geometry in terms of the "flower of life" depicted here, we should point out that all of these polyhedrons are implicit in that diagram; they are just not obvious when it is viewed in its flat form. The jitterbug polyhedrons arise when you intersect a number of those flowers to form a 3-dimensional structure. In every case, the initial lines of connection form into 60 degree angles. More complex shapes and angles are derived when these are extended and transformed. This process is simple and well explained in Robert Gray's paper referred to above.

We are laying out some basic principles of sacred geometry, also following the insights of that earlier iconoclast and outlier, Buckminster Fuller. In order to feel the shape of creation we need to extend and combine these steps:-

- Bring together the many elements –

    o the flower of life,

    o the pulse of the jitterbug,

    o the balance between tension and compression

    o the ways that spheres pack together.

- Scale these elements, seeing them as components of the creation fractal.

- See the wave components of creation in three dimensions, not the two of a paper representation and then encompass spin, because everything spins.

- To all of this we should add the fundamental knowledge that energy cannot be created or destroyed, only transformed and that there is an essential symmetry to this relationship.

When we put all of these elements together we will have an alternative picture of material creation that shifts the viewpoint away from a big bang, and delivers a cyclical universe of infinite life and death, ever-balancing yin-yang, in-breath and out-breath. Rather than one big bang we will find trillions of little bangs, within a living process of creation and re-creation.

In a previous chapter we listed a number of choices that physics has adopted over the decades. The result of those choices has been to reduce the richness. Newton presented an isolated frame of reference that excludes elements of rotation (spin). Wave diagrams are drawn in 2D (even if physicists know they are 3-dimensional). Forces and particles are repeatedly conjured up and adjusted to eliminate discrepancies. The inconvenient infinite is curtailed and then the adjustment is conveniently forgotten. Space is described as a vacuum when it is filled with energy. Dark matter is invented when we have no idea how it might be found. We look entirely at the forces of expansion and are unable to see the where the contractive balance is happening. Something is not adding up.

It may turn out and probably will that some – even much – of what has been created in these theories will still hold true. But we need to reframe the whole of it. When we looked at paradoxes we could see that the resolution to some of them is to widen the frame of reference, find a higher perspective and bring in the missing dimension. This is what we must do here, so that we provide cosmology with a new framework.

Accordingly we will try to present an image of Universe which meets the fundamental principles and at the same time resolves some of the anomalies.

- It will need to be fractal – building from the small to the large.

- It will need to be 3-dimensional.

- It must allow information to have effects which appear to operate faster than light-speed.

- It will need to find how to include the infinitely small and the infinitely large – regardless of our human inability to comprehend the infinite and without making concessions to our discomfort or to the awkwardness of the mathematics.

Perhaps human mathematics is not adequate for infinities and paradoxes, and maybe it is not the best way for us to understand every feature of creation. This image will need to show how energy is never created nor destroyed, but retains a symmetry of form and dissolution that produces newness on a continuous basis. Where energy becomes matter it will need to balance being and becoming and also to allow death and rebirth.

What we present does not need to be totally new. Instead it is a re-assembling of known pieces to form a new pattern. We will see what others have seen, but we will do so with new eyes that support humans to think what we have not previously thought. Amidst that new thought will be the part played by information and consciousness. There will be a place for that to fit.

Building Universe from the bottom up

We began this process above in our description of sacred geometry. Let's try in as short a space as possible, to build the new picture just as creation itself did. Let us start from the smallest elements of the fractal and grow.

We started by describing triangles. The closed equilateral triangle is the most fundamental stable pattern of forces in two dimensions (in the plane of the triangle) but it can be broken by a force from the third dimension (at right angles to that plane). It locks into 3-dimensional stability when a tetrahedron is formed; then it can resist distortion from any direction. The tetrahedron provides the minimum possible form in which to enclose volume – the most elementary 3D structure to have both an inside and an outside – the fewest corners and lines of force to accomplish this, and the first and simplest to be inherently stable.

The tetrahedron is thus the first form that delivers "being" – the state of sustainability over time - the first that shifts beyond evanescence, that locks the movement of particles into stasis, the first that rises above the randomness and true chaos of raw energy potentiality. It is most certainly fractal, because it stacks and scales up so easily and elegantly. The first picture to get is that of twenty tetrahedrons stacked together. This is what they look like if you place 10 as a base, 6 above them, three at the next layer and one on top. You get a new, larger tetrahedron. This is the start of a fractal process.

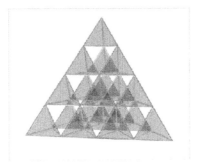

It is easy to see that the stack of twenty is shaped as a tetrahedron, but is not fully solid. It has gaps, white spaces, and is not in equilibrium. The symmetry can be increased, and the gaps filled, when you stack another set of twenty in the opposite direction, interleaved with the first set. This delivers the star tetrahedron which encloses the space more completely and has greater symmetry. It is getting closer to a sphere. At the centre of the tetrahedron though not obvious in this illustration is the shape of the Vector Equilibrium.

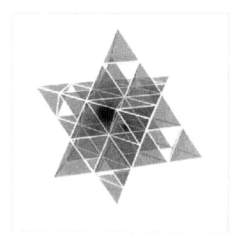

The next stage of addition is to fill in the gaps around the star where there is still negative (white) space showing. There are eight points to the star and each can be "surrounded" by a group of three tetrahedrons. When this happens, the resulting

shape is comprised of 64 tetrahedrons in total and looks like this.

The 64-tetrahedron structure is also composed of 8 individual star tetrahedrons, creating a second layer of potential fractal development. Energetically it is internally balanced with a mixture of radiative forces and contractive forces and it is a structure that can be seen as a building block such that it scales up in an unlimited way to infinite (or Universe) size.

And then, crucially, it can spin. When it spins it can do so simultaneously in all three spatial dimensions and when it does will describe the shape of a sphere which is the smallest surface area that will enclose a given volume. Anything that has fluidity, if it has equal pressure from all external directions or consistent attractive tension across the surface of the fluid, will naturally form a sphere. Even raw energy in motion has this potential to behave as a fluid.

The picture we are presenting is simplified and we merely skim the surface. There are larger geometries which have other possibilities of symmetry, spin and internal jitterbugging. Examples of these can be seen in the R. W Gray paper that we referred to previously and in the work of Lynnclaire Dennis among others. Readers who follow the links below[35] will find an abundance of fascinating exploration, some of it animated. The sequence to 64 tetrahedrons (and much more of what follows here) can be seen on Nassim Haramein's excellent video "Black Whole".

For the sake of the following descriptions, please leave very flexible any concept that you might have or any image that you might be using of energy being a "wave", and any previous description of the relationship between that and a particle. Bear in mind even the conventional QED view that it requires both descriptions to be complete. We will deal with this transitory reality in due course.

Meanwhile, we refer back to our earlier description of triangles and our observation that two lines cannot occupy the same space. When lines of force[36] meet, as in the closed triangle, they terminate at the meeting point (in fact they never quite meet). As described earlier, when they do not come that close they create two possibilities. One is that two open triangles will merge and form a tetrahedron and the other is that two or more open triangles of energy in motion will combine to form a vortex, an energy spiral. As just pointed out, closed triangles are vulnerable to be broken by force from outside their plane. A closed tetrahedron will resist forces, and so will a vortex which can accommodate pressure from other directions. Between vortices (dynamically sustainable motion) and tetrahedrons (structural stability) we have the building blocks for a living Universe.

These building blocks and the principles which formed them provide a base to create more complex geometric structures. There are so many that we cannot possibly deal with them here. Many of the developed structures have significance in the natural material world and in the esoteric and energy-relationship worlds. In

---

35   www.rwgrayprojects.com and www.mereon.org

36   Those with technical understanding will recognise that "force" is not necessarily a technically correct description, but it is easier to think of them in this way. Force is an influence on a body with mass, and there are none of those present here. What is really happening is a line of energy in motion, a vector. A vector has a direction, a magnitude and a point of origin (though the latter is often ignored).

addition to the websites just referred to Keith Critchlow[37] has shown how many forms there are in nature. Bruce Rawles[38] has a very rich book of diagrams, which those who enjoy patterns can colour in and meditate on (or both at the same time). If you would like your children to develop a feel for this kind of structural mathematics, Mary Laycock's very usable "Bucky for Beginners" is a great way in for them and for adults alike. There is nothing quite like exploring this physically, being able to touch the forms and turn them in your hands.

---

37   The Hidden Geometry of Flowers. Floris Books 2011

38   www.brucerawles.com

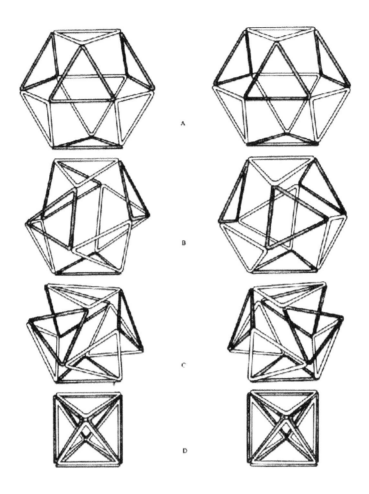

Avoiding the richness of these multiple possible structures, we will focus on the Vector Equilibrium. If you didn't view our video of the Vector Equilibrium in its cycle of contraction and re-opening[39] we encourage you to do so now. It will make what follows easier to visualise. When the Vector Equilibrium is open, it represents self-balance. There are 24 struts in the VE, four times six hexagons interconnected. When, starting with A left the top triangle is pressed down it does not rotate and neither does the base. All of the triangles, being structurally stable, retain their shape, but the ones at the side rotate around one corner. The square sides are not stable though, and collapse. Midway down the shape is such that there could be a further strut across each of the diagonals in the collapsing square, forming triangles of the same size (visualise this in the front face of B left image).

_____

39 www.scienceofpossibility.net/resources   (QR link on P 268)

The structure this creates is that of an icosahedron.

From six triangular and six square faces we have formed eighteen triangular ones. As we compress further the square faces collapse completely and we form an octahedron – eight triangular faces with twelve edges - the tension vectors are doubled up. There is insufficient inherent compression dynamic to maintain this structure. Once the hand is lifted it releases the stored energy and resumes the open VE structure of the cube icosahedron.

The VE has its stability because the distances from the centre are all equal in size to the outer edges. We need to think of the sticks that make up the structure as lines of force made visible. They all cancel each other out. But now notice that all the distances to the corners (vertices) are equal – not just in one plane but in the dimensions of all four hexagons. So now see if you can imagine the entire structure spinning – not just on one axis as it would if held up by a string, but in all three – top-bottom, side-side and front-back. What you get is a sphere, very much like the panelled structure of a soccer ball, but completely smooth and even. (In fact you can form spheres easily by spinning evenly any shape which has a regularity in 3 dimensions, so the tetrahedron despite its apparent spikiness can also generate a sphere, as noted previously.)

We can look at the same transformation in another way. Look at this flower of life diagram. The traditional flat-surface flower of life is a design which appears carved on or etched into the rock in sacred sites such as the temple of Osiris in Egypt (in some cases mysteriously burned in using techniques modern humans cannot replicate) and is created by drawing first one circle, and then subsequent circles centred at the edge of that one, and at points where intersections occur. This too is a fractal, where an element repeats and grows – potentially to infinity. None of this is surprising; the form of the flower lives within the stack of 64 tetrahedrons. The flower is a two-dimensional projection of the rotating spheres that can be generated from spinning those geometries (i.e. when each of the tetrahedrons in the stack is rotating in place).

The traditional view of the 2D flower of life stems from those original sacred sites, and from our communication now in writing, on paper and screen. Real life is three-dimensional and a better image is as conveyed in the picture of a real-life model that you can see below. (With thanks to Jiva Carter)

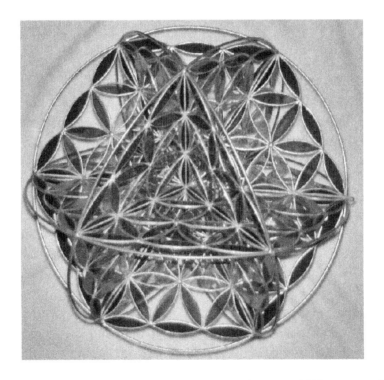

As well as being fractal in nature as you scale the flower of life outwards, it is also fractal going in and this gives a very important and interesting feature. If you consider a circle as defining Universe, and conceive of it as infinite in nature a new possibility opens up. The following diagrams illustrate this. If you begin with a star tetrahedron occupying the entire circle, it looks like the traditional and mystical "star of David". Remaining inside the boundary you can then draw circles around each of the triangles and then draw a further tetrahedron inside that new circle. We hope that you can visualise that this process can carry on at smaller and smaller scales, with no end.

If you see this clearly then you can understand that there is no theoretical limit to the infinitely small. And if you conceive of the first, outermost circle as defining the boundaries to Universe you can also see that there is no limit to what is possible inside of an infinitely large boundary. You can also see that there is no incompatibility between the idea of the infinitely large and the infinitely small.

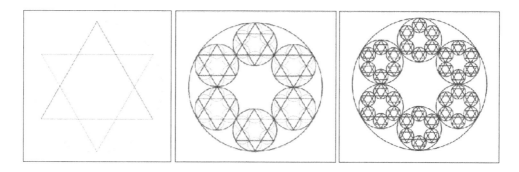

The scaling is limitless in either direction.

Do the locomotion

We have considered three forms of motion thus far. There was the jitterbug dynamic within the Vector Equilibrium. This potentially exists within others of the geometries – either as wholesale transformations, or as oscillations of interior lines of energy. Many of the forms have potential subdivisions and energy fluctuations of the possible configurations. This was visible in the various jitterbug motions defined by Robert Gray in the "Shape and dynamics of space" paper we referred to earlier[40]. Secondly there was the spin dynamic by which rotation of a polyhedron in three dimensions simultaneously would create a spherical form to the energy movement. We had a third motion offered by the dynamics of the vortex, an expanding and self-balancing developmental spiral that sits at the heart of this book, and we still need to say more about vortices.

We haven't yet exhausted the possibilities for movement. When you rotate an

---

40 www.rwgrayprojects.com

object around its centre you create spin and the potential for spherical shape. What happens when you rotate it around its edge? Imagine a ring-donut, or a bagel. Rather than slicing it horizontally as if to put in a filling, think of it sliced vertically. The cross-section that you see is a circle, or rather two circles which show how one circle can travel around the shape. This shape, called a torus, is what happens when you rotate a circle around a single point on its circumference.

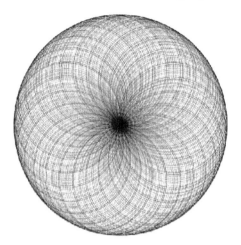

You may also be able to visualize that lines of energy might circulate around the torus in a spiral like fashion. This is like the child's "slinky" toy – a helical spring that has been bent around so that the two ends meet. However, the torus becomes a little more complex in the real world because of coriolis.

Coriolis is closely connected with rotation or spin. Coriolis is not a force (though sometimes it is called one) but an effect of relative motion in connection with a rotating body. One way to visualise it is in terms of an object in Earth's gravity. We think of a dropped object as travelling in a straight line drawn by the force which has its effective centre at the centre of the Earth.

Now imagine an object dropped from an extreme height – say 5 miles above the equator. At the time the object is dropped, the straight line of attractive force to the centre of the earth passes through one particular point on the Earth's surface. But since the Earth is turning, by the time the object is half-way to the ground, the surface has moved and the straight line to centre passes through a new point. By the time it hits the ground it will be at a third point. What this means is that the isolated frame of reference that is assumed in Newton's laws is not strictly correct when considering the Earth, the solar system or the galaxy, all of which are rotating. We referred to this in describing **Choice 1** earlier.

For bodies which turn slowly this effect is often negligible in relation to small masses, but the impact on our living conditions is considerable. Since the movement of the surface is faster at the equator than nearer the poles, and since there is a large mass of air and water vapour (clouds) surrounding the planet, the relative motion affects the behaviour of air masses that are already rising and falling, or flowing due to the influences of the sun's heat, the oceans and mountain ranges. The results are a significant factor in how our weather is determined. They

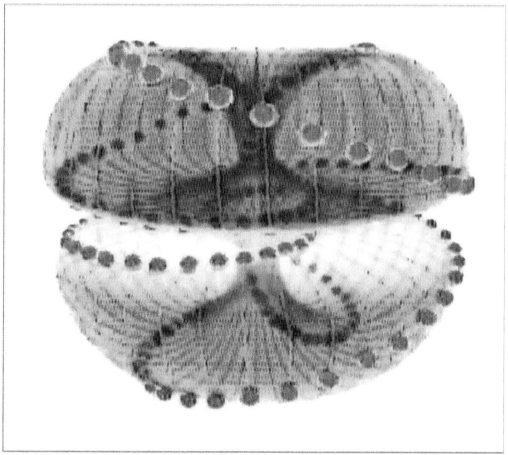

cause hurricanes to spin opposite ways North or South of the equator and to a degree divide the Earth in two, with movements either between the North Pole or South Pole and the equator, but not between the two poles. And where the body is rotating rapidly as it can be with spinning particles the effect of coriolis can become more significant.

Diagramatically, the combination of torus and coriolis looks something like the above, but coriolis is not all that is happening. The effects of gravity and rotation supply real forces such that the whole energetic structure is subject to torque, or rotational force. What happens when we scale up such effects to galactic or even Universe levels? And what happens in the realms of the infinitely small, where energy forces are spinning themselves into matter?

Before we address that question, one more type of motion is required to complete the set, and it is one that we have been discussing right from the beginning. We need to talk about waves and to see them in relation to the formation of particles. We are used to thinking of waves as drawn in 2D, or even as depicted with two 2D waves at right angles to each other. (See page 235.) But what if you take a 2D wave, and rotate it to occupy the 3D space? This will be a truer depiction of the real world, since there is no reason why energy would operate in only two dimensions.

The next step is to visualize that the wave of energy shifts in such a fashion that each of the nodes in the wave becomes distinct. The five nodes depicted in the second of the diagrams here can then separate from each other to form one or more spheres. You might imagine all of the energy in the wave being "pulled" into the rotation of the particle being formed. While not necessarily a literal depiction of the event, this gives us a visual sense of what a transition between wave and particle might involve. We are describing a deeply fluid world, in which all of the shapes and motions that have been laid out are interactive and interchangeable.

The Black Hole in the Donut

Every one of us has an intrinsically spherical frame of reference. Your view of the world extends equally in every direction. You may typically classify that view as "in front of me" or "behind me", and likewise in terms of left-right or above-below. But if you were a fighter pilot you might say "bandits at four o'clock high" and if you point to something it may be anywhere within the sphere. It's not a square world.

Scientific frames of reference are conditioned by a choice to look at the sphere via co-ordinates in 3-dimensional axes of reference. The world that we have just described is inherently and dynamically curvaceous. Yet the trigonometry you will have learned in school (if any) will not have been spherical. Scientific convention constrains our frame of reference.

Consider then, the possibility that the shape of Universe may be donut or hamburger-like – that we may be located anywhere in the cosmos, imagining that we are within a sphere. Consider as a result that we may be interpreting everything we see via our telescopes through a filtered mindset. Imagine if we are somewhere on or in the torus, sitting as a speck of jam in the donut or relish in the hamburger, and quite unable to conceptualise the movements that we see in appropriate terms. Consider too that in the torus depiction gravitational effects may alter the path of light itself and that tracking the movement of entities on the periphery may yield accurate results only over extended time-frames.

What kind of perceptual window are we looking through? At the moment all the data from our telescopes is interpreted through the big-bang model. So when we observe different and distant galaxies, receding from us at different rates according to their age, and come to the conclusion that the expansion of the universe is accelerating, is it possible that we are seeing something quite different, like a distant spin or a coriolis effect on bodies at the extremities of the torus?

We don't claim that this toroidal model is the certain answer or that we know what the answer is, but we must pose the question of what might be happening if there wasn't a big bang? We return to the question of symmetry – since energy cannot be created or destroyed, how do we produce a model for a symmetrical Universe? We can see an expanding creative dynamic which dominates our attention, yet we know also that something else is happening. Not only do we face it in the life/death polarity and in our perennial questions about soul, spirit, past-lives, life-after-death. In recent decades we have also faced it through physics because we have all heard of black holes and probably have some perception of these as parts of the cosmos where there is so much gravitational pull that even photons cannot escape; black holes are black because we cannot see any light where they are. We also have a notion of black holes as a destructive zone, the opposite of an expansive dynamic; in their vicinity whole star systems are being sucked in, never to be seen again. Here we may be coming to the least-developed part of current thinking. Certainly this is the view that Nassim Haramein[41] presents and we would like to give an indication of where this leads.

---

41 http://resonance.is/wp-content/uploads/2013/05/1367405491-Haramein342013PRRI3363.pdf

What happens inside a Black Hole?

The original notion of a gravitational mass big enough to prevent light escaping dates back to the 18th century, but began to seem more real as an outcome of Einstein's relativity field equations. Even then, it was not until the 1960s that such phenomena were studied in depth and the label itself probably first appeared in 1964, to be made mainstream by John Wheeler's use of it in 1967. In the strictest sense of course, we have never actually seen a black hole. We can only detect one by the absence of light from that location, and by the effects it has on whatever stars or galaxies are near to it. This is not a trivial or pedantic observation but emphasises that information is even more inherently restricted than for other galactic phenomena. We cannot tell what is happening inside.

Then once again we discover as we look more deeply that things are not as simple as they first appear. The theory and conjecture in this information-poor realm is that there are different types of black hole with differing characteristics. There are spherical black holes which are viewed as stationary. There are rotating black holes and there are flattened out ring-like black holes. There are super-massive black holes which are believed to be at the centre of galaxies (including ours) and which might have a mass equivalent to hundreds of suns. Then as a theoretical possibility there are tiny black holes that would be too small for us to detect under normal conditions. In conventional theory it is suggested that the smallest might be around the size of the Planck length and have a mass of 22 millionths of a gram.

According to the theory, a spherical, stationary black hole would keep sucking everything around it in, and continuously increase its mass. Around such black holes is something called an "event horizon" – denoting a boundary beyond which events do not affect external observers because it is like a one-way gate through which everything can pass inwards and nothing passes out. However, there is also conviction in favour of Stephen Hawking's prediction that all black holes will produce radiation. Unfortunately that radiation would not necessarily be above the level of the cosmic background and so it remains hard to discern, but given enough time would this allow all energy to dissipate. It is also suggested that the smaller black holes would be hotter than the large cosmic bodies, would radiate very rapidly and not remain in existence for long. Possibly the smaller they are, the briefer their existence.

Rotating black holes are thought to be different. It is from these entities that it is conjectured that a new spacetime continuum might be generated, and this produces the possibility of wormholes, which is very helpful to science fiction, with its potential to traverse galactic distances in an instant. If you have seen the film "Contact" you will have a notion of what might happen if we had the technology to create a temporary wormhole, and you may recall that the device that did so was spinning at extremely high speed.

Most of these possibilities exist in the minds of physicists, just as "Contact" emerged from the mind of astrophysicist Carl Sagan. They may or may not exist anywhere else. That is not to say that we don't believe in black holes; the evidence that something of this nature is happening is very strong because of the visible effects. The uncertainty concerns whether the theories themselves will hold true in their current form.

Predictably this area throws up its paradoxes too. It is a fundamental tenet of science that in principle the complete information about a system will determine its state at another time. It is equally fundamental to quantum theory that the information about a system is encoded in its wave function equations (one of the reasons why "virtual" particles were needed). Unfortunately, the problem with black holes is that elements in the equations tend to become infinite. They are dealing with infinitely large amounts of mass or energy compressed into infinitely small space – zero volume and infinite density. There are conjectures that time and space cease to obey the rules of relativity under these conditions (called a "singularity"). But possibly worse (for science) is that nobody knows what happens to anything that passes the event horizon. In particular there is no knowing whether the information about that system, the fundamental building block of scientific thinking and quantum theory, is preserved. When everything else is breaking all known boundaries, how could it be?

The solution to this problem has been to propose that information about a system is stored according to the "holographic principle" which says that the description of a volume of space can be thought of as encoded on a light-like boundary on the system's gravitational horizon. This is an interesting notion. There are some very advanced theories in this area, such as that of Beckenstein who has proposed that there is a maximum amount of information that can be potentially stored in a given finite region of space which has a finite amount of energy. The idea also led to the venerated John Archibald Wheeler's[42] suggestion, known as "it from bit" that all things physical are information-theoretic in origin and that this is a participatory

---

42 Wheeler was author of the seminal work on gravitation, worked closely with Einstein, was Professor at Princeton, tutor to Feynman (among many) and was active into his nineties.

universe. That is "the physical world is made of information, with energy and matter as accidentals". This is perhaps the closest that physics comes to the central theme of this book, but we should note that they mean something different and narrower than we do by the term "information". This will become even more apparent in our final section.

Here too, we are not able to resolve the many questions that are presented. We are not even sure how many of the questions are real ones, and how many are purely intellectual constructs and artefacts of some very clever theories produced by some amazing minds and which may ultimately prove not to be real. We don't mean this as a critical comment. There is a difficult line to be walked. At its best such creativity produces great insights like those of Einstein that open new horizons. Enough of the insights turn out to be predictive, later supported by the evidence from ever-improving observational equipment. At the same time, anyone who cares to Google around string theory, membranes and the like will find speculations involving 10 or more dimensions, amidst some extremely abstract notions. The difficulty for ordinary mortals is that neither we nor they know where the boundary lies between what is "real" and what is a construct of human imagination, possibly verified with technology that will only see what it is constructed to see. This 10-dimensional string world with 17 types of particle doesn't look like nature, it looks like something that humans make up. So what would be a more elegant and natural solution? Could it be equally successful in explaining known measurements and can it unite large and small?

Nassim Haramein brings together a number of perceptions that we have been listing and describing here.

- If space is filled with energy and is not simply a vacuum then space has the potential not to separate out but rather to connect the objects of the visible universe.

- If the way that energy and forces move is generated from first principles by the tetrahedron, the star tetrahedron, the vector equilibrium and all of the fractal scaled-up versions of these and more complex geometries, then these are the forms that fill space. Because they are in equilibrium they are not obvious to our detection systems.

- If black holes behave like twin-hemisphere toroidal topologies with the torque and coriolis forces that are missing from Einstein's field equations then there are radiative effects that alter their characteristics from those described by the standard model.

These characteristics would completely transform the nature of black holes from what we currently conceive them to be. They alter the speculations regarding what is happening on the other side of the event horizon. At the large scale black holes become vehicles of transformation. At the small scale, rather than being unable due to their high temperature to sustain form, black holes may find stability and be a central component in the formation of matter.

Black Wholes

A theory that meets all of these requirements, however elegant, is bound to be complex. We will try in the following paragraphs to describe the simplexity that emerges on its far side. Let's start from some very basic principles.

How much energy does the Universe contain (as energy or as matter)? The answer that scientists use is derived by starting from the Planck length, which is the smallest distance that they conceive to exist, about one-twentieth of the diameter of a proton. The Planck length is the distance above which it is believed that matter might be formed and is the boundary below which the rules of Einsteinian relativity cease to apply. If there is any reality smaller than this it exists only in the quantum realm. The smallest element of mass that could be formed at that size is $1/100,000$ $(1 * 10^{-5})$ grams.

So if the Planck length would represent the smallest particle conceivable, how many of those might you fit into one cubic centimeter of space (about a fifth of a teaspoonful)? Because the Planck length is so tiny, the number you can pack in is remarkably large, of the order of 10^{99} and the mass that this would deliver is of the order 10^{94} grams. That would be the maximum possible and we are inclined to borrow Gary Douglas's[43] humorous term "godzillion" – a number so big only God can conceive of it.

So if it is theoretically possible to fit 10^{94} grams of mass-energy in less than a teaspoon of space how much matter is there in fact? Well, as best may be estimated – bearing in mind that you can't put a single galaxy on a set of scales, still less several billion of them –the amount of actual matter in the universe totals something like 10^{39} grams. That means that there is about 10^{55} grams left over that might have been formed but has not. That potential (energy density) exists as energy fluctuations in space (or the vacuum as it is still called, since it contains no matter). So we know what we are looking for – about half a godzillion tons of

---

43 Founder of "Access Consciousness"

mass-energy distributed throughout the universe.

Nassim Haramein[44] together with his colleague Elisabeth Rauscher has proposed that there are black holes everywhere in the universe. Effectively these singularities are the invisible glue, the cosmic thread that connects Universe. We are to imagine that not only galaxies but stars and planets too have singularities at their core, and even further than that, that singularities are at the centre of every atom in the form of protons.

What defines a black hole technically? The answer to this was derived in 1916 by Karl Schwarzschild from Einstein's field equations. For any given amount of mass, what is the distance at which its gravitational pull would be sufficient to prevent photons from escaping? If all the mass is inside that distance, which is called the Schwarzschild radius, then singularity conditions are created. Thus for our Sun the Schwarzschild radius is approximately 3 kilometres; it is bigger than that, so we can see it. The Earth would need to compress to the size of a pea before it exhibited black hole conditions.

The Schwarzchild conditions apply to just one type of black hole, but as mentioned above there are others. In fact there are five general categories of black hole solutions. They are: (1) an uncharged, non-rotating black hole which is described by the Schwarzschild solution field equations, (2) a charged, non- rotating black hole which is described by the Reisner-Nordstrom solution, (3) an uncharged but rotating black hole which is described by the Kerr solution, (4) a rotating, charged black hole which is described by the Kerr-Newman solution and 5) the Haramein-Rauscher solution with the inclusion of torque and Coriolis forces to define the origin of spin.

Haramein and Rauscher have produced the graph below depicting the relationship between vibrational frequency and gravitational radius for all scales of existence in our Universe. The point labelled BB refers to the notional conditions of the big-bang singularity according to conventional theory. At the other end is the scale of the entire Universe, with two different sizes of galaxy (G1, G2) and our Sun in between. Also shown in between is a value "A" for the atomic scale. This is not the conventional value, which would not fit the scaling line relationship.

---

44   http://www.resonance.is

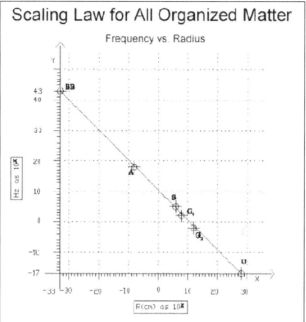

Figure 2a. A scaling law for organized matter of frequency vs. radius. The black hole system is presented in this figure. Plotted from the top left is the mini black hole at the Planck distance of $10^{-33} cm$ through to the stellar-sized black holes, larger black holes, galactic center black holes and at the lower right is a Universe-sized black hole. Note that in between the stellar size and the Planck distance mini black hole we have included a data point for the atomic size which we as well calculate a new value for its mass that includes the energy available in the vacuum space of a nuclei and yields the correct radius to describe an atomic resolution as mini black holes (see equation (5) to (18)). It is of interest that the microtubules of eukaryotic cells, which have a typical length of $2 \times 10^{-5} cm$ and an estimated vibrational frequency of $10^{9} to 10^{14}$ H~ lie quite close to the line specified by the scaling law and intermediate between the stellar and atomic scales [19].

The point A shown in the graph is that which applies when the Haramein-Rauscher[45] solution is used. They postulate a condition where all of the vacuum energy is available to the protons in an atomic nucleus. If a proton can attain a mass sufficient to produce Schwarzschild conditions then it exerts a gravitational pull which is <u>exactly</u> that which quantum theory indicates as being the size of the strong nuclear force.

The strong nuclear force was introduced to quantum theory because two protons which have like charge should not be able to come together. The postulated "strong force" is that which is necessary to overcome the Coulomb force of that repulsion. The Haramein solution would mean that the strong force is no longer required. The mass of the Schwarzschild protons would produce a gravitational

45   SCALE UNIFICATION – A UNIVERSAL SCALING LAW FOR ORGANIZED MATTER   Nassim Haramein, Michael Hyson, E. A. Rauscher :   From Proceedings of the Unified Theories Conference (2008)

attraction of the required magnitude. Eliminating the strong force and applying gravity at the atomic level brings unification between Einstein and quantum, between infinitely large and infinitesimally small, much closer.

It does something else too. It places black holes at the centre of every atom of the material universe. You could conceive of this as creating breakdown conditions at the edge of the Einsteinian universe. Within the theory is an implicit access point to vacuum energy from everywhere in the universe, a connectivity not unlike that which is offered by wormholes, but omnipresent. Beyond that there is the possibility that there may be singularities within the Earth, the sun and the stars themselves. It is highly speculative but not inappropriate to suggest that if we are seeking a dimension in which all of the information about the universe might be carried, including physical information about matter and "pure" information about events, thoughts and knowledge of conditions, the vacuum would be capable of doing so. The connectedness of inner (atomic) and outer (cosmic) space would seem to make this possible.

To repeat a regular refrain, our thesis does not require that we supply all of the answers to how the connectivity of the Universe works. We have produced a mass of evidence that the connectivity exists. The Haramein solution may be the answer or it may be a step towards an answer. From the standpoint of this book it serves to show how big are the gaps in our current theories, and how much space exists for new scientific answers to emerge. It does not even matter if we don't find those answers. We only need to stop believing that the non-visible world is of less significance than the visible one so that we can find a more connected and coherent way to live life.

And if Haramein is correct, or if a solution similar to his were found to fit, we would see several of the difficult areas of current theories fade away. There would be no need for the mysterious dark energy and dark matter. Einstein's desire for there to be a dynamic balance of expansion and contraction would be satisfied. Rather than conjecturing a long-time-future "big crunch" where the elastic of the big bang snaps back, we have a dynamic and perpetual lifecycle for Universe that reflects the rest of nature. Creation is continuous.

The seemingly endless subdivision of virtual particles and attributes at the energy / matter boundary would no longer be needed. We could learn to accept that the energy/matter boundary is not a "digital" event with a mathematically defined description. We could live within the bounds of Gödel's theorem, safe in the knowledge that life will always contain "undecidable propositions". We might even accept life's mysteries, fundamental uncertainties and inherent paradoxes and learn to live with nature. We know that what can't be measured, cannot be controlled. This would mean we could abandon our futile attempts to control life in favour of co-creating with life. This could free us from the hubris which leads humanity into such a damaging and seemingly endless stream of unintended consequences. Instead we might cease giving our authority to "science", have it be our servant rather than our master and find our way to co-create with nature in a world which is, and always has been "analogue".

The human mind finds binary thinking useful. It is part of our survival apparatus to label things as hostile or safe, edible or toxic. But most of life is not binary. We do not choose between breathing in and breathing out. We manage the polarity and live in the dynamic interplay of alternatives.[46] Most of us are raised to believe that the world is characterized by right and wrong, to judge it, others or ourselves as good or bad. This fixed way of being is a comfort-blanket for fearful minds. It squeezes our creative potential and causes us to be less than we can be.

What if every one of our cells and atoms contains a gateway to the infinity of time and space? What if some of our being oscillates in and out of the information realm, as the electrons that form us shift back and forth between matter and energy. What if the limits that we imagine as constraining us – the size and shape of our biological forms – are an external illusion that dupes our thinking into the belief that we are not infinitely connected to all that is? What if our conscious minds are a device for experiencing the biological form but are also only the tip of an iceberg? What if, beneath the surface, sits infinitude of conscious connectivity with the rest of creation through the singularities at the centre of every atom of creation? What would life be like if we were to open the boundaries of our conscious mind, let it all in, behold and see that it is good?

In the model of creation just described we have a way of understanding why energy forms the material patterns that it does. The answer to what causes it to do so at all is unknown, but might best be described as "because it can", and that when it did, the pattern was sustainable and repeatable. The forms that matter takes were determined by a kind of trial and error. The ones that we see are the ones that worked.

This is the nature of our cosmos – the outworking of a self-exploration by all that the universe is. We humans, in our turn, are the most recent outcome in the world of living beings – the product of 4 billion years in which matter, having been

---

46 See Barry Johnson's "Polarity Management" for an interesting exposition of this.

created, produced forms that would replicate themselves – the amino-acids, proteins and cells that would eventually emerge as multi-cell organisms.

Every part of this journey, every element of this creation is built from the patterns of tetrahedra that span and coalesced into particles. The fractal nature of emergence allowed pattern to form on pattern to produce increasing complexity. That complexity eventually became sufficient to produce an element of self-awareness, for us to be conscious that we are conscious. How we then use that consciousness to become who we are individually and collectively, and what we can do with that consciousness in relation to the universe is the subject matter for our final section.

> "We are participators in bringing into being not only the near and here but the far away and long ago. We are in this sense, participators in bringing about something of the universe in the distant past and if we have one explanation for what's happening in the distant past why should we need more?"
>
> *John Archibald Wheeler – radio broadcast 2006 aged 94 on the "Participatory Anthropic Principle"*

Review

In this chapter we have presented a picture of an alternative to quantum mechanical solutions to the understanding of cosmology. This picture is coherent from smallest to largest scales and avoids having to invent forms of matter and energy that we cannot detect. The forces that shape reality in this point of view are real ones, ones that we can recognise as fundamental to the geometries and patternings of the physical world. While this picture too may yet evolve into something better, it serves to support our viewpoint, providing a mechanism by which Physics would allow the connectedness through all of creation that is central to our presentation.

Section 4: The Science of Human Experience

I have said that the soul is not more than the body,
And I have said that the body is not more than the soul,
And nothing, not God, is greater to one than one's self is,
And whoever walks a furlong without sympathy walks to his own funeral
drest in his shroud,
And I or you pocketless of a dime may purchase the pick of the earth,
And to glance with an eye or show a bean in its pod confounds the learning of all times,
And there is no trade or employment but the young man following it may
become a hero,
And there is no object so soft but it makes a hub for the wheel'd universe,
And I say to any man or woman, Let your soul stand cool and composed
before a million universes.

And I say to mankind, Be not curious about God,
For I who am curious about each am not curious about God,
(No array of terms can say how much I am at peace about God and about death.)

I hear and behold God in every object, yet understand God not in the least,
Nor do I understand who there can be more wonderful than myself.
Why should I wish to see God better than this day?
I see something of God each hour of the twenty-four, and each moment then,
In the faces of men and women I see God, and in my own face in the glass,
I find letters from God dropt in the street, and every one is signed by God's name,
And I leave them where they are, for I know that wheresoe'er I go
Others will punctually come for ever and ever.

Walt Whitman   *Song of myself*

"Meaning is invisible; but the invisible is not contradictory of the visible: the visible itself has an invisible inner framework, and the in-visible is the secret counterpart of the visible.

M. Merleau-Ponty   *"Working Notes"*

Human Experience

The fourth section of this book will come as a relief to those readers who do not lean towards hard science. We acknowledge that Sections 2 and 3 contained demanding content and called for shifts in mindset. With the biology and physics behind us we now have a basis for exploring the ways in which humans thus far have framed the world. We can then take the trend-line forward and anticipate what is coming towards humanity, and what is possible for us.

We hope that the story so far has persuaded you that universe is a conscious co-creation. All that exists has been a flowering of energetic geometry, particle physics and chemical building blocks that eventually combined to produce first inorganic and then organic life. Organic life has constructed itself in myriad forms, principally based on combinations of cellular templates assembled into vegetable and biological complexes, interacting and co-adapting in multiple ecologies.

The field of self-knowledge which holds every outcome of every exploration that our Universe has taken, now includes several billion human beings consciously enquiring into all that is, self-aware in ways that are seemingly unique to our species and now engaging actively in shaping the future. We are creating new materials and forms by ourselves – a meta-nature world that rests on the natural world, but is additional to it. The systems within our 50-trillion cell bodies are of huge sophistication in their ability to produce coherent action and to contain complex thought structures that map across both time and space. In addition those systems are open to the field of consciousness and able to detect information from it – apparently without boundary of time or space. Those systems are also able to influence the content of the field of universal information. We can move objects and events with our minds. We have an impact on the fractal unfolding of creation.

Our meta-natural world is not always harmonious with the natural one. The co-creative experiment is in great tension where human technological power outstrips human understanding of the consequences of our actions. We do not yet know whether the human experiment will prove to be a success; we have the capacity to destroy all that sustains planetary life. Less extreme, we face the possibility that humans could be an evolutionary dead-end that did not even last as long as the Tyrannosaurus. It is also possible that the dynamics in the rest of our living co-creation will produce a new balance that in the short term means huge reduction of human numbers. If seven billion is unsustainable, the law of unintended consequences may return us catastrophically to smaller numbers.

On the way it may damage beyond repair all of the systems that we currently enjoy.

It takes large-scale activity to sustain international shipping and air transportation, oil extraction and refining and a world-wide web. Even the alternative technologies for small-scale sustainability are threatened in this scenario. To produce a solar panel demands an intricate web of mineral extraction and refining, technical manufacturing process and wide-scale distribution. Using solar panels is locally sustainable but they are not produced by cottage industry. Even the washing machine requires a factory supported by energy supply, steel and copper production, components and machine tools. Few in the West would willingly revert to an existence too primitive to provide the washing machine.

So our familiar question recurs. What else is possible? What is the nature of our co-creative process at the psychological and social levels? How have we evolved our social systems and our thinking processes; how have we built our relationships with the planet, its creatures, its ecologies and each other? Most of all, what can we do now? What do we need to do now? If we have brought ourselves to the edge, what is the shift that we must make in order to find the new ways of thinking and being that will solve rather than compound the problems we have created?

The first three sections dealt with the big questions; who are we and where did we come from. They provided something of the answer to the question "why are we here?" The remainder of that answer is in a field of choice because the answer is within us. The next phase of the co-creation experiment is influenced by who we decide that we wish to be. What kind of life do we aspire to create? It should be self-evident that this is not an independent, willful and selfish choice since such choices have brought us to the edge. However we cannot be passive either. We cannot uncreate our technologies, cannot put the genie back in the bottle. And we all know that the secret with genies is to ask the right questions.

18.   Reality and the Social Order

"The mental and the material are two sides of one overall process that are (like form and content) separated only in thought and not in actuality. Rather there is one energy that is the basis of all reality ... There is never any real division between mental and material sides at any stage of the overall process."

David Bohm   *Wholeness and the implicate order*

Said Plato:"The things that we feel
Are not ontologically real
But just the excrescence
Of numinous essence
Our senses can never reveal"

Anon

Theme

If consciousness is fundamental in creating and defining Universe, what is "real"? We have already discovered that reality is not only material. We have seen how the metaphysical is as important to creation as the physical. Since Universe is a scenario, where is humanity placed within the field of consciousness. Is reality outside of us, waiting to be perceived? Is it within us, awaiting manifestation? This chapter explores the nature of the human mind and the way in which we construct the various layers of reality that we share.

Inner and outer mind

We all experience imagination, dreaming, illusions of the waking senses and errors of perception. Sooner or later, we all notice that other people experience reality differently than we do. For those who do not think this simply means everyone else is wrong and stupid this inevitably leads into fundamental questions of ourselves and the universe. We want to know what is real. We want to know what it is that we truly know.

The view of creation expressed in the last chapter and the presentation throughout this book of multiple levels of awareness and knowing have profound implications for our understanding of reality. In the universe we are presenting it is hard to distinguish Descartes' "I think therefore I am" from its reverse "I am, therefore I think", from Popeye's "I am what I am". Or is that Buddha?

All definitions are up for grabs. The very fabric of the universe is maintained in the information dimension, the zero-point field, the continuum of black holes beyond time and space. Our biological being is the outcome of a number of vibrations in the quantum soup, responding to and interacting with multiple stimuli beyond our conscious awareness. Our consciousness is everywhere, and nowhere. Information about ourselves that we think of as personal, is held somewhere beyond what we think of as our physical body.

More than this, the information that is available in this external field or dimension is also beyond present time. Both the past and the future also seem to be available judging both by common personal accounts and by J.W. Dunne's documentation of his and other's experiences.

The fact that everything "known" is somehow co-ordinated by a field of information puts a very odd slant on what we have considered to be "real". The Oxford Dictionary has its first definition of the word as "Actually existing as a thing or occurring in fact, objective, genuine". Most people are more comfortable with "real" having something of solid substance behind it, but the blackbird song from outside our window is also regarded as real. The description of the universe presented in this book greatly blurs the boundary between subjective and objective and places the determining features of the material world partly or completely in a non-material realm. Arguably this has been known ever since wave-particle duality, but we suggest that it is now somehow more personal. It doesn't just affect what you and I know of the world around us; it goes to the very root of who we see ourselves to be.

The end to realism and idealism

This book is not intended as a history of philosophy. But there are some core strands of thinking which need finally to resolve and find a new place in human consciousness. Most of philosophy, most of science and most of human thought has operated for millennia on the basis of dualities; the either / or of subject-object, mind-matter, good-bad, true-false, reason-emotion, body-spirit distinctions. None of these work effectively in a consciousness-driven universe, except when we apply them within quite specific boundaries. They are useful distinctions which make conversation easier and support an analytical process. But they are not reality itself. As Robert Pirsig puts it, in "Zen and the Art of Motorcycle Maintenance"

> "This eternally dualistic subject-object way of describing the motorcycle sounds right to us because we're used to it. But it's not right. It's always been an artificial interpretation imposed on reality. It's never been reality itself."

It is for this reason that we have repeatedly emphasized the requirement for an additional dimension or perspective beyond these dualities as depicted in our 'target' image and as represented in the spiral vortex picture.

Philosophical investigation into what can and cannot be known and into what is real is ancient and deep. The idea that things can be real only when they are observed was most famously epitomized in Bishop Berkeley's view that the tree in the quadrangle ceased to exist when no-one was there to observe it. Ronald Knox satirized this view:-

> There was once a man who said `God
> Must think it exceedingly odd
> If he finds that this tree
> Continues to be
> When there's no one about in the Quad.'

But was answered by an anonymous respondent

> Dear Sir, Your astonishment's odd:
> I am always about in the Quad.
> And that's why the tree
> Will continue to be,
> Since observed by Yours faithfully, God.

The central notion of idealism can be summarised as saying that all things are an expression of mind, or dependent in some way upon mind. You might see a reflection of Berkeley's view in the notion that only humans can collapse the wave-function that determines the fate of Schrödinger's cat. There are many subtle variations on the theme of idealism, but they tend to have a central agreement that some kind of observer is required (one mind, many minds, the mind of God etc..)

Plato is seen as the architect of the "realist" view, in which the existence of objects has nothing to do with humans or observers of any kind. The objects are deemed to have a reality of their own, in which forms are related to particulars – the actual instances of objects and properties. A particular is regarded as a copy of its form as in the example that a particular apple is said to be a copy of the form of Applehood and the apple's redness is a copy of the form of Redness.

The resonance with Plato is more like the concept of a morphic field. The realist view has things existing independent of any sense of an observer. Even a reference to "the mind of God" is meaningful only in the sense that particulars could be held there; actual forms do not depend on God as a perceiver for their existence.

Humberto Maturana and Francisco Varela have taken a different approach to this question. Their theory of what can be known is developed from the roots of biology. Where philosophy has argued its various points of view based largely on pure reasoning and analysis, these authors have examined the development of cognition and behaviour from the most primitive organisms to humankind.

In their beautifully argued book "The Tree of Knowledge" – far too detailed and subtle to summarise – they reach the conclusion that cognition is a "bringing forth of the world through the process of living itself". Their view like ours is of a world which is self-creative; in common with Elisabet Sahtouris they use the term "autopoietic". They provide thorough evidence and analysis that human reality is not "out there", and neither is it simply an internal construct that individuals develop. Rather it is a world in which they suggest that any individual who understands what they have said

> "will be impelled to look at everything he does – smelling, seeing, building, preferring, rejecting, conversing – as a world brought forth in co-existence with other people."

In consequence, reality has no independent point of reference. There is no external certainty available to validate our descriptions and assertions. The reality that humans live in is something we have constructed together and there is no absolute Truth. In their presentation, our neighbour's certainty is as legitimate and valid as our own; the act of recognition that accepts this fact is the act of love. It is the biological basis of social life, a biological dynamic with deep roots leading to operational coherence in the social realm.

Reality is socially constructed

Since people do not simply construct their knowledge of reality inside them unaided and uninfluenced, and neither does reality consist solely of the "things" of the world, the underlying fabric of our universe is created at the social level. As we have seen, in the zero-point field or the akashic vacuum, information is held outside of the things or the beings which experience them. In a similar way in this social realm the information exists in a realm of its own, independent of us. Even science is a social construct and not an absolute truth; scientific objectivity also relies on consensus between its practitioners. It even insists on it. In its absence there is no agreement about facts and no scientific "truth". You might give a cheer at this point.

To some degree this way of thinking is present in older work by Berger and Luckmann. The title of their book "The Social Construction of Reality" sums up their viewpoint concisely. They present the way in which knowledge is held socially. Not only is society itself constructed by a process of social legitimization, but the individual's view of personal identity is framed within society's terms and his / her view of the world is framed by concepts which society itself provides.

You can see this for yourself in your view of identity. Please answer the following question. Who are you?

Now please answer it again, differently. And again. You can play this game for as long as you keep coming up with new responses, or until you realize that none of the answers truly define who you are.

It is almost certain that your responses will include items such as your name, your gender, your job, your family status, your age and your relationships to others. You may possibly have reached the point where personal preferences or inner qualities began to be listed. Apart from a few of the more objective statements such as gender and age, the typical responses one gives to this question have a social context and are, in effect, given to us by the way in which society views us. And even when it comes to inner qualities, these are not free from social construction.

Another theorist, George Kelly, working in the arena of personality came to the view that our individuality is formed according to the constructs we find to be significant, and where we measure ourselves and others to be on those constructs. George Kelly regarded each human being as a scientist attempting to predict the behaviour of the world around them. In his words

> *"A person anticipates events by construing their replications".*

These predictions would take place on the basis of a repertory of measures that are personal to the individual. That is

> "A person's construction system is composed of a finite number of dichotomous constructs."

For one person, an important measure applied to other people might be based on the construct "for me / against me". To another person that measure might be irrelevant, whereas the construct "communicates / does not communicate" might be very important. We recently heard of an American businessman who was called "evil" by a Chinese contact because he recruited new hires on merit, rather than giving the position to a family member. The Chinese have a much stronger "family – not family" construct than most Western cultures"

Kelly predicted and showed that the level of empathy between one person and another would be based on the similarity in constructs that they use to predict or evaluate the world. It is clear that constructs might arise both as a result of personal experience and choice and by virtue of being provided by a shared culture. For example, to a person brought up among the Amish, the constructs of "Behaves like Amish / Not like Amish" might be important in their attitude to people. Their attitude to things might be similarly culture-related as "Referred to in the Bible / not….". A Tibetan Buddhist would not share these constructs. Nor would he/she start from the same concept of what is "real".

Eastern religions have traditionally taken a different approach to reality. As Bede Griffiths describes it

> "There is no objective reality outside us as opposed to a subjective world within. There is one reality, which manifests itself objectively outside us and subjectively within, but which itself is beyond the distinction of subject and object, and is known when the human mind transcends both sense (by which we perceive the "outside" world) and reason (by which we conceive the mental world of science and philosophy) and discovers the Reality itself, which is both being and consciousness in an indivisible unity."

No doubt Griffiths, from his background as a Catholic Priest, also saw this as a parallel to the notion of restoration to oneness through Christ. But the Eastern formulation of this also comes closer to that of Robert Pirsig. Our divided view and our experience are seen to derive from ignorance and illusion; that division exists only in us, not in Reality. Restoration to oneness with God is the underlying solution expressed as Buddhist "Nirvana" and Hinduism's "Moksha". Pirsig looked to the "Quality", to excellence, dharma and arête.

> "It is Quality, not dialectic, which is the generator of everything we know".[47]

From the above brief examples it is evident that the approaches we are taking have been pre-figured by thinkers and present in some religious perspectives on the spiritual world down the ages. The scientific evidence may only be coming together now, but there are thinkers who had the analytical understanding to see it coming and / or the spiritual experience to feel it and there are other cultures in the world whose approach starts from a much more helpful standpoint than the now-

---

47 "Zen and the Art of Motorcycle maintenance" P. 385.

embedded Western scientific paradigm. The views being presented in this book pull these strands together and provide additional cohesion. We do not claim to have invented them. What interests us now and in the remainder of this book, is what do we now do with the knowledge?

A new way to frame reality

The emphasis that science has placed on one particular form of observation has made any kind of unified view quite impossible. As said here before and often complained of elsewhere, the dualistic view is self-justifying. It makes a (quite spurious) claim to "objectivity" in such a way as to exclude human experience and knowing from any possibility of delivering "truth". The objectivity is spurious because that choice of perspective is a subjective one. Perhaps more importantly, it is quite simply out of touch with the scientific nature of reality. Remember, we have used science itself both in our method and in our reasoning, to establish that the "mind" and the "matter" are not in any way separable. Put another way, we have provided scientific evidence for the existence of those things which people describe as "spirit", "essence" and "soul".

Throughout our text we have documented the scientific evidence supporting the view which many others also express, that consciousness is an inherent part of material reality and totally inseparable from it. As we have shown, there is a significant level of understanding of the foundations for this relationship. While it may lack detail it is sufficiently advanced for us to work with in practice. With this established, it is time for the western world and the scientific fraternity to embrace the new reality and to let go of the old one. It is no longer fruitful for there to be philosophical debates about realism and idealism, when both are equally true (or untrue). It is no longer of any value to maintain the pretence of a meaningless objectivity in science beyond the boundaries where it has justification. As Bede Griffiths argues (echoed by many others through titles such as "The Tao of Physics" and "Dancing Wu-li Masters") there is a need for a Marriage of East and West.

We took this apparent diversion into the territory of knowledge construction, of quality and value systems (constructs could also be described as values) for two big reasons. The first, which we will examine in a moment, concerns the way in which the spiral of development continues beyond the biological realm. Since our proposal is that material existence is a process by which the "All" experiences itself, the existence of culture and large-scale society is a significant part of its development and of our own. This is taking place in a small blue-green planet at the outer edge of just one galaxy among billions, but that fact does not reduce its significance, not even if humans are the first to develop in such ways. If that were the case you would still potentially be an influence in the establishment of a morphic field for the rest of creation.

More crucially, that large-scale social construction is also of considerable interest to human beings. A large part of human evolution is now taking place, not through changes in our biology, but by means of the cultures that we create. As all of us can see, that human social and intellectual emergence has a huge impact on the whole of planetary life. To understand who you are in a scientific way requires a model for this cultural development and the growth of human knowledge. We will see shortly that the human aspect of spirituality and belief has its own evolutionary path.

Secondly, our knowledge, our values and the perceptions we carry of what is "real" are secondly of major significance when it comes to our individual and collective choices around religion, or around non-religious spiritual orientations. We will deal with this further as we complete this journey. The available choices are the very stuff of life and the way in which humans may now see our own potential are exciting both personally and collectively. This will be the climax of our story.

Review

In this brief exploration of the nature of reality we have laid the ground for understanding that reality is something that humans hold as a shared perception, something that we construct socially and hold within us as chosen patterns. In the next chapter we will discuss how those choices change and evolve.

19.   Values and Memes: Evolving societal dynamics

Everything is metamorphosis in life – in plants, animals to man and in him too.

J.W. von Goethe   Spruche

If you're going to have a worldview that is sophisticated, you have to take into account that dirt got up and started writing poetry.

Ken Wilber

Theme

In this chapter we take the idea that humans construct their reality a step further and examine a theoretical basis for understanding the evolutionary nature of human thinking systems in terms of the ways that the constructs people hold are adaptive systems that enable us to meet life's changing demands, and to increase the capacity and range of our collective and individual possibilities.

We would like to start this piece by acknowledging Dr Richard Dawkins. Having been critical of his views we owe him recognition for a brilliant contribution to the following discussion. Observing that he is unable to provide any genetic explanation for human culture, he has introduced into our language a fresh and potent concept, the "meme". The meme is a unit of cultural information that propagates itself across the ecologies of mind and produces the content of belief systems. The memes that we are particularly interested in are those that drive the Values governing how people think, which we will call Vmemes. You may detect some resonance in what follows to George Kelly's Construct Theory. Why do people value some things and not others?

The theoretical structure that we are about to introduce has the name Spiral Dynamics integral (SDi). It is based on theories developed by Prof. Clare W. Graves and articulated in this form by Prof. Don Beck with Christopher Cowan and enhanced by the Integral perspective of Ken Wilber. SDi is a wonderful example of simplexity; a theory in which simplicity emerges elegantly on the far side of complexity.

Many scientists have difficulty in accepting as a science any discipline that has to do with society as a whole or with the psychology of its members. While it is clearly not an arena for the kind of experimentation that a chemist might perform, it has many of the elements that govern other accepted sciences – study of form, breakdown to components, collection of data, statistical analysis, verification against real-life examples. Here we will draw out some parallels between the way that genetics and evolutionary theory reveal dynamics in biological development, and the features of cultural development as it affects individuals and all sizes of social group. The balance of forces governing individual vs group and competition vs co-operation dynamics occur again here. So too does the progressive development of life through dynamic co-operation, together with increasing levels of behavioural sophistication that we saw in the Sahtouris loop. All of these are in play.

SDi is also one of the most scientific of approaches to society because of the way in which Clare W. Graves developed it. Where many theories emerge from academics thinking and speculating about society, Clare Graves started by collecting data. Based on questioning huge numbers of people about the Values they hold and the conditions governing their lives, his theory emerged from the data itself. It is in the finest traditions of first observing, then measuring, then analyzing the results. At the same time it satisfies all the demands we can make for explanation of high-level patterns and of an intuitive feeling of rightness to a living process, with high predictive success and many effective toolkits for supporting change.

A theory which does all of these things will inevitably have much complexity within it. Those who seek full technical understanding are encouraged towards Beck and Cowan's authoritative exposition. For a narrative overview, Don Beck's CD set[48] offers a more accessible approach. A shorter, one-hour introduction is available here.[49] Our presentation inevitably must be high-level and simplified. No change there, then.

The codes and principles that underlie spiral social evolution can be viewed as a form of psychosocial genetic material. Just as a biological being is required to respond to its context and its surrounding ecology in order to thrive, so too are individuals and social organisms. In the same way that changes occur in biological organisms over generations as adaptations are incorporated into the form and repertoire of the species, human thought systems also change and elaborate over time. Where a biological life-form deals with variability in climate, habitat, predation or food source, the human individual too, responds to life conditions. Those life conditions are partly grounded in physical survival - the difference

---

48 "Spiral Dynamics" www.soundstrue.com
49 www.che-hub.com and www.theintegralvillage.org

between tundra and rain-forest - but increasingly relate to survival in the social context itself as societies grow in size and complexity from village to megalopolis. The social genetic material is in the value codes, the Vmemes which live within individuals and which govern their individual and collective management of change. We are not dealing with different types of people, but with thinking systems in people that express their way of coping with varying life conditions.

The codes and principles also have to deal with the in-built tension that exists between social cohesion forces and individualization forces. (Anthroposophical readers might recognize here Rudolf Steiner's Social and Antisocial forces in the human being.) Both forces are necessary and healthy. On one side are the gains that accrue from stable social structures; this is associated with society's tendency to require conformity to shared values. On the other is the need for individual creativity and adaptation. Society needs those who can lead or pioneer responses to conditions of change and there is a natural impulse in human individuals for self-expression and self-actualisation. As Steiner would say, the healthiest societies exist when the community is reflected in the individual and when the creative strength of each individual manifests within the society.

The balance between these pressures is far from static though. It varies with the external life conditions. It also creates its own push-pull. Each stage of development reaches a threshold at which the internal social conditions themselves demand a fresh response. As a result, it does not swing like a pendulum between the two, but expands and grows, shifting to a new level as it does so.

We can make this more intelligible by describing the levels of the spiral. The developmental process we are charting reflects the evolution of human society over time. It also reflects developmental changes in the human psyche from the simplest and most infant-like, towards the most fully developed levels of human capability. However, please be vigilant as you read these descriptions, for the reasons we are about to give.

What Spiral Dynamics isn't

Many psychological systems seek to establish typologies – different classifications of humans. These are the introvert / extrovert types of model in which people are unchangeably one or the other. The SDi stages are not types of people. They are thinking systems, coping mechanisms and value-sets that people may adopt or relinquish in relation to their life conditions.

The view of evolution in which humans are seen as the pinnacle of creation could also be mirrored in the perception that the different layers or stages of the spiral represent a hierarchy, and a ladder which humans should aspire to climb. Where the movement through the spiral might be seen as representing progress, this is only the case where the life conditions themselves are developing that way. Where they are not, it is entirely right for individuals and societies to remain in layers that are adapted to those conditions, or even to revert to earlier stages when conditions deteriorate. Each coping system and behavioural repertoire remains in the population and has value in supporting the stages beyond. There is no hierarchy.

Similarly, since each of us will read the description that follows from our own perspective we cannot avoid a tendency to filter. We might unconsciously assess each layer according to our own value set. You are encouraged to read the description from a "value-neutral" perspective if you can. You can learn a lot about your own thinking if you notice your view of the descriptions as you read them. Try to appreciate the positive qualities of them all.

The last recommendation is that you should not confuse any of these descriptions with anything that might relate to a particular race, skin pigment, creed, gender or political orientation. Human beings in any of these classifications can and do operate in any of the Vmeme systems that we will describe. Indeed, the original motivation for introducing the use of colours to denote the various value systems was in order to provide a more meaningful way to understand the complexities of the South African transition than that which was offered by skin-colour and racial designations, which actually obscured the real issues and inhibited creative solutions. So having established that this is not a typology, not a hierarchy, not a ladder, and does not intend judgements about certain categories of human, let's look at what Graves' theory actually says about human emergence. You may find it helpful to look at the diagram to be found on www.scienceofpossibility.net[50] as you do so, as well as referring to the table that follows.

www.scienceofpossibility.net/sdispiral.html

Colour	Life Conditions	Adaptive Capacities	Core values	Pinch points
Beige	A. Survival bands	N. Instinctive hunter-gatherer	Keeping Alive	Conflict and insecurity
Purple	B. Safety in numbers and collective wisdoms	O. Interpreting nature's ways. Kinship and elder wisdoms	Bonding, making sense of the world and safety	Lack of freedom for self-expression
Red	C. Self-expression and strength	P. Taking power over the world, Heroism	Impulsive, pleasing self, attention and respect	Bullying, exploitation and slave-empires
Blue	D. Seeking purpose and order	Q. Legal and governance systems	Finding a right way to live. Order and purpose	Constricts innovation. Control through guilt
Orange	E. Strategies to prosper by	R. Striving, analytical, technological, scientific	Success, status, capability, autonomy	Manipulation, greed, inequality, lack of care
Green	F. Seeking equality and fairness	S. Sensitivity, sharing and caring, community	Inner peace, equality, consensus	Over-conformity, limits to development
Second Tier Systems				
Yellow	G. Integrating and aligning systems in flexible ways	T. Knowledge, overview & integration	Competence, Functional flexibility, Natural flow	Material focus, lack of collective feel
Turquoise	H. Seeking synergies and holistic viewpoints	U. Intuitive and spiritual awareness, seeing self and whole together	Supporting the whole Making spiritual sense	? Individual mastery ? ? One size fits all ?

What Spiral Dynamics is

Spiral Dynamics integral provides us with a way of viewing the whole of human psychosocial development from our earliest days to now. It provides a lens through which to view our own individual development since birth. It offers a descriptive framework for the many ways in which nations and societies are currently living, for what they might expect to be next in their development and for how we might support that healthy emergence. It gives us a way to look at organisations and groups so that we can improve how effectively they function. It does all of these things because it reveals some very fundamental patterns of who people are and how they think.

With biological evolution we are very used to the idea of changes that enable organisms to adapt to their environment and function more effectively in it. Our Values are codes to live by, generalised answers to the question "what does it take for us to live healthily and effectively?" The answers to that question are not static because they vary according to the conditions that we are living in. The story that follows is the story of that dance between life conditions and response priorities. Without being conscious of it, human beings are continually adapting to that underlying, unspoken question – "What's enables us to thrive now?" The story takes us through stages of development. The journey is a continuum just as a spiral is, but the stages identify landmarks in that journey, broad categories of Values systems.

To begin the journey we invite you to imagine yourself as one of the earliest humans, part of a small band living as hunter-gatherers. You could be on the savannahs of Africa, or in the forests of Borneo. You may have the most primitive of tools, you may have learned how to make or keep fire. In this, **Stage 1** of the spiral, the priority code is stark and simple; it is to **survive**. How do you find food and water? What will shelter you from the elements? How do you keep safe from predators or from other bands that may be hostile to you, jealous of their territory? Beyond this day to day existence the only long-term thought that you have is to reproduce and bring your children to an age of independent survival. We colour-code this stage **Beige**. This stage of human existence is above that of the animals, but not far above. Of necessity individual, personal survival comes first. Beige humanity relies on its physical senses, its instincts, its sixth sense and its vigilance. It knows its territory well but its capacity for retaining and passing on knowledge is likely to be limited and dependent on one or two individuals.

There is a corresponding stage in our personal development. Humans are all born helpless and depend for our survival on someone else taking care of our basic needs. Those needs are absolute. Our food, warmth, safety and other biological requirements are met by others and we see those others through the lens of our needs because our survival depends on them.

Each stage of the spiral has visible limitations and for Beige the fragility of existence creates a need for greater safety. As humanity's overall numbers increased it became possible to solve this problem by coming together in larger groups, bringing **Stage 2,** the tribal stage, into existence.

The priority at Stage 2 is **safety**. As well as deriving from increased numbers, the greater stability that numbers provide enables more ancestral wisdom to be carried. The ancestral ways become ritualised as a way to build stability and encode beliefs. The tribe is typically led by its elders, the ones who carry the greatest knowledge of what will benefit the tribe. It is likely that there will be some structures – customs about who may marry whom, who will take care of the widowed or orphaned, defined property relationships such as dowries.

This stage, colour-coded **Purple** is also very aware of its relationship with its natural world. On top of the Beige instincts and intuitions, relationships with the natural world are established. The spirits of the ancestors are recognised as part of life's continuum; the spirits of plants and animals are perceived as a reality that creates a possibility for relationships to be made with those spirits. This relationship may be active in many tribe members, or may be a special function performed by a small number of shamans. In many cases the shamanistic experience will be part of the rite of passage, particularly for adolescent males. Whatever the form, the relationship with spirit is a direct one. The Divine realm is to be found in the relationship with the natural and ancestral worlds and there is no singular God.

All of the "cool" colours – ones with blue in them – are collective Values systems. The warm, red-based colours are individual, so there is an oscillation between the I and the We. That dynamic, that polarity management, is a permanent feature of the spiral since it is a permanent feature of life. In our personal development the equivalent of the Purple stage occurs in infancy as the child begins to crawl and to interact with its world. At this time the relationship with parents is central. The child sees through their eyes, copies their behaviour and absorbs the accepted way to be. There is little by way of cognitive engagement and so no questions are asked. While we grow out of Purple in that sense it is retained deeply and forever as a fundamental desire to belong. This may show up as family and ethnic loyalty but we may also adopt a tribe; Manchester United is one of the biggest tribes on the planet and companies like Coca Cola and Nike spend millions in order to create a tribalistic brand loyalty.

Each stage in the spiral contains the seeds of its successor because each creates conditions with some element of tension. The tension in Purple is the expectation for conformity, stasis and suppression of individuality. Going against tribal traditions can be a painful or fatal choice, an offence against the ancestor spirits. Since it is natural for humans to seek to express their creative spark and their urge for individual distinction, Purple may be felt by some individuals to be restrictive. If the conditions around the tribe change, the old ways may become maladapted, requiring a heroic leader who will overturn the accepted order to take the tribe somewhere new, or who will lead a breakaway group.

From Survival and Safety to Power and Purpose

Stage 3 with its individuation, creative exploration and potentially heroic leadership is an opportunity to bring about the new. This stage, colour-coded **Red** contains an impulse to take **Power**. That power may be "power to..." an empowered choice to carve our new ways. It can also express as a wish to take power over others and this can be the downside of Red. The Red values stage remains visibly active across the planet in warlord cultures such as Afghanistan and Somalia. You can find it in inner cities as street gangs and mafia-style criminal organisations. The power relationship in Red is active whether individuals have power or not. Thriving in Red means finding a personal place in the pecking order. It may be a more successful Red strategy to find a safe place half-way down than to risk competing to be number one. This does however require some control of the naturally impulsive quality that Red has. Red wants what it wants and it wants it now. Red will take the shortest route to satisfy its desires.

You can take a little time to examine your relationship with all these qualities because they all live within you. The heroic side of Red can also be present in the individual sportsman or the explorer. Its negative expressions can bring disapproval from those who consider themselves more civilised, but without Red humanity would never have advanced. This is obvious when we look at the Red stage of individual development. An infant cannot remain forever in its parent's reality and we know that there is a point in time at which all children discover their individual being. The long and comfortable stage of compliance changes when they utter the magic and scary word "no". The child's desire to meet their own needs rises strongly. The child in Red is impulsive and when it doesn't get what it wants, there may be trouble. We don't refer to this stage as the "terrible two's" for nothing. Yet it is an essential process, without which a child doesn't discover their own will force, a beginning of self-actualisation without which they will achieve less as adults than they might have.

The limitations and pressures of the Red stage lie in its impulsive and power-driven qualities. Its individualism is a stress for the collective and a glance at Afghanistan or an episode of the Sopranos will tell you that it can be full of conflict and chaos. For individuals who achieve power through Red dominance it is not safe to get old. There is always someone who is ready to fight them for what they have and leaving any form of legacy is difficult.

This tension calls for **order**, a form of organisation that is stable, a set of codes to live by, a structure that assists longevity. That is what the **Blue** stage brings. **Stage 4** in history brought the rise of the established religions, the introduction of writing systems and accountancy. Warlord bands morph into established armies with allegiance to a state or an ideology rather than an individual. Blue puts boundaries around the Red chaos, brings it more into line, channelling its heroism for a **purpose**, requiring the individual will to be in service of something bigger. Blue creates legal systems, commercial rules and supplies the means of civil enforcement – police forces and law-courts for both the criminal and civil administration. Blue also creates governmental structures and public institutions. It is the foundation for settled living in towns and cities.

In the Red system, the relationship with the Divine was more distant than Purple's direct engagement. Red places power outside and sacrifices to the Gods that it creates in the hope that they will be benevolent. Blue takes this distancing one stage further. The Judaeo-christian religions have priesthoods to interpret the word of God, which is now written down (for many centuries making it only accessible to elites) and even when the majority can read it is assumed that a direct experience of or relationship with the Divine is beyond ordinary mortals. Divine power is no longer sacrificed to, but it is worshipped and it is placed where it must be invoked by prayer and in the hope that the Divine will favour us with love.

The transition to Blue can be difficult. Rome exemplified this in its ability to produce great works of engineering and its attempt to move from god-emperors towards a republic. The over-reaching ambitions of Red can rebound – trying for more power than it can sustain, taking on one enemy too many. Rome's boundaries were too stretched, Napoleon and Hitler foundered in Russia. Imposing

Blue values from outside is hard if the culture is not ready internally and modern countries have conspicuously failed to overcome the embedded power of warlords whether in Afghanistan or Somalia. Red always sits within and societal weakness will be exploited by ambitious Red-driven leaders. Hitler rose to power in an economically stressed Germany, channeling Blue from Red within. Red and Blue can operate alongside each other for long periods. Elizabeth I had a Blue internal order and a Christian religious base, but encouraged piracy on the high seas. Her period had laws, but the monarch was above the law.

We emphasise this relationship because it also sits strongly in our individual development. The tantrum-prone two-year-old needs boundaries. They hate having their impulses denied but will become insecure if they do not meet with some structure. As with the big historical picture, it takes a long time to take on Blue. In the West, primary / grade school is a Blue-centred experience lasting many years where children learn how to live with rules, how to use the Blue tools of writing and number and how to express their individuality in a context of a structured collective. They also fight in the schoolyard to establish the pecking order.

Blue lays the foundation stones for what we think of as civilisation and everything that is to come rests on those foundations. You cannot operate cities without Blue values. But as with Purple, the collective enforcements can become a constraint. The challenge of Blue is that it can become restrictive. Its rules and its processes may inhibit or suppress innovation. Blue structures like things to stay in their assigned place, so class boundaries operate against new intelligence from "below". A society operating in Blue can become obsessive about how things are done, creating a box-ticking culture of compliance regardless of whether the outcomes are successful. It can turn into a police state, enforcing its "one right way" with rigidity, showing up as state or religious forms like the KGB and the inquisition. In a similar tension as existed between Purple and Red, the individuality, creativity and impulse to expand must break free of the constraints. Blue paralysis looks like communist Russian economy, like British Leyland cars and like every "jobsworth"[51] and rigid functionary you have ever encountered.

Mastering materiality, mastering ourselves

Stage 5 arises in response to this tension. It is another "I" response to an excess of "We" thinking and a reaction to over-dependence on rules. It is by nature

---

51 Response to all procedural challenge - "It's more than my job's worth mate!"

entrepreneurial, expressive of the self that seeks now horizons. It exploits new ideas and seeks ways to turn new scientific discoveries into products that benefit us all. The **Orange"** stage of development is expansive. It is driven, striving for excellence, hungry to show "look what I can do". It is status-driven (as distinct from power-driven Red) and measures its success both by what it can do, and how visibly it is seen to have done it. In Orange, he who dies with the most toys, wins.

Orange has brought huge blessings to humanity. Its **Strive and Drive** quality has an impulse to improve things for their own sake as well as for any commercial reasons. It will challenge the authority of Blue if that is in the way of those improvements and it will create strategies to get around the rules if the rules are in the way of creating a better life for itself. Individuals with strong Orange are driven to express themselves. This does not have the "at all costs" qualities of Red impulsiveness which is inclined to destroy whatever is in its way before even considering consequences. But it resembles Red in being determined to find a way round limitations. People with strong core Orange values will do everything to find strategies to subvert the Blue restrictions.

In recent times we can all see this aspect of Orange in the way that the financial industry created products that were too complex for regulators to understand, defeating both the legal frameworks and the credit ratings agencies. This sits squarely in the weakness of Orange thinking – it is profoundly materialistic in every sense. It becomes greedy, sees money as an end in itself and not as the tool for commercial relationship that it was originally created to be. It is equally focussed on the material in the science that it produces, which is unable to acknowledge any form of spiritual experience – can't even view it as real. The Divine, which has receded from human view ever since Purple, is no longer relevant at all. Humans have made themselves into technological Gods, imposing their will upon the Earth with minimal understanding of the consequences. The focus on material existence drives out not only the spiritual but also the personal. People are only valued for the functions they perform. Human connection and care is diminished.

For the developing child, the Orange Values system drives the process of establishing who they are to be as an individual. It takes place as teenage years are entered, as the hormonal surges wake up fresh impulses and break down previous thought-patterns. The emerging adult must draw back from parents because parents know how the world was, not how it is. The young person wants to map out her own future, find his own identity and opportunities, not repeat what a previous generation has done, or says should be done. This is a creative and essential process towards establishing healthy adult autonomy, but may involve breaking through in any area where their earlier Red attempts to establish independence were excessively suppressed. They may also challenge the Blue rules that they have been accepting until now in order to develop their own view of what is right or wrong. There may be strong and contradictory polarisations as they do this; what was absolutely right yesterday is absolutely wrong today.

The teenager can easily find him or herself alienated, mirroring the loss of connection for societal Orange as a whole. In both cases the response to the disconnected quality of Orange individuality calls forth the next cool colour, the next expression of "We". **Stage 6** in the spiral of human development is colour-coded **Green** and is concerned above all with the re-establishment of the **Human Bond.** Green was the leading edge of human emergence through the twentieth century – sitting on top of the continuing surge of technological Orange and bringing in addition an interest in who we are. The academic disciplines of psychology and sociology, the social expression of care through pensions systems, unemployment benefits and socialised medical support are all indications of emerging Green values systems.[52] The teenager may explore Green at the same time as Orange, finding support in their peer group and possibly also a collective response to generational change.

Where Green values become predominant in the individual or the group there is a strong desire for consensus. This can have the positive effect of inclusion but can also show up as intolerance of hierarchy and as an inability to make independent decisions which leads to personal or organisational paralysis. Green is also very concerned with fairness and may emphatically reject the prior Orange Values system as being competitive and hostile. Green Values can turn against the technological and commercial successes in a way that is understandable when looking at the ecological problems and the credit crunch, but would throw the baby out with the bathwater. Green values as expressed through social care systems rest on economic surplus and it is technology that has liberated western society from drudgery, an advance which has particularly freed women to take on work roles, many of which are in the health, teaching and caring professions. The washing machine is one of the keys to gender equality. Technology also made 40-hour working weeks viable with its higher productivity.

Just as Red power did not often deliver happiness or peace, Orange achievement likewise fails to satisfy some of the deeper human needs. Where Blue compensated

---

52 Lynne McTaggart has documented the science of the Green system in great detail in her book "The Bond"

for Red's turbulence with order, Green meets the empty materialism of Orange with inclusivity, caring and the search for community. Where the Tribe was ethnic and geographical and Blue collectives form around "isms" the Green community is at root voluntary, inclusive and intentional. At best it retains this openness. At worst it turns into a form of consensual enforcement that can be intolerant of divergent opinions. The intolerant form of consensus in Green can echo the "one-right way" of Blue and some call it "mean-Green".

We hope that a pattern is revealing itself in these descriptions. To repeat with emphasis, the dynamics and not the stages themselves are the core of the Spiral. At every stage, even between the stages, we are in the dance of individual and collective and we are in a Goldilocks world where too much or too little of a Values system can become dysfunctional. Every Values system fulfils a purpose. Each of them meets a need that humanity has found as it developed, because that development is a steady increase in capacity and size / scale, from the small Beige bands to today's cities with their many millions of citizens.

Conflicts are Values-based

We must also emphasise again that these stages are systems in people, not types of people. We hope that as you read these descriptions you will have had moments of recognition where you saw that you hold, or have held some of these Values yourselves. Perhaps you have also seen the developmental unfolding in your children. And maybe you can look through your window or through your daily news feed and see the various colours being expressed in events, debates and conversations. The US healthcare debate for example, revealed many flavours of Blue, Orange and Green. It seems on the surface as if the debate is about how best to manage an issue. This is rarely the case. Typically the conflict is between what people hold as their primary Values. Whatever your own stance, it is an interesting exercise to peer beneath and examine how the views you hold are expression of those deeper priority codes.

We would also hope that the evolutionary power of Values system development is visible. It is not only that scale and numbers are growing. Since the various Values systems are present within us like a set of Russian babushka dolls and since they show up in our societies all operating together we are creating ecologies, networks of interdependent systems in competition and collaboration with each other. This complexity is rich and fertile. It is also a challenge to manage.

Some conflicts appear to be between different forms of a single Values system, but this is rarely the whole story. The war between Catholic and Protestant in Ireland is only partly religious. There are also Values discrepancies relating to economic status. Orange Values are reflected in protestant land ownership (the association with the Orange order is a historical coincidence) The Catholic population were historically more Blue, or even Red-Blue. Some of the shift towards peaceful co-existence derives not from the negotiations themselves, but from the rise of an Orange values system with economic advancement south of the border, reducing the Values differences between the two communities. This holds a lesson for what is needed in the Middle-East, and to understand how a similar reconciliation might be brought about there, the work of Don Beck with Elza Maalouf holds more possibility than any of the repetitively abortive diplomatic peace strategies.[53]

This may appear to be a diversion, a tangent from the core of our story of consciousness. In fact it is central to it. What we are describing is the way that humans think. **How** people think is more important than **what** they think. We have described how the physical structure of the planet emerged through a dance of co-creative consciousness and how evolution follows the same dynamics in its achievement of dynamic balance in ecological niches. The evolution of societies and the nature of the conflicts that these dynamics reveal are likewise rooted in the field of consciousness. Our collective beliefs make us who we are and underpin our similarities and our diversity. We can follow similar threads in the way organisations work. When we understand these we can find better ways of managing many of our current challenges.

Green was not the end of Values evolution. There is no end because each turn of the spiral generates new conditions which call forth new Values adaptations to deal with them. Clare Graves called this "the never-ending quest". Two more systems have already emerged since Green, and one more is on the horizon. But in order to explore these, we need to recognise that there is a different level of shift that occurs after Green.

A different kind of change: Integrating diversity and complexity

There are a few critical factors in the conditions that arise with Green. The first is that Green Values in the individual cause them to look at human existence in a new way. Spiral Dynamics theory exists now because psychology has arrived, because we are examining ourselves. Although the earlier stages are nested and all remain within each of us, we have not been aware of them. Each of those stages contains a

---

conscious rejection of what went before. Red Values bring rejection of the Purple constraints and Blue wants to contain the Red impulses. The Blue system may occasionally be grateful for the heroic acts on its behalf and it may even reward them, but it doesn't trust them. The Green impulse in us begins in its self-conscious investigation to catch glimpses of what lies beneath. Core Green still rejects soulless Orange and rigid Blue, but leading-edge Green begins to see something more in the diversity.

A second feature of human planetary existence is that we are all becoming increasingly aware of the complexity and the speed of change. We referred on Page 8 to the 2011 survey of 1500 top corporate CEO's in which IBM discovered that the number one concern, expressed by approximately two-thirds of respondents, is unpredictability. The world has become chaotic and non-linear. There is talk of emergent systems, of fractals, of tipping points and black swans.

A third and related aspect of life as we can see it today is that the complexity and speed are compounding the negative outcomes of Orange blind-spots. Our impact on the planet, the visible damage that we are doing to ecology, material resource base, energy sources, species diversity and probably climate (where we seem at least to be compounding a natural cycle) appear to take us closer day by day to catastrophe. We have been treating money as something that has intrinsic value when it is only a representation of the relative value of what is truly real (land, water, houses, machinery). We have come to believe that by creating more figures in bank accounts we have somehow added to the wealth of the world. This belief system has distorted the way that we manage our economies, run our industries and make our life choices. It brought us close to financial meltdown and threatens to take us all the way there next time around.[54]

Added to the complexity, speed, non-linearity and urgency with which solutions to these problems are required, is their global scale. The combination of Green diversity awareness and Orange technological connectivity means that humanity can see all of this across the planet. Beyond the subtle image of a butterfly wing-flap in Hawaii causing a hurricane in Tokyo is something more concrete; a terrorist act in Tokyo might trigger a financial tipping point in London. An inventive idea in Mumbai could wipe out an entire industry in Philadelphia or Houston. Osama bin Laden's actions rooted in his one-right-way Blue Values, his Red expression of power and his strategic Orange thinking changed the world in a few years.

We are describing this complexity at the global scale, but the corporate CEOs

---

54 Our relationship with finance, and the changes that are needed are dealt with in depth in Jon's book "Future Money". Our systems reflect the Values that they were built from. Our Values are shifting; systems must too.

experience the same phenomenon. Some of their unpredictability is in the external world of competition, resource prices and political or economic shifts but they also face the challenge of internal diversity, of niche technical expertise which breaks hierarchical systems, of internal complexity within large-scale organisations and of the need to create the responsiveness of an athlete in organisations that move like Giant Tortoises.

Making the momentous leap

Welcome to the Green-Yellow shift, a stage which Clare Graves characterised as "a momentous leap". In the first six stages (which he called "First Tier") humankind has been in subsistence. In Beige there is limited future orientation because survival is day to day and even in Orange our ability to take on board long-term consequences of our actions show obvious deficiencies. In Graves' words

> "As man moves from the sixth or personalistic level – the level of being with self and other men – to the seventh level – the cognitive level of existence – a chasm of unbelievable depth of meaning is crossed. The gap between the sixth level and the seventh level is the gap between getting and giving, taking and contributing, destroying and constructing. It is the gap between deficiency or deficit motivation and growth or abundance motivations. It is the gap between similarity to animals and dissimilarity to animals, because only man is possessed of future orientation."

In **Stage 7**, which is colour-coded **Yellow**, we take a leap towards integration. The first-tier levels do not disappear and they do not become irrelevant, far from it. They are the platform on which our being depends. The task is not to jettison the first six stages but to see them all, to mobilise them within ourselves and within society to the best possible effect. We do not cease to need Blue order or Orange entrepreneurialism or Green caring. We need to integrate, balance and use them all purposefully. We need the inbuilt balances of the whole spiral to-date.

The leap that is required has another more important characteristic. In the first tier it has been possible to operate with a linear thinking, cause-and-effect model. People were able to disentangle the strands and separate out the variables. We can no longer expect to do that. We have to find a new way that is more agile, flexible and immediate. We have to be able to make decisions now, for now, and be willing to adjust them as necessary in light of outcomes or the external changes as they arise.

This does not mean that we can let go of first-stage Values. They each are applicable in the appropriate conditions. It does however mean that we must let go of earlier decisions that we made, habits that we embedded or rules that we articulated from those Values systems. We cannot leap to second tier carrying the baggage of the first. When we apply the first-tier Values in the second-tier world we must apply them fresh, we must presence them in the moment. In order to respond to fractal, chaotic complexity we must be **flexing and flowing**, surfing the wave of existence.

Can you imagine what this shift in thinking means for governments? You may have noticed what a hard time they are given when they change policy – it is treated as a weakness, characterised as a failure and a U-turn. This is understandable because humans like predictability; it makes them feel safe to be in the known and to be managed by people who appear to have certainty. Politicians are adept at projecting such sureness even when we all know that they do not and cannot know. To some degree the psychological desire is to treat them like our all-knowing parents or wise elders and cocoon ourselves in that safety. How ready is each one of us to take responsibility as adults, to support them, even sympathise with them in facing that challenge? We may all be about to discover the answer to that question.

The same will apply at all levels of societal organisation. Fortunately local conditions will not change drastically for many people. Those who are well adapted to their Red, Blue or Orange life conditions do not have to change substantially. Can we create communications media that no longer see their success as based on scaring people with the unknown? Can we step out of our addiction to being scared or angry, to having our unproductive emotions stirred up?

The Yellow response is integrative and systemic. The companies of the future and the governments of the future will need to find new blends of central control with distributed autonomy. The shift, as Don Beck puts it, is from "live and let live" to "thrive and help thrive". This new level of thinking will not buy the destructive myth that all of evolution was competitive because it will fully recognise the equal contribution that collaboration makes to successful co-existence. Indeed, learning how to collaborate and compete at the same time will become a key to Yellow existence.

In seeing clearly the damage that humans have inflicted through the unaware parts of Orange Values, Yellow will apply its awareness of how much humanity is dependent on the planet, its resources, its other inhabitants and each other for continued collective well-being. Yellow-driven thinking doesn't embrace the knee-jerk "all change is dangerous" reactivity of Green. It cannot afford to because some of the problems technology has created may be best solved by new technology. But it will be much more cautious about what it does not know and much more keen to take decisions in such a way that they can be undone when unintended consequences show up. When Yellow Values are present in our leaders they will ensure that when decisions are taken they build in the monitoring that will detect those consequences early and will place the economic cost of that monitoring and rectification with those who are responsible, not with external agencies. Society will expect such accountability to be built in to commercial considerations.

As individuals we must become adults. We are no longer children or adolescents and we are now responsible for our lives. Not only does this mean that we cannot blame our parents any more, it means that we cannot make governments or others surrogates to carry that blame. Second tier is a leap into personal accountability. We will have to ensure that the accountability, the responsibility for who we are is balanced by empowerment to take our own choices. Autonomous decision-making will be essential because the centralised awareness cannot see everything at the periphery. Command and control expires, along with linear thinking, because it cannot achieve the flexibility and responsiveness that are needed.

In the shift to second tier, the non-linear and flexible response and the integrated view of what has gone before, it becomes possible to act on the basis of what we know and not of what we feel. Humans with Yellow values are more capable of stepping free in themselves of the survival fears, the impulsiveness and greed of what has gone before and less in need of reacting to those aspects in others with control and constraint. In Yellow we also regain visibility of the non-material and the science that this book presents opens the door into a new perception of connectedness. Yellow is still a warm-colour system but it's "I" aspect transfers in some degree to the human species in its awareness that people depend on the planetary system and on each other. Our species survives together or falls together. However the science that this book proposes goes a stage further than this. In the world of the shaman we are consciously connected to the other beings.

The science presented here is the science of the **eighth stage**, known as **Turquoise,** reconnecting spirit through both physics and biology. Overlaying on the meshwork of Yellow interconnection in the practical and systemic realms, Turquoise brings a level of **holistic** awareness. It is a revisiting of Purple spirituality at a higher level where we re-engage both personally and collectively with the spiritual life in all things. The planet as a whole takes priority over or at least alongside humanity because we know and experience that the two are not separate.

Turquoise is the leading edge of human development at this point. SDi theory does anticipate a next stage which it has colour-code as **Coral.** It is likely that this stage will echo the Red individuation that rose out of tribal Purple, so one might expect that new forms of individual empowerment will develop as humans increase their mastery of the new levels of connectedness, and their ability to engage with the creative aspects of a conscious universe. To say even that much is speculation and we will say no more.

This presentation simplifies SDi greatly. Describing the stages can obscure the fact that each one has stages within, which are the transitional times as cultures enter the new band, as they become centred in it, and as they reach the exit conditions of moving to the next phase. It can sometimes be difficult to understand the difference between underlying Values and the content with which they may be expressed. And in truth, this is a science still in its early stages, with a great deal more of the subtlety to learn. The technology of application of this theory to individuals, organizations and cultures, while already powerful and urgently needed in the world, is nonetheless still in its early development.

For those who are curious to explore this rich territory we recommend the publications referred to earlier, and the website www.spiraldynamics.net.[55] There are many powerful practical applications of the theory that assist individuals, groups, companies and societies to manage the complex transitions that they face[56]. Regrettably these are also beyond our scope here. We should make it clear that the description we have given addresses perhaps 20% of the explanatory depth that SDi offers and which is probably only approached in the full Beck and Cowan text. The full power of Graves' theory is still waiting to emerge. Our view is that when it does, it will eventually be seen as a paradigm that is as important to our understanding of humans and society as Darwinism has been for biology. One illustration of its power is to be found in the book "The Crucible" by Don Beck and Graham Linscott which describes the application to South Africa's transition from Apartheid. Copies are hard to find but we are hopeful that a new printing will be available in the near future.

One purpose in describing SDi was to show how the principles that we saw in

---

[55] My introductory video can be found on www.scienceofpossibility.net/sdintro.html
[56] See the work of the Centres for Human Emergence (www.CHE-hub.com and others)

biological development are repeated at the level of maturing individual psychology and in the development of social organisms. Although we have given space to describing the layers, the underlying dynamics are of greater significance. But we hope to have done enough to indicate the way in which the balance of individual and collective success is held within the dynamics of changing societal life. We also hope that it is highly visible that the repeating cycles of co-operation and increased complexity that Elisabet Sahtouris describes are operating again here.

The other feature which some readers may by now have anticipated is to recognise that we again reflect the particle level and the biological levels in the way that information is held both within and independently of the human individuals. This section is about the way that we construct and create "reality". The values at each level of SDi are held by people, but also become encoded in rituals and forms, in doctrines and accepted wisdoms and in shared views. Some are encoded in the language like the Red "it's a dog eat dog world". The reality is socially constructed and the Value memes become a part of a collective consciousness.

Whether held in language, habit, shared experience or collective folk wisdom these parts of consciousness are all in the realm of information rather than material things. We would also suggest that it is consistent with every other facet of existence to expect that the codification is also "in the ether"; this would reflect both Jung's phrase "the collective unconscious" and our presentation that we are embedded in the field of the All. As we have reached the most recent levels of the evolutionary spiral, the emergence of a global information network technology is facilitating increasingly rapid transition through the levels and adding a conscious, externally embedded level to the collective unconscious. The other feature that web-like technology increasingly shares with the "All" is that the flow is in both directions. Just as we influence the "All" through our own existence as part of it and through choices such as prayer, the technological information field has also democratized and globalised our ability to influence and share with one another. The planet's driving stories are now on Television and in Facebook and YouTube.

Spirals: past into future

There is a reason why the image of spiral form occurs regularly in nature and is appropriate also to a dynamic of social development. There is extensive literature on the significance of the Fibonacci series and the golden mean as providing a mathematical reason why spirals are a recurrent natural form. The concept boils down to the fact that as a spiral develops, it retains its shape as it is added to at its open end. Nature builds on itself and layer builds upon layer. Nothing is discarded. That which is already there becomes fixed and embedded, held within the development of the whole; it carries the record of its own past. What is newly added is influenced by the shape of what exists. The spiral conveys the image of the fractal; the space for its growth is partly preconditioned and partly open.

The opening part of this section examined the history of attempts to understand what "reality" is, and to discover boundaries to what can be regarded as Truth. Throughout this book we have repeatedly encountered the brick walls that science has constructed around the search for Truth. We have seen that arbitrary choices have brought a focus on components at the expense of wholes, and on the aspects of life which can only be seen when the organism is dead. The frequent question that people ask about the existence of a soul, is whether it has material reality? Does it have mass, and where does it exist? One suspects that the question is much more often asked by those who believe there is no such thing, in order to present challenges to those who know there is.

The Nature of Things

An alternative approach to science (and reality) has been in existence all along. Wolfgang von Goethe is remembered as a poet and playwright, but only because his scientific work, which he regarded as his major achievement, has been treated as an irrelevance rather than accorded the respect it deserves. Goethe was concerned to observe the nature of things, but then to find the true essence of natural phenomena which is not given in immediate sensory observation, but appears after painstaking research within the observables as a "higher nature within nature" It is the "ideal in the real...the idea in the phenomenon," which, according to Goethe, it is the researcher's task "to seek out". He too was looking for the underlying Quality.

Goethe was responsible for introducing the term "morphology" into biology and developed theories of colour (his understanding of wave theory was unfairly ignored as particle physics developed) and of plant and animal form. These theories are concerned with the whole rather than the part, and so approach knowledge from the opposite direction to conventional science.

The Goethian approach to form has been continued by such notables as Rudolf Steiner (Harmony of the Creative Word) and Wolfgang Schad (Man and Mammals). These viewpoints restore the perspective that science requires an understanding not just of mechanisms, but of the patterns and energy relationships that determine the forms that life takes. Mechanisms are not explanations and the story that they tell is incomplete.

Gregory Bateson, in his book "Mind and Nature" makes a very persuasive case for the need to understand "the patterns that connect". Where much of science looks at things and processes and expects that the patterns will emerge from them, Bateson like Goethe recognizes that there is an underlying unity which is not revealed by comparative anatomy or description of form. It requires an understanding of relationship. By definition relationships can never be encoded as properties of things. They have to be external and exist between things. The intuitive understanding of such connectivity is just as important to our true knowledge of the world. Einstein, Gödel and Feynman all recognized the part intuition played in their ability to develop theory from observation and the scientific literature abounds with stories of intuition and dreaming as a source of breakthrough. Alfred Russell Wallace and Charles Darwin were first and foremost naturalists who observed.

Bateson tells a story of a very powerful computer, to which a man asks a question about its mind – "Do you compute that you will ever think like a human being?" After a period of self-analysis, the machine produces its answer. "THAT REMINDS ME OF A STORY". As Bateson points out, stories are a way in which humans encapsulate knowledge of relevance. The connecting strands between events and phenomena are in multiple realms of connectedness. As he puts it

> *"If I am at all fundamentally right in what I am saying, then thinking in terms of stories must be shared by all mind or minds, whether ours or those of redwood forests or sea anemones."*

The world of science needs to understand not just the events that happen, but the story that links those events. In Steiner's "Harmony of the Creative Word" he too refers to Max Planck's statement

> *"Anything that can be measured is real; anything that cannot be measured is not real."*

It should by now be abundantly plain that there is much that is real in any normal understanding of that word, and yet not measurable. Steiner's presentation of the world of form is startling in its contrast with modern science. His language seems flowery and ill-defined. A scientific skeptic will tear its subjectivity apart in seconds, usually with great subjectivity one might add. And yet the overall picture of life which emerges from his methodology and likewise from Wolfgang Schad has an elegance and consistency which adds considerable meaning to the "patterns that connect". We suspect that there is a further spiral that reflects development of form as seen through the three dynamics that they identify of Metabolism / digestion, Circulation / Rhythm and Nervous system / Brain. This is a realm for other experts, but is an indication of the "story" to be told, the story science does not yet tell.

Much of our text has examined the complexity of living systems. The current scientific view seeks to understand complexity by finding simple components. Biologist and mathematician Brian Goodwin, author of "How the Leopard changed its spots" argues the need for a science which can encompass complexity and echoes the views of Ian Stewart and Jack Cohen in "The Collapse of Chaos", an excellent and insightful book which guided some of our descriptions of genetics. Goodwin's view is that a new scientific methodology is required which can incorporate the intuitive and holistic in its way of knowing. Echoing Pirsig, his term for this is a "Science of Qualities".

There are so many levels in the mechanisms we have examined. The spiral of life from fundamental particles through chemical elements, cells and organisms to societies and cultures has to be understood in its entirety, and this is not an easy task. But we hope that what has emerged provides some of the fundamental relationships that become visible when the whole is viewed together, and that some simplicity and tentative sense of "qualities" is visible in the recurring patterns.

Qualities, in our view, would include the recurring balance between stability and creative flexibility, the interplay between chaos and order, and the interaction of the material world and the world of information. They would include the balance and tension between the forces that support individuality and uniqueness, and those which encourage consistency and stability across groups. They would include the existence of an energy connection that mediates the interaction between material and information realms, recognizes that these are completely embedded in each other and thus locates each organism interactively within the field of creation, each with its own "address". In the image of the spiral, the place at which creation and stability interact is at the open, emerging end of the spiral form. It is the place where new elements are added to the building blocks of the old.

In our early chapters we presented the evidence for intuitive capability. We showed the consistency of world-view through the cultures which have maintained an intuitive relationship with the world, and among those from our culture who have entered into that realm. It should be obvious that viewing the world in pieces has led directly to our inability to understand the consequences that will follow from our many technological actions. It is not science itself which brings this about, but an attitude that underpins science and narrows its viewpoint. That attitude is a choice to define reality in ways which deny the intuitive, the whole and the patterns that connect, reducing our capability as a culture to understand ourselves and our relationships with the world. The results in ecological damage are self-evident.

There is a corresponding impact to our understanding of who we are and of our place in creation. If human beings were less robust and more inclined to believe science above their own experience the sense of alienation would be much worse than it is. But we would hope it is as clear to others as it is to us that both science and society would benefit greatly from a more accurate framing of reality. If there were full recognition of the extent to which our understandings exist beyond the things we look at, and often independent of the material of our own bodies, we could begin a fresh and more rich engagement with every aspect of life.

In our closing chapters we will return to the central theme of spirituality and the ways we have available to restore meaning and empowerment to our relationship with the world. Religion may seem like a victim to the growth of science, but it too has been complicit in putting power outside of human beings and human realms. Science created disconnection through its mechanistic views. Some religious forms have discouraged connection and personal authority by placing rules, historical texts and hierarchies of priesthoods between people and their spiritual experience. Our last chapter will seek to re-establish the values which are present at the centre of all the great faiths and to find a unifying sense of empowerment for all in the ways that they choose to interact with the realm of consciousness.

Review

In the last two chapters we have presented the basis on which we understand reality. At both a personal and social level, human beings construct their reality in terms of the things that they value, the thinking structures which they see as enabling their psychological and social survival as well as their physical well-being. Kelly's constructs and the Gravesian value systems are both elements in the way that we work together to co-create what is regarded as "real". In either case, the information exists between people, in a personal and cultural field.

The outcome is that just as the entanglement of particles places information beyond the physical boundaries of the particles themselves and just as morphic fields affect the development of biological form, our perceptual systems have an external existence that exists in the information realm. The spiral continues at this level as at all others to be an interaction between the physical, the energetic and the informational. Some information is cultural and explicit. Some is intuitive and held in the space of photon coherence and subtle effects. All of these work together to create the "mind" of the all and our experience of it.

20.   Holistic Spirituality

One of the things that today's science lacks is an effective theory of complexity.

Jack Cohen and Ian Stewart   *The Collapse of Chaos*

We have lost the core of Christianity. We have lost Shiva, the dancer of Hinduism whose dance at the trivial level is both creation and destruction but in whole is beauty. We have lost Abraxas, the terrible and beautiful god of both day and night in Gnosticism. We have lost Totemism, the sense of parallelism between man's organisation and that of the animals and plants. We have even lost the Dying God.

Gregory Bateson   *Mind and Nature*

Theme

This chapter reviews our journey so far in order to lead into our concluding observations. We have presented our material in the context of this particular time, framed in the indications that a significant evolutionary shift is taking place. Perhaps for the first time on Earth this change is not primarily biological. What can we do now and what may be available to us when we complete the transition?

We have called this book "The Science of Possibility" and have given plenty of clues to why we chose that title. In these final chapters we would like to draw together the strands of thinking and the patterns of information that led us here. We would then like to close with some elaboration of what we think it all might mean, deepening the sense of what "possibility" involves and most importantly what humankind might be able to do as a result.

In describing conscious evolution through the lens of Spiral Dynamics theory we indicated that there is a qualitative shift between Green and Yellow, or between first and second tier consciousness. That shift is underway right now. Most of the world's people still live in Red, Red-Blue and Blue life conditions and are expressing the Values systems that fit their world even while using the results of Orange commerce and technology, drinking Coca Cola and leapfrogging our copper-wire telephony infrastructure with mobile communications.

Simultaneously the global reality, driven by Western influence, evidences a greater level of complexity. The world is combating problems of habitat, climate, financial, political and resource challenges that are serious, urgent and compound. The rate of change is faster than we can predict and you probably feel this in your own life.

While unpicking the strands of scientific thinking and revealing the gaps in presentation of human biology, genetics, evolution and quantum physics in order to create space for an expanded perception of how the universe works, we have introduced a fundamentally new way of seeing. If you still think that reality consists only of matter, if you think that metaphysics cannot be part of science, if you still doubt that it possible for people to be intuitive and psychic and if you disbelieve all that we have offered in the way of evidence then you will not like our conclusions. But in case you are hovering on the edge, perhaps lost in the breadth, depth and complexity of all this data and argument, let us try to put all of the pieces on the table in the hope that seeing the whole picture in one place will help you share our excitement for the world that is revealed.

Humanity needs this point of view. We cannot hope to deal with the problems, the complexity, the urgency and the unpredictability using the thinking systems of the past. The purely material view got us into this mess and we need a new level of thinking if we are to enter the second tier world. The following summary describes the world of everything else that makes the universe function and recapitulates the journey of this book so that you can see all of it in one place.

Section 1. The non-material world of connection and information

At the core of our challenge lies two thousand years of flawed philosophy, which chose to view the world as if we are separate from it, and as if it consists solely of things. The world also contains qualities, information and values. All of these are real and all have an effect on the real world but are excluded from scientific discourse.

From that philosophy, further viewpoints have developed which are designed to exclude humans and their experience from any evaluation of what may be regarded as "true" or "proven". This is claimed to increase objectivity, which it doesn't, and it excludes any possibility of there being unseen connections between humans and other aspects of the world. In this and other ways science forces itself to deny what turns out to be real and meaningful.

In contrast, the reality which we have shown is that humans are connected to the natural world. We can know things intuitively that are not possible through our standard senses.

Within that connected reality we have shown that the usual view of present and future is misleading and that information is available to us from the future. We have shown that minds can access information over considerable distances of space, without apparent limit. We have shown that they can direct enquiries with apparent precision towards other individuals, even when they do not know their location. We can know beyond the accepted boundaries of time and space.

Plants also have the capability to sense their environment and so do bacteria. The evidence indicates that all living things possess this capability. Humans are not the only "minds" inhabiting the Earth.

All of this evidence points to the presence of an information network or field of consciousness that is available to all living things and through which we can both receive information and send influence. We find it helpful to think of the information as the data that the universe holds about itself, and of consciousness as being the awareness that we (or other conscious entities) have of that data.

Within the realm of information that we cannot see using our biological senses, beings exist which are part of the natural world and which influence it. There are also fields of consciousness within plants and animals which govern the growth of organisms, can be communicated with individually and can mediate in healing.

Within that realm of information lies the capability to influence those organisms by means of energy "vibrations" and resonance effects. One means of doing so is homeopathy, which was developed using entirely scientific methods. As well as bringing about healing through controlled and targeted application of specific remedies, homeopathy has demonstrated that there are other links between generations than genetic DNA. There is evidence to support all of the major approaches to spiritual and energy-based medicine.

Section 2: The biological realm and the human organism

Genes do not have a one-to-one correspondence to characteristics.

There aren't enough genes to explain by themselves the way that an organism develops.

The conventional presentation of what "genes" are is misleading.

Influences are passed from generation to generation by other means besides chromosomal DNA.

The determination of development is not only controlled by chromosomal genes, but by other elements of genetic material.

The determination of development is subject to environmental influences and is flexible rather than fixed.

The conventional presentation of the role that genes perform is also misleading.

Layers and complexity

There are many levels of function and complexity in a typical higher mammal such as a human being. These lead to multi-faceted and multimodal functional co-ordination. There is much more than brains and nerves involved.

Complex functionality is present at the level of individual cells – cells are not blobs.

Cells contain organelles performing parts of the cellular function. These have probably developed via the accumulation of simpler functions in component bacteria.

Many aspects of cell functionality are mediated through the cell membrane, which is a liquid crystal.

The development of life progressed from simple (prokaryote) bacteria through complex (eukaryote) bacteria towards bacterial colonies functioning in a co-ordinated way. Co-ordination was mediated by photon communication and chemical messengers and was responsive to environmental triggers.

Further development came via the development of multi-cellular organisms of increasing complexity, with increasing differentiation of cellular function, and with formation of organs and structures.

Co-ordination has been built upon the earlier layers with the addition of synchronising rhythms, nervous system pathways, electrochemical connection and connective tissue with jump-conduction of protons through the liquid-crystal continuum.

Organisms maintain a degree of quantum-level coherence that makes them rapidly responsive to small-scale triggers.

The small-scale triggers may include the detection of subtle energies including Zero-point / thought-field influences and vibrational energies such as those provided by homeopathy and other forms of "energy medicine".

The heart is a driver for rhythmic co-ordination and in large measure influences the brain rather than being controlled by it

Some of the interaction between mind and body is mediated by neuro-chemical messengers which distribute control and emotional experience throughout the body. Emotions are triggered by chemicals which arise from many areas of the body to influence the organism and are not simply side-effects of brain activity.

Multiple levels of synchronisation and control are interleaved in the organism, creating a holistic and indivisible totality which cannot be understood merely as a combination of its parts.

Subtle energies influence the body throughout, supported by such self-organising phenomena as pumped phase coherence and Bose-Einstein condensates which mediate the relationship between the physical body, internal consciousness and the inherent fields / dimension of information.

There is external evidence of long-range order, efficient and symmetrical energy transduction and high sensitivity to external cues combined with "noiseless" communication.

Every cell in the body and the organism as a holon are embedded together in the field of consciousness, influenced by it and influencing it.

Evolution

The evolutionary process has been a self-actualising (autopoietic) creative process in which all species of organisms have balanced their individual creativity with the creative growth of the totality, the primary creative consciousness, in a blend of competition and co-operation.

The Darwinian model for evolution is not wrong, but its origin in Malthusian thinking has over-emphasised the significance of scarcity and competition and has not allowed for a sufficient degree of interactive or environmental influence on the creative dynamic.

The autopoietic nature of the universe provides an inherent drift towards increasing levels of complexity as all possibilities are explored.

Developing complexity is supported by a cycle of individuation and mutuality which spirals through successive levels of biology and speciation.

Morphogenetic fields, or other such qualities in the "information field" bring repetitive use of elements and patterns, and variations in the blend of informational (sensory), rhythmic (breath / heartbeat) and action metabolism elements among differentiated species.

There is no evidence that humans are the peak of the evolutionary process, nor that it is complete. There is equally no evidence that we are not. It would be entirely consistent with this evidence for human beings to be the source of the entire field of creation and to have created everything we experience as a collective dreaming. How would we know? While not our view, this possibility should not be dismissed.

Section 3: Cosmology, Physics and philosophy of science

The underlying mathematics of Kantor, Boltzmann, Gödel and Turing show a flexible and paradoxical universe with no bounds to creative possibility.

The subject-object dichotomy that we inherited from Greek thinking is false and leads to spurious choices of what is regarded as "knowledge" or "proof".

A scientific method which views the world excessively through its components and by taking life apart, leads to inadequate perceptions of the whole. It is incapable by itself of delivering a complete understanding of life, and needs to be complemented by the "Goethian" approach.

As it becomes clear that the information which connects all of creation is held somewhere in the "spaces between" material particles or living entities, our science must find ways to observe these spaces.

Quantum physics shows that our natural view of reality is inadequate. There is much taking place that we cannot detect with our senses.

There are in-built limits to all rules and predictive models as shown by Gödel's theorem and Heisenberg's Uncertainty principle.

At the lowest level of matter, everything is a form of energy which started in an indeterminate state and might be described as either particles or waves.

Everything that we see is the result of a process by which some energy has been "determined" as taking a stable form. We, and the matter that we interact with, are parts of what has already been determined.

The "determining" process is rooted in natural patterns of energy, in the simplest forms that are sustainable and in the geometric building blocks that these patterns generate. These forms can be understood in modern and mathematical ways and have also been known for millennia as "sacred" geometries.

The "determining" process is inherent in creation. From the outset (e.g. "big bang") that which becomes stable has its defining information in the field of "what is known".

The relationship between matter and consciousness (and between organisms and the whole of creation) takes place outside of the four dimensions of space and time that we normally work in.

One popular possibility for the way this works is through the "zero-point" energy-field (ZPF). Information is thus held in the "environment" and is instantaneously available to everything, everywhere. Much of the energy potential of the universe is undetectable, existing within singularities (black holes). The alternative possibility, that there are additional dimensions, is also capable of explaining a world of stored information. Either model will support our case for connected consciousness.

Human awareness of the information realm may well be mediated through the presence of pumped systems or Bose-Einstein condensates in the brain or body-mind.

There is evidence that coherence is supported by alpha brain-rhythms and heart-rate patterns and that human detection systems are at least partly mediated by the heart. Experiments also show that this detection can take place pre-cognitively.

Information storage, both within the brain and in the field of consciousness has properties which are holographic and holistic.

Human ability to use thought to significantly affect physical reality uses these mechanisms to provide a small trigger which creates an "attractor" within a system of unfolding fractal chaos.

Morphic field resonances may also operate at global or universal levels as attractors within such systems and could be intensified by larger numbers of people.

Section 4: The psycho-social realm

Human beings build their view of reality based on external information, most of which is socially defined.

Human beings construct predictive and adaptive models in the form of constructs (polarities) and Values (Vmemes); these guide their choices and behaviour.

In a parallel to biological evolution, patterns are present in social evolution which balance individual and collective interests.

In a parallel to the world of physics, our social reality arises and is held outside of individual organisms. It too inhabits the "spaces between".

As with biological development, social forms are adaptive responses which reflect the life conditions around them. These evolutionary coping strategies are based on identifiable Values systems.

The development of social systems reflects the spiral patterns of individuation and co-operation observed to arise when Sahtouris cycles are repeated at higher levels

Just as complex biological organisms contain within them the earliest biological forms (bacterial components and single cells) and are built in a way that transcends and includes previous developmental levels, so too is the psychosocial realm. Earlier Vmeme systems remain available and can re-emerge if required.

Connecting consciousness and spirituality

A previous book containing portions of this material described itself as a scientific guide to the spiritual habitat. Titled "God's Ecology" it framed itself as an attempt to restore the perceptual ecosystem in which "God" or "Spirit" might continue to live in human beings and societies. Mindful too of Gregory Bateson's "Steps to an Ecology of Mind" we also have the sense that what is emerging takes that impulse further. We are perhaps looking at an Ecology of Being. That might also have been an appropriate title.

There are many interpretations of what "spirit" might mean, and there are as many possible views of "spirituality". You will probably have your own and we do not believe any particular form needs to be defined. Broadly, we treat spirituality as the way that most humans relate to the non-material universe, and particularly to the awareness of the realms of information and consciousness, including "Divine" consciousness.

When describing the colours of Spiral development we noted that at each stage, human relationships with "the Divine" or "God" or "spiritual realms" takes on a different character. We see all of these viewpoints as reflections of other aspects of the Values systems for their stage. Since we wish to reflect the 2nd tier Yellow and Turquoise systems, our perspective has both the integrative view of Yellow, which would see all previous viewpoints as functional for their time and then of Turquoise, which is aware of the connectedness of the whole. Thus our view of spirituality is that it requires a holistic presence of the non-material, non-visible, non-ordinary reality to pervade the visible and material realm.

To some degree the use of words like "spiritual" and "Divine" reflected the earlier stages where both are seen to be outside of us. Our use of the word consciousness recognises that there are no boundaries. Since the information is everywhere in this holistic perspective there is nowhere that is not spiritual or Divine. The choice to speak of consciousness has a similar inclusivity. As individuals we have a portion of the whole of consciousness – our unique perspective that is what we "see" of the whole. Universe can be viewed as having consciousness of itself since it has access to all of the information that is. It is not significant for our purposes whether or not you choose to see the universe as "self-aware" in a similar way to humans.

More significant is the relationship that people typically have to this non-material or spiritual context. Dr. William Bloom, among the most lucid, compassionate and authentic of the pioneering voices for new spirituality has been championing the holistic agenda for decades now. In his book "Soulution: The holistic manifesto" he argues the case for a coming-of-age in the trend towards alternative views of religion, spirituality and health. He draws out the many strands of commonality and coherence in what sometimes appears from the outside to be a mixed bag of practices and belief systems and shows the maturity of the world-view that emerges.

He quotes a British survey in 2000 which indicated that 70% of people believe that there is "something there" – personal God, life force or whatever. Only 9% of people believe that their own path to God is the only way. 33% believe that there is a way to God outside organised religion and a similar number believe that all religions offer a path to God. A 2007 US survey indicates that even in a country often seen as strongly fundamentalist, 70 percent of Americans affiliated with a religion or denomination said they agreed that "many religions can lead to eternal life," including majorities among Protestants and Catholics. Among evangelical

Christians, 57 percent agreed with the statement, and among Catholics, 79 percent did. Among minority faiths, more than 80 percent of Jews, Hindus and Buddhists agreed with the statement, and more than half of Muslims also did.

Evidence exists to suggest that other parts of Europe are similar to Britain and if this finding is anywhere near representative of humanity as a whole, it suggests that the sense of spiritual belief in general is strong, even if formalized religions are declining in some areas. Fundamentalists typically make the most noise but they do not appear to be even close to a majority in any major religion. Overall this data indicates that there are many people who are in some sense unfulfilled, undirected or "closet" believers in a generally spiritual reality. We hope that this book might help change the climate so that more of them will feel safe to "come out".

The next level: Emerging into Possibility

When describing the "Momentous Leap" that human consciousness must take in its transition to Second Tier existence, we wrote of complexity and of non-linear thinking. The leap begins in integration; it starts with our collective ability to embrace a wider bandwidth of thinking, to see the virtues and vices of all of the Values systems, in learning how to bring them into internal healthy expression and into balance with one another. Powerful as this will be, and however great the tools and understandings that it will give us for how to reshape our communities, our organisations, our nations and our global systems, it is only the platform for the next phase of our journey.

The new scientific perspective presented here has huge implications. The leap out of survival emotions, out of existence and into being is more than a psychological change – at least in our old notion of psychology. Among the changes may be a new understanding of what psychology is. That study has until now been concerned with our internal experience, with the structures in human brains and with the ways that our logic, our emotions and our stored experiences generate our thoughts and behaviours. For humans to express new Values systems leading to new behaviours is far from trivial, but embracing a consciously creative reality is a bigger step and enters a different dimension.

Who are you, who can you be if you take on the full implications of a universe in which you engage as a small piece of an almost infinite whole, knowing as you do so that you are simultaneously and inexplicably both a finite individual and in unbounded connection with that whole?

We will rephrase that and suggest that you consider it carefully.

You now know that you are simultaneously and inexplicably both a finite individual and in unbounded connection with that whole. If you take on the full implications of a universe in which you engage as a small piece of an effectively infinite whole, who are you and who can you be?

Merely to grasp that the question is real is a stretch for most of us. Even to write it calls for expansion in the moment. And yet this is the reality that we are presenting, one in which we all face a conceptual chasm. In our pasts are the material beings that we have conceived ourselves to be. In our future are the beings that we might become, the ones that emerge when we embrace every aspect of what it means to be matter that is formed from consciousness.

It has been conventional to conceive of ourselves as human beings having conscious or spiritual experiences. The evidence we have assembled here describes entities of a quite different kind. We are beings of consciousness having a human experience. We are spirit made flesh. In any moment we are who we think we are. Jon sits at a computer typing words, having a conventional human interaction, being the Jon who appears to be a normal person with a house, a family and a car – possibly known by some to have unusual points of view, but in general seeming to be as other men are. You, whoever and wherever you are, are reading this book. That by itself is the action of a normal person.

In contrast, if you were all that this scientific perspective implies that you could be, you would not need this book at all. You would already have access to all of the information that it contains. That you are reading it indicates that you too think of yourself in a relatively conventional way, as a primarily biological being, a member of a species that has developed its consciousness to the point of reading books that explore who we are.

The very first paragraph in this book read "What if almost everything that you have ever learned, experienced or been told about our world has been skewed, misaligned and misunderstood? What if our ways of perceiving who we are, what the world is, and how we relate to it are somehow inside-out and back-to-front?" Our first chapter title asked the question "What else is possible?" What does it really mean for humans to engage with the world the right way around?

Of course we, the authors, don't know – not in any specific sense. What we do know is that we have been presenting evolution as a co-creative activity, taking place in a universe where "reality" is whatever the field of information says that it is. If the field of information defines an electron as having a particular form then that is what is "real" – or as near as we will get to it. As we have presented it, the field of information developed without there being an intention. We are assuming that to start with there was only a kind of inevitable accident. Sooner or later the patterns that we now recognise as the primary source of energy stability – the tetrahedrons and vector equilibrium geometries spinning into forms that could sustain themselves – were bound to happen.

Of course it is entirely possible that there is still some form of "being", an elemental consciousness that was "before everything" and that was the source of creation akin to the Old Testament Yahweh. We don't know; it would seem that it is unknowable and we don't know that it matters. It is also possible that there came a point at which creation's self-exploration became self-aware. There may have been no intention operating to begin with, but the universe may have reached a stage when it was more than the sum total of all that had happened so far and was capable of doing what we humans can now do, holding the past in conscious awareness and projecting ahead in time to see what could possibly happen next. Your guess is as good as ours.

We don't know that this happened but we do know that it is possible. We know this because it is what happened to humanity. Seemingly we were once creatures that like insects and lizards could not and did not project their awareness into the future. At some point in evolution creatures came into being that can and do. We don't know if we are the only ones but we do know that we are its leading exponents.

Whether or not we are the first, and whether or not the universe or any other part of it has intention in what it creates, two facts tell us that we can. The first fact is the successful results in the PEAR psychokinesis data, and experiments such as that of Dr. Larissa Cheran[57] who has demonstrated a strong influence of human intention on cultures of neuron cells. Our minds can influence the field of consciousness with materially measurable outcomes. The second is our capability as creatures that model and anticipate the world to come. Other life-forms react and respond. Some do so in the moment based on hardwired programs. The moth will go back to the lamp again and again. Mammals typically have memory and will modify their responses based on stored experience. The human life-form anticipates, replacing reaction with creation. This is a huge evolutionary step.

---

57 Winner of the 2012 Mind-Matter mapping prize – Journal of Non-locality Vol I No. 1

From prayer to creative visualisation to unlimited possibility

Prayer has been a part of the human repertoire for a long time. Some experts argue that the most ancient of cave-paintings are more than an artistic depiction of how those people lived, that they are an invocation intended to call forth success in the hunt. Cultures which sacrificed animals to their gods had perhaps overlaid their prayers with a bargain and the Islamic "Insha'Allah" or Christian expression *Deo Volente* also hoped to influence outcomes even if the balance of power lies definitely with an external entity.

The New Age equivalent of prayer is often known as "creative visualisation". Some versions of this would place all the power within you as an individual and many of the sales hypes from those who propagate techniques of this kind would seem to offer something close to omnipotence. If you visualise it clearly enough, call for it strongly enough and have no doubts or contradictory thoughts, any desired outcome can be yours.

So where does the truth of our capability sit between supplication and omnipotence? Despite decades of personal work on our clarity, connectedness and internal alignment we have not as yet found a consistent secret of miraculous manifestation although it has happened several times. At the same time our experience of life has been one of continuous improvement. Much of what has happened to us has not been planned or visualised consciously; some of it has come in forms that surprised us and in general it has felt as if there is a bigger and better plan that we don't know of consciously – at least not yet. .

We do believe that the more we acknowledge and recognise our own potency, the more we clear out conflicting belief systems, the more we manifest what would previously have seemed miraculous. Thus we don't think that human beings have been stupid or delusional in their widespread underlying belief – whatever form it has taken – that there is something beyond our physical actions that moulds our future. We think that all along there has been a deep instinctual knowing of the non-material world. All of our evidence and the picture that we have constructed around it supports that knowing.

That said, the question arises - what are the boundaries to individual human capacity in relation to our collective creation, our planetary reality or the outworking of the entire universe? Whilst one answer has to be "how could we possibly know?" we can at the same time ask "what else is possible?" and continue to explore what is beyond our rational knowing. And there lies the conundrum for us all if we begin to embrace our position as an intentional part of the creator consciousness.

If you are not an arch-sceptic this is not an intellectual game or a trick question. During our narrative we have given examples of people with extraordinary abilities, as for instance with the psychic surgeon Stephen Turoff. We have not included the many stories of avatars like Sai Baba who is said to have manifested jewels or like Babaji who was said to be capable of teleportation. What if we are all avatars in the making? What if we all have the potential to heal as Jesus did? Was it a trick that he turned water into wine?

Millions of people have believed such things to be possible – possible for others, possible for demigods and somehow "special" individuals. Some of us may believe that it takes extraordinary spiritual development or requires us to have God as our progenitor. Humans routinely attribute special powers to fantasy characters in movies. What would it take to believe of such possibilities for ourselves, for each other, for our children? What if humans truly believed the message of Jesus – that it is only our lack of faith that makes us less?

We, Juliana and Jon, were brought up as most others have been – Jon to doubt anything that was not visible or audible, Juliana to believe only in the Catholic God. Both of us have spent our lives exploring places that our cultural backgrounds would have caused us to reject. Both of us, in very different ways have spent several decades experiencing a world that seemed impossible, that science tells us does not exist. Whenever we think that we have encountered the boundaries – gone as far as we possibly can, something else comes along. This book merely puts a toe into the water of those experiences.

So we don't know what is possible. We don't know where the boundaries are. We don't know who we can be – still less who you truly are or who you might be. And since who we are as individuals arises and operates in a field that is also the creation – the co-creation – of others, we have no idea how the constraints on that field of belief may be affecting what is possible for us. So we have even less of an idea regarding what humanity might be capable of collectively if we were to free ourselves of the current limitations to what we conceive to be possible.

Our purpose

This book has only one real purpose and that is to crack open the pervading belief system. It could have been a very short book if we could simply have said "see what we have seen and accept what these others have experienced." We only needed two hundred pages of biology and physics because there is such a well-established structure which we are all taught to believe as fact when it is actually a limited point of view – one with the power to take us to the stars but which will prevent us from truly experiencing what is there when we reach them. That limited point of view is flawed, a building that has grown crooked from misaligned foundations. It is upright because it has been buttressed and given counterbalancing corrections with successive storeys. Its seeming robustness derives from the amount of effort and material that has been put in to shoring it up.

Have we persuaded you that this building is no longer fit for human habitation? What might cause you even now to doubt the evidence that we have provided? And if you are persuaded, is there some hesitation to leave the apparent security of what you have known? In cracking open the current belief system we are unlocking the door of that building so that you can go into the garden. We do not need a new building because the garden has been there all along and is blissfully habitable.

The garden you enter is the same one that you were ejected from at the beginning of biblical time. It is the one that you left, the moment that you accepted any thought or belief that you were separated from the whole and less than Divine. It is the one that we all leave every moment that we judge ourselves – set ourselves in the image of the punitive Old Testament God – the one who says what is right and good, lays down that there are aspects to life which are bad and wrong and which must not be known. When we adopt that judgemental way of being, we expel ourselves from the garden. No external agent is required.

By letting go of that flawed material belief system and of the disconnection that it has encouraged us all to have from the oneness of creation, from the consciousness of the All, you have an opportunity to re-embrace that original innocence. You have the possibility of finding out who you truly be. You have the opportunity of empowerment as a full participant in co-creation. That is our purpose. That has been our purpose all along.

21.   So what else is possible?

"Dwell in possibility..."

Emily Dickinson

We are in a different time in History. We have turned the page on possibility. But at the same time we're not simply local; we are also nonlocal. There is an underlying connectedness that exists in everything.

Jean Houston

Given that we do not know what might be possible for you, for us or for the collective, what follows will be speculative. Your guess is as good as ours, but we would like to suggest some trains of thought that we see as interesting and productive.

We would like to preface any specific areas of application by exploring further what life is like following the leap into 2nd tier. We described the integration of 1st Tier systems a few pages ago as a platform for a new way of thinking. Four aspects of the shift we are describing now come together and it is more than a new way of thinking. It is a new way of Being.

1. We have described the complexity and unpredictability of today's life conditions and indicated that they will call for non-linear responses. In earlier, slower-paced and less complex times we could attempt to predict the outcome of our actions. Our singularly human ability to project the future worked moderately well. It seemed effective to apply cause-and-effect thinking and make plans. If we got it wrong that was because the Gods had other ideas, because we didn't know enough or because Murphy's Law lives; we didn't think it was because those linear processes themselves were flawed. Now they have passed their use-by date; we can no longer sit on the beach and analyse the waves. We will have to be up on our surfboards, sensing our position, adjusting our balance and the direction and tilt of the board continuously as the wind and water change. There is no final "right" adjustment. The dynamics are living systems. Living systems are dynamic.

2. We have described a transition from ways of thought based on past decisions (ours or other people's) about what is good or bad, right or wrong. It will be necessary to integrate into our shared awareness all of the Values systems that have been before and to recognise what they contribute to

human existence. We will see how and where they are working well, and where they are out of balance in order to adjust conditions and behaviours according to what seems best for the whole. But that must take place in the here and now so that our adjustments are based on what we can see to be needed, not on a preconception or rule of what worked before. In leaping the chasm to the new world we can all bring our competence, instinct, understanding and intelligence. Our previous judgements and old solutions will weigh us down. What is right or wrong today will be whatever we assess and feel as delivering the best outcomes. What is good or bad tomorrow may adjust that in the light of what has changed, assessing the vital signs as we might with an intensive care patient.

3. The adjustments that we can make may be in the material, external world – altering systems and applying resources. They may be in the human realm – engaging our capacities and talents, learning new skills or improving our relationships with ourselves and each other. We have had those possibilities all along. We have described a new reality framework in which we have the potential to be co-creative – to make our mark on emerging reality directly through conscious engagement with the field of information. Now we have a new possibility, that we can change our inner being with the intention to create something different through consciousness. We can call forth new systems and relationships from the ground of who we are; engaging in the information field with what already is in order to bring forth what is next.

4. As a consequence of carrying our past decisions with us as if they will be right for next time (2 above) we have been in the habit of looking for answers. We have treated those answers as if they are permanently reliable rather than contingent on the conditions in which they were arrived at. And since those answers were seen as what we need to drive our engagement with the field (3 above), what we have contributed to the unfolding of a creative conscious reality has consisted of expectations based on the past. That is to say, sometimes we have not been creative at all.

In every moment we are at a fork in the road. One path will be towards repetition of what went before. It will be a contraction, a constriction to the fractal, an inhibition on the dynamic of creation. The other path will be towards the new. Instead of giving instruction to "the field" that it should give us more of the same, we have the possibility of drawing something new forth from reality. To do this we need fewer answers and more questions.

Until now, human answers have generally been specific and our intellectual history has been to give them as much definition as we are capable of. Not only have they been restricted in their openness to anything new, they have kept even the old within tight boundaries. Naturally we have struggled with paradox! Our questions must do the opposite. This means that they must be framed in an open-ended way. The difference is like that between "Will we have pizza for dinner?" and "What would be really nice to have for dinner?" When we think that we know what we want, the knowing places a limitation on what the creative unfolding might deliver. What if the unfolding has more to offer us than our minds have imagined being possible? Our creative input into the universe needs to be less in the form of concrete thought-forms and more in the formlessness of wishes, Values, qualitative choices and images of outcome.

Until now each shift in Values has been a change in the priority codes that people are applying in their thinking, a partial shift in HOW they think which affects the content of WHAT they think. Here we are describing a much more substantial shift in how to think. You might even see it as changing what "to think" means. Or you might view it that choices and decisions will come to be based less in thinking and more in a combination of thinking, feeling, sensing, intuiting and Values-based choosing. The HOW is expanding and the WHAT is rooted not in things but in Qualities.

It is no accident that this chapter has the title "So what else is possible?" We must again acknowledge Gary Douglas, the founder of Access Consciousness for his formulation of this phrase and of the way of thinking that it encapsulates. We have asked this question pointedly at various points in our text. This was not merely a rhetorical device; it was central to our purpose and the reason why the strapline to this book is "The Science of Possibility". And even while acknowledging that we recognise that there are long-standing spiritual and mystical traditions that have had the same open-endedness. Some forms of Buddhism do so as does the Vigyan Bhairav Tantra tradition taught by Osho.

Embracing a non-linear future

This is why the shift from linear thinking into creative, emergent, fractal, chaotic non-linear responsiveness is so dramatic. The leap into possibility is transformational and revolutionary. It is not merely more radical than we have imagined, it is more radical than we are able to imagine. That is the blessing it brings, that the future is bigger than the individual mind can make it. However, the consequences of a shift of this magnitude for the way humanity lives are also bigger than we are able to imagine.

You might argue that humanity's faith in that linear predictability was always behind the game but either way it is what we have done and it is still how most of us think and how most of our systems attempt to operate. Governments still make plans and formulate policies, and we as citizens expect that they should get that right.

Hitherto it has also been necessary to justify policies with rational argument. This has never been quite what it seems or pretends to be. As described previously the choices are generally driven by the underlying Values that people hold. If you know what these are the arguments are predictable and the policy results are what their proponents want them to be. One definition of an intelligent person is someone who can rationalize their prejudices. According to that our governments are filled with intelligent people. In practice people tend to lay out all the arguments as rationally as they are able or feel obligated to do; then they make a subjective decision, the one they wanted all along.

Our future now requires a different level of intelligence and different types of intelligence. That future calls us to step beyond the intellect and its attempts to pin everything down. It will not be viable to make decisions according to narrow Values frameworks, blind to the bandwidth of a multi-faceted human existence and we will not be able to afford the coarse and wasteful policy swings that take place on administration life-span cycles. Nor will we be able to afford the inadequacy of sensing systems that we have been accustomed to. Until now if it has been present at all, intuition and gut-feel has had to be masked. We will need broadband rational, systemic, emotional and spiritual intelligences but all of these will need to be harmonized with an embodied wisdom and deeper knowing. It is a move from intellect and conscious cognition into embodied awareness.

Humanity will need a different attitude to research as well. Research will tend to tell us what has been, not what is coming. Change may happen faster than research can be carried out. Societal research can only control for a limited number of variables, too limited for the complexity of scenarios we are dealing with. Perhaps we will be able to construct better models with which computers can project the interactions into the future. No doubt we will use our new mindsets to develop new tools, as humanity always has done. Yet although data is essential and we should use every tool at our disposal we are handicapped if we believe that we can always figure everything out. Research and modelling are based on the knowledge of the past; they do not take account of the potential for our creative choices to shift the underlying conditions. Research and modelling take existence as it has been and do not allow for a newly emergent reality. This will call for courage and trust – a leap out of fear that we will get it wrong.

Apps for human futures

Writing this, we are in the same position as IBM's CEOs, if of course you ignore earnings and power. We cannot predict the future either. It is one thing to write something in a transitory blog and quite another to do so in a book that is intended for a long life-span. So what follows is not intended as prediction. It is more an indication of emergent trends that we see and directions that we believe might be followed that will be more compatible with our reframed view of reality.

These trends are already being anticipated by others. While presented separately they are inevitably interdependent. All will depend on the choices we make, both at the material level and at the creative consciousness level we are suggesting to add. All will be influenced in ways that we cannot anticipate. Even though we have talked of scientific limitations, the technical breakthroughs will continue to arrive. We do not know what the impact of fracking or 3D printing or graphene based technologies will be, still less what will emerge from a laboratory tomorrow. What we do know is that the problems to be solved remain urgent and that most of the systems we are addressing are unstable.

Evolving our relationship with finance

All the signs we see are that the financial crisis of 2008 failed to do what was needed – a system reset. If the fundamental principles of free-market economics had been adhered to, many institutions would have collapsed and much corporate and national indebtedness wiped out, taking us all down with it. That was much too scary, both for governments and for the people with most to lose. Self-correcting market economics was thrown out of the window by the very leaders of capitalism who so deeply defend the free market. Self-interest in some areas conspired with political panic in others and yards of band-aid were applied. Some of this was with the best of intentions (paving the road to hell) so that the system has been propped up with Quantitative Easing ever since. Major nations are competing to drive their currencies down in value (similar behaviour to that following the 1929 Wall St. Crash) but cannot do so fast enough to compensate for the over-valuations before 2007. Stock markets rise but they do so with lower-value money. That is a recipe for societal conflict.

The above is being written in June 2013. The conditions being described are highly unstable. Some radical economists predict imminent collapse. We see the last four years as a juggling act with several jugglers in a high wind, none of whom entirely trust each other. We don't know which ball will be dropped or by whom but believe that it is just a matter of time.

The way that this relates to our overall perspective is that the system we are operating was designed in accordance with the Values of the time (SDi Blue and early Orange). Many features have been added such as computerised speed of transactions, the ability to generate "virtual" money (90% of new money does not come from government currency issue[58]) and the invention of derivative financial products that defeated regulatory functions.

The system remains fear-driven, competitive and globally exploitative. The Values systems which would call for us to design in fairness or to represent the economics of ecological consequences have not yet impacted the systemic design. Green social Values have been clumsily bolted on. The political costs of economic dominance strategies which then lead to costly military demands (both offensive and defensive) are outside of the corporate balance sheet. And even the military becomes an industry that we count as part of national product that citizens are taxed to pay for, viewed by some nations as the cost of doing business.

What this calls for is an entirely new systemic design based on the Values that we now hold, or aspire to hold. The reasons and parameters for this are set out in Jon's book "Future Money" and are summarised in the "11/11/11 manifesto"[59]. Others such as Charles Eisenstein, Jordan Bruce Macleod and Louis Bohtlingk[60] have written on aspects of this area of change. Many of the elements of a new design are present. Currently lacking are the will and intention to apply the changes, and the adequate global governance to bring it about.[61] Some changes might also be accelerated if individuals were to disconnect from current systems and support the emerging ones, like micro-finance and alternative currencies.

We have presented finance first because the materialist perception in scientism is paralleled by a deep misunderstanding about the reality of money, leading to wide societal blindness to the importance of anything else. Finance is perhaps the keystone in the arch that sustains the world in its current state and prevents healthy emergence.

---

58 For more information see www.positivemoney.org
59 Can be downloaded from www.emergenthuman.com
60 See bibliography
61 See also www.simpol.org

Love, Fear and the Destiny of Nations

The subtitle above is taken from the book by Richard Barrett. Like SDi, Barrett takes a Values-based approach and he has done more than anyone in practice to measure national mindsets[62]. His strapline to Volume 1 of this book is "The Impact of Human Evolution on World Affairs" and his presentation of evolution (Foreword) closely mirrors that of SDi.

Barrett has long argued for and written of a new leadership paradigm. In this book he develops that argument and in addition builds evidence of both the need for and the inevitability of a system of global governance.

Globalisation in practical terms is a manifest reality. It may be far from complete, patchy, fragmented and subject to many forms of internal conflict but its existence is undeniable. Our challenge is that the conversations which govern that reality are not representative of the human collective. When global conversations do take place (e.g. climate summits) governments are present. They are heavily influenced or bought outright by corporate interests, many of which are bigger than smaller nations. When combined with national self-interest the result is that the planet and its human inhabitants have no seat at the table.

Think Globally, Act Locally

Paradoxically we ask whether global governance will be the answer. For sure, the domination of large nations and corporations is not. But many solutions will not be global. We have a balance to maintain. The large systems, both man-made (Sea and air transportation, global communications, food and mineral distribution) and natural (CO_2, acid rain, water distribution, climate shifts) are all essential to our future. At the same time we need to cut our dependence on these systems, reduce our food-miles and make our local economies self-sustaining. Our cities are large organisms in their own right and most are only sustainable on the basis of external food-supply, waste removal, energy import etc. Dr Marilyn Hamilton has written brilliantly on the challenges this presents and the routes to solutions.[63]

Many of our cities are also too big to be thought of as local solutions. They are internally complex and many of the global imbalances are mirrored there.

In order to respond to complexity the shift out of linear thinking also demands a shift from command and control centralization. Our description of the human body

---

62 www.valuescentre.com
63 Integral City: Evolutionary Intelligences for the Human Hive

offers an image of the change that is required. Remember - the brain is our most important organ – according to the brain! It is conventional to think that the brain makes decisions. In our presentation of biological coordination we have encouraged you instead to view the body as "mind all over". Our planet, our cities, our organizations large and small will also need to reform around this image because the intelligence of the collective needs to operate responsively. For this to happen, frameworks and guiding Values need to be present throughout, but the ability to detect and respond must devolve to the level of the organs and the cells. The subtitle above is "think globally, act locally". If we unpack that we have a call for global awareness, planetary Values and coordinated high-level policy. But locally operating life conditions and Values systems demand that the implementation, the action decisions must be made at the "coal-face".

To achieve this we have to be willing to empower individuals more than ever. Fear, and the predominance of Blue-Orange mindsets have brought an intense focus on process and organizations are now often driven by process. For sure we need processes, but their dominance has been damaging. An example that is current in the UK as we write is that the attempt to raise standards in the Health Service has been built on the premise that you cannot manage what you cannot measure. The outcome has been hospitals which have failed to deliver care because a) care is not easily measurable and thus is absent from process and paperwork b) managers have driven staff according to those care-absent targets with the result that c) many staff whose motivation was to care have been driven out of the organization. The spirit of the organization has been systematically designed out and driven out.

Arthur Costa, Emeritus Professor at California State University sums up a similar challenge in education thus:-

> "What was educationally significant and hard to measure has been replaced with what is educationally insignificant and easy to measure. So now we measure how well we taught what isn't worth learning."

Such examples generalize widely. Companies spend a lot on advertising and brand awareness, only to destroy it because employees have little discretion and cannot deliver the experience that produces happy customers. The policies, designed in a vacuum by lawyers, process engineers and accountants and backed up by inflexible IT and HR systems make it more difficult to deliver the desired outcomes. The evidence, as documented in books like Raj Sisodia's "Firms of Endearment" and others[64] is that the companies which are able to get the balance right are both more admired and more profitable. Some views place these experiential intangibles as constituting above 80% of corporate value.

The human organism

When the skin is damaged, resources rush to the site of the injury. The brain has nothing to say about this. Nor does the brain make a conscious decision regarding whether the liver, the kidneys or the spleen should receive resources. Evolution has spent millions of years creating a system of sufficient inbuilt intelligence that 50 trillion cells manage the complexities of biological existence by having responsiveness at the level of the cells and the organs themselves.

The body manages its resources. It manages all of its resources simultaneously. Humanity is not yet capable of this. In an emergency the body will divert resources. For example it will shift blood supply to the muscles to ensure oxygen and fuels are available for fight or flight. When the crisis is passed it will ensure that balance is restored. It responds to conditions and ensures that all needs are met.

It does not do this by putting the oxygen first. Nor is it supply driven. There is an overarching principle that each part of the organism must survive and the whole must survive. It will not normally sacrifice a part and there is inbuilt recognition that most parts are essential. Nor does it manage by process. Each part makes its demands for resources and there is a system-wide (whole body) response that is demand driven. Supply is a constraint, not a driver.

The materialist focus of our culture falls short of this capability in two ways with the result that the complexity of the body and its ability to respond to its many environmental demands exceed human capability. The materialist focus causes us to lose the necessary awareness of each individual cell – the humans in the collective. Current reality permits cells (humans) to be wasted. That materialist focus causes societal awareness and management systems to focus on money – as if ensuring that the money supply is thriving will ensure that the human collective thrives. We will assume that you find it is as visibly dysfunctional as we do, and if not then we recommend researching the human happiness index.

The materialist focus also allows imbalances that natural systems would not allow. If a part of the body stored excess oxygen it would experience that as toxicity. Yet

---

[64] John Mackey "Conscious Capitalism", Colin Mayer "Firm commitment".

our system sees it as healthy for individuals to accumulate vast amounts of wealth – not at the proportionate level that provides security against emergencies or ensures reserves to support a coming generation but at a level which means that its circulation is cut off. Even in an emergency (as you may view our current crisis to be) it is the exception and not the rule for those who have the huge surpluses to ensure that these are released for the good of the whole.[65]

Redistributing intelligence

Whether we are speaking of commercial corporations or public health and other services, whether we are talking of planet-wide organisations or medium-size enterprises, whether we are talking of big countries or small ones, the principles that we are describing here will be essential if we are to transition to a sustainable future for 7 billion humans and above. The intelligence that pulls your hand from a flame begins in the cells at the surface. We need that responsiveness at the human level. We don't yet have it. As E. O. Wilson has said:-

"The real problem of humanity is the following; We have Paleolithic emotions; Medieval institutions; and god-like technology."

Human history and organisational models alike demonstrate that it has been normal to believe that people must be controlled and directed. There are of course times when that holds true. The Spiral describes several stages when the individual can get out of step with the needs of the collective. In the first tier systems that imbalance has been within humanity. As humanity approaches second tier it is learning that the imbalance is also between one species (ours) and all the others.

One possible outcome of this imbalance is that the need of the planetary collective will be expressed through the failure of ecologies that we depend upon – the mass death of pollinating species for example. But we are at this point because of our species' phenomenal capacities. Humanity is an exploring child – the toddler that pulls the dog's tail to see what happens. If we do not hear and interpret the growl correctly we will get bitten.

Any species other than ours has only responded to such pressures through biology. When the ecological niche changes an adaptive response is called for; the species expands, dwindles or goes extinct according to that adaptation. Humanity has grown from thousands to billions through socio-psychological adaptation. Moving out from the savannahs of Africa we have explored all of Earth's niches and

---

65 Warren Buffett and Bill Gates providing the examples that are exceptions

colonised most of them. We can conceive of extending into Space and have begun to do so[66]. At the same time we know that such expansion requires stable foundations here. Forget science fiction stories where humanity escapes Earth's problems and sends colonies to the stars. If we do not solve our problems here we will not have that capacity.

That capacity cannot emerge when society's operating assumptions are that people will behave badly if not controlled and directed. We will not know what people are capable of contributing if we do not ask them. We will not get the best of that contribution if we do not make space for it to come forth or if we starve it of resources. We will not see any creative contribution if we think that we know best and place unnecessary boundaries around what is possible for others.

Many fears that are projected on to others are fears that people have of themselves. They may have been told as youngsters that they were selfish and greedy, programmed with the belief that they and other humans are bad and sinful and that only severe constraints will save us all from our inherently wicked natures. This is not only about "original sin"; it is a social convention that pervades parenting and education. There is no reason to accept that convention, much evidence that when ignored great success is possible and it is probable that in almost all of the instances where it appears to be true that this has been a self-fulfilling prophecy, an outcome that stems directly from those fear-based belief systems.

Perversely, the way that many schools operate is the obverse of what research states will support learning and creativity. Criticism stimulates the frontal lobes which are the gatekeepers that inhibit divergent thinking. Teaching methods are left-brain oriented when it is the right brain that comes up with insightful ideas. Those insights are preceded by a voluntary inhibition in the visual cortex to allow space for the new connections to be made. This too is inhibited when concentration is demanded. Creativity is also supported both by unpredictable pattern-breaking and by periods of undemanding repetitive activity (like doodling). Neither of these is supported in a typical classroom scenario. Yet these are the capacities which our future will call for if we are to meet the needs of a non-linear world.

What if we were to adopt different assumptions, learn to parent and educate children to develop a core sense of who they are as a unique individual so that they will find what they are able to contribute to the world? The Steiner (Waldorf) education system has been successfully doing this for a century. In addition,

---

66 Visionary example: Howard Bloom's "Space Development Steering Committee"

humans are naturally curious and will educate themselves. Prof. Sugata Mitra has demonstrated this very elegantly in India.[67]

All of the above indicates why we have to redistribute human intelligence and human empowerment. This is true at every scale: nations, corporations, enterprises, care systems, communities and all the way to individuals. We will need to raise and educate citizens to carry and accept responsibility. This means ensuring that we raise personal capacity and eliminate blame cultures which discourage individuals from stepping forward. It will take empowered and emotionally intelligent individuals to hold such responsibility. It will take emotionally and spiritually intelligent individuals to lead them[68].

Redistribution of wealth and resources will help – not if based on blind equality, not responding merely to desire and not responding to psychological or emotional need, but ensuring that each has the resources that will maximise their contribution to the whole. There is much misunderstanding in this area. Our social care systems have encouraged dependency. The excessive remuneration of top executives with its huge differentials from other staff does not only arise from greed, it also reflects the ludicrous imbalance in how responsibility is perceived – the outdated view of people as automata. Redistribution cannot be at the cost of incentivising entrepreneurial activity. On the other hand there are other rewards than money, even for those with core SD "Orange" values systems, such as status and recognition.

The blend of global thinking with local empowerment is also reflected in an equally paradoxical need to adopt perspectives of long-term sustainability even while we are suggesting that our decision-making cannot be based on "answers" and must be flexible and responsive. It is only to the linear mind that these apparent paradoxes present a problem. Our natural intelligence knows better.

The reality is that humans behave like this all the time. When you decide that you will drive to another city 3 hours away you do not time your departure to the second, do not analyse every judgement that you will make on the way. Each junction, overtaking manoeuvre and even pit-stop is likely to be made during the journey. Your arrival time will be whenever you get there and may reflect your intention or may have been subject to road-works, tail-backs and the like.

Our future requires the same balance; it takes the same principles and applies them at a higher scale of our existence particularly in our collective undertakings. The

---

67 www.ted.com/talks/sugata_mitra_shows_how_kids_teach_themselves.html
68 For deeper understanding of this comment see Cindy Wigglesworth: "SQ21"

corporation will need its vision of what it seeks to achieve and it will need to articulate the balances it desires – happy customers, employees, shareholders, suppliers, society and environment. These will be images and aspirations that they will live by. Sometimes these choices will be encoded in detailed central policies but in more cases employees and other partners in the enterprise will be expected to make decisions or set local policies informed by these images. They will be accountable to Values and Qualities, measured by outcomes more than processes.

Thus the transition to distributed intelligence is driven by Values in several senses of the word : Personal and collective choices that guide behaviours, priority codes that assist individuals and collectives to thrive, aspirations for the qualities that we want in our lives, moral and ethical frameworks that guide all of these. In place of quantitative and materially driven rules, commandments, processes, conventions and consensual orthodoxies we will make our decisions according to higher order images and guidelines. Our Values will represent a qualitative view of the world. While there is some truth still in the view that we cannot manage what we cannot measure, we will increasingly measure by outcomes. Those outcomes can be qualitative and the assessments can often be subjective.

The transition to possibility

All that we have described above fits within the space of our current frames for what is possible. None of it rests on the non-material aspects of scientific reality. For sure it engages with subjective responses and with our emotional intelligence but it does not demand prayer or creative visualization. It calls for greater freedom of thought, for a more open-ended view of our material and human interactions but it does not live in the space of pure emergent possibility that is the most radical implication of the scientific worldview presented by this book.

So what else becomes possible when we also engage with that more mysterious, magical, quantum-uncertain and unbounded infinite view of who humans are? While repeating that we do not and cannot know, our experience is that previous limitations begin to look like illusions – mindsets and thought systems that we have constructed for ourselves without apparent justification.

The science and the stories in this book indicate that humans have barely scratched the surface of who we can be. The authors' personal experiences so greatly exceed what we have described here that we would deeply urge readers to accept that we have understated and not exaggerated what we know at every level of our being. From the outset we have been aware that this new perception of reality might seem to you like a fantasy, like magic, like a dream, like anything other than the new, practical, functional existence that it is capable of being.

The implications of this new existence are transformative. They bring a fundamental reframe of how the maintenance of health and the treatment of ill-health can be achieved. We have some very powerful technological and pharmaceutical capabilities, but what would happen if these were our backup systems, not our primary care? What would happen if individuals would know how to maintain health using their mind-body connection supported by alternative practitioners and were educated to do so? If that isn't a good enough idea for its own sake, think how much we could reduce the costs of healthcare and perhaps increase healthspan to match lifespan?

The way corporations work in the future has the potential to be a net contributor, to find the territory of win-win-win. Humans have achieved material progress at the expense of the planet and its other inhabitants. It has become the norm to accept the "side-effects" as if they are unavoidable when this is merely a failure to design them out even though the necessary design principles exist. What more might be achieved if humans bring deeper and more intuitive awareness to that design and engage in the connection with life's other co-creators when doing so? This is a step further than finding out how not to do harm. It introduces the possibility of employing the shamanistic intelligence to work together.

There is also the possibility to develop a new relationship with politics and government. All of the reasons why this is too difficult and all of the statements that might be made about the state of humanity right now will inevitably surface when such a suggestion is made. We are not describing a flick of a switch. But if we all shift the mindsets we can step out of the either-or and left-right polarizations in order to deliver more flexible responses. More than that, if we are employing the "field" awareness, connecting to the collective intelligence and avoiding the rigidities of yesterday's rights and wrongs we can emerge choices and solutions that represent a different level of knowing.

Those who put forward the new perspectives are frequently dismissed as "woo-woo" or given other insulting labels that suggest that we are a few sandwiches short of a picnic. We are not fantasists, we are not ungrounded and we are not out of touch with the material world. We are not suggesting that we give up technology in order to run the world on magic. We are not suggesting that we abandon the use of the rational intelligence in favour of prayer. Our philosophy is

"Trust in Allah, but first tether thy camel."[69] And it is more than that.

Engaging with the non-material aspects of reality is not an alternative to science. It is additional to science and expands science. It adds to the range of possible tools in our kit. It amplifies the knowledge that science has brought already and lifts it to a higher level. We have made clear how our approach unlocks some deep knots in current theory but our interest is not theoretical. The theories are a means to an end and that end is the application of greater knowledge and wisdom to the problems facing humanity.

You do not have to look far or work hard to find others who will tell you how many problems humans have created, how critical some of them are, how urgent it is that we find new ways to address them and how deep is our need for new ways of dealing with the oncoming crisis. There are some who would tell you that the tipping point has already been passed; the damage is no longer reversible. It is possible that this could be so.

We don't believe it has to be. For sure it is something of a last-minute rescue, but that seems to be something that humans like to do; there is a vast film and TV industry which is built on that tension – how exciting it is to wait until the very last moment for the cavalry to come riding over the hill. We don't believe that we have come all of this way in order to snatch defeat from the jaws of Victory. We generally create stories with happy endings.

There is a very large collective on Earth who share the reality that we are describing. Most have no idea how it works or the science that might describe it. Many could hardly care less; they know what they know. There are many others who don't need to know but who might engage in what they are doing with greater confidence. This collective is very fragmented and there are many, many belief systems about what the spiritual reality consists of. Gurus and guides abound, there are teachers of techniques and there are many channels who speak the thoughts of beings that are in the field of consciousness but not incarnated. Many bring wisdom, though we are reminded of the comment from Ram Dass that the fact that a being doesn't have a body does not necessarily make it wise. You know enough to be discriminating and one useful guideline that if something makes you feel a lightness of being then it is more likely to be true for you. If it feels heavy then it almost certainly is not.

It might seem to some, particularly if you are used to the traditional religions, that

---

69 Variously attributed to the Prophet Mohammed or to later Arab proverb

there must be one right answer, one voice that has the Truth. What if that voice is yours? What if the Truth needs only to be the Truth for you and there is no greater Truth? We suggest that the essence of co-creation is that we are discovering how to blend our individual voices in the choir, to be one and to be different at the same instant. We feel that is richer and more exciting than we have words for. It is huge, epic and joyous because it is without limit.

It is not a problem that the collective is fragmented in its belief systems. Each of us encounters our own reality on a daily basis. If you want to know what your beliefs are you only have to look at what you are creating. Changing that may not be an overnight job, but it is certainly possible to change. And each of us who engages with that change finds their own journey. Because there are many of us and because we are all so different, starting from unique places, no two journeys can be alike. By definition, cults that know what is right for everybody have lost the plot.

Science has spent hundreds, thousands of years attempting to "know the mind of God". The "God" that it sought to know was outside of humans, outside of material creation so doomed to incompleteness. The attributes of God are omnipresence, omniscience and omnipotence. God is everywhere, knows everything and has power over everything. God can only be omnipresent if it is as much within each of us as it is outside of us. It can only be omniscient if it contains the knowing of everything that is in creation – including you. It can only be omnipotent if it encompasses and includes the power of every part of creation. That's you too.

The science we have described leads to the recognition of an all-pervading consciousness, an emergent field of knowing and creation in which all of universe is engaged. Since there are no boundaries, each of us has the capability to be aware of the whole, whether we can fit that knowing inside our cognitive minds or not. There is nothing more that we can tell you.

Each of us has a choice. You can know that this is who you are. You can accept that anything that limits your ability to know it is a habit, a belief system; a program that you have been given that is not your Truth. Each of us can then work to clear ourselves of everything that we have unconsciously taken on so that we spend more of our time in the potency of our creative capacity. We can undo everything that we have been told that would have us be less than that. We don't have to make anybody else be anything. Our freedom is not dependent on other's permission and their freedom to be who they are requires no approval from us.

For sure, the more of us who become empowered and conscious in this way, and whose creativity is directed toward all the things that represent human ideals – ease, peace, justice, harmony, joy, grace, glory, pleasure, togetherness, love – the more and faster we will move towards the manifestation of that in the world as a whole. At the same time you are not dependent on others to create that local reality, to have that be the nature of your own life. The more you do that, the more obvious it will become to others that they too have a choice. We are all leaders if we choose to be, leading ourselves and each other into the Promised Land – i.e. the land that we have promised ourselves.

And that is all it is, a choice to be who we are. When we enter into that place of being we don't need to understand God and you may already have heard that the peace of God is beyond understanding. So our destiny is not about knowing, it is about being, about being and becoming, joining our visions and aspirations with each other and with the remainder of creation. Together we can become what we have always been but not allowed ourselves to believe; we can become the mind of God.

Jon Freeman

Juliana Freeman

Summer 2013

Politics and Religion are obsolete.

The time has come for Science and Spirituality.

Jawaharlal Nehru

Acknowledgements

Written at Oscar time, with some sympathy for those who are recognising that it is impossible to acknowledge the hundreds of people who touch our lives and without whom things would be different, but whose gratitude impels them to try, to the embarrassment of all.

Jon wishes to acknowledge his father Dr. Jack Massey for always asking questions and for his dedication to making life better, together with his mother for the love of music and for showing that the world is full of feeling. Acknowledgements to those who have lived around me and had to bear with unpredictable life paths, obsession, brain-on-a-stalk behaviour and other flaws. Jack, Yannis, Joby – this is my excuse and my apology.

To Charles Paul Brown – the first to say "Write your book". Sorry it took so long, Chuck. To the creators of the Oxford University Human Sciences degree, to Vernon Reynolds, Bryan Little, Godfrey Lienhardt and other tutors way back when who wouldn't even recognise my name – thank you anyway for intellectual inspiration. Along the way, in addition to those whose work is mentioned elsewhere, Leonard Orr, Jacob Bronowski, Jose Silva, Ram Dass, Marshall Rosenberg, and Rudolf Steiner. Among those mentioned elsewhere, particular personal acknowledgement to Jean Houston, Don Beck, William Bloom and Gary Douglas / Dain Heer of Access Consciousness plus Juliet and Jiva Carter at The Template . Thanks also to the Prophets Conference – Robin and the late Cody Johnson – for bringing great people within my reach. Thanks to other readers whose constructive comments were very helpful, Dr. Howard Smith, Michiel Doorn, Dr Vinicio Sergo, Ian McDonald, Sophia Pairman, Pam Gregory and Marcos Frangos. Blessings on Stephen and Ondrea Levine for their poem, and their permission. Thank you to guiding spirits, guardian angels and nature beings. It is awesome how many different forms consciousness may take. For musical inspiration while writing, special thanks to Bela and the Flecktones, Gary Burton and John McLaughlin – all exemplars of the creative coherence and dynamic equilibrium that Mae-Wan Ho calls quantum jazz - plus gratitude also to Johann Sebastian and Ludwig v B. for order and impulse.

Finally and most obviously to Juliana – co-creator, supporter, healer, friend, and beloved former life partner. More than words can say.

Juliana would like to acknowledge all the beings that have supported her growth and unfolding – her mother for the family sensitivities to the world of spirit and her father for his enthusiasm and love. To my early teachers – Raymond Armin, Peter and Simon who opened me to the unseen worlds; Betty Balcome for her training of my psychic abilities; To Eliot Cowan, for introducing me to the plant spirits. To Gary Douglas for the incredible tools of Access Consciousness. There have been many other people who have influenced and assisted me and I thank you all. My most profound thanks and appreciation to all the wondrous beings that have graced my life up to this point and who I deeply desire will continue to walk beside me: the angelic beings, the nature spirits, plants and crystal beings, dragons, my spiritual teachers and mentors and all those whose presence is felt even when not seen. This world would be far less sweet without you all.

To Jon – for the courage and integrity to always speak your truth even in the most difficult circumstances and for daring to write this book. For your constant loving support and your willingness to give your time and energies for the benefit of others. For all that you are. With my love and devotion.

Bibliography, references and resources.

All quotations used are copyright of the authors and publishers listed. We are most grateful for the permission to include their material, where used. In other cases, works are listed because they were inspirational or influential. Further suggestions appear below.

Norman Shealey and Caroline Myss from "Anatomy of the Spirit", Bantam Books 0553-50527-0.

MonaLisa Shulz from "Awakening Intuition", Bantam Books 0-553-81212-2

Candace Pert from "Molecules of Emotion", Simon and Schuster UK Pocket Books 0671-03397-2

J.W. Dunne from "An experiment with Time". Recently re-published by Russell Targ with Hampton Roads Publishing Company. 1-57174-234-4

Peter Tompkins and Christopher Bird from "The Secret Life of Plants", Penguin 01400-3930-9

Bob Jahn and Brenda Dunne from the Princeton Engineering Anomalies Research www.princeton.edu/~pear   More up-to-date work and DVD's available from www.icrl.org

Malidoma Somé from "Of Water and the Spirit". Penguin Arkana 0-14-019496-7. Copyright ©1994 by Malidoma Patrice Somé. Used by permission of Jeremy P. Tarcher, an imprint of Penguin Group (USA) Inc.

Machaelle Small Wright from "Behaving as if the God in All Life Mattered", Perelandra books 0-9617713-0-5

Peter Caddy and Dorothy McLean from the Findhorn website, www.findhorn.org

Eliot Cowan from "Plant Spirit Medicine", Swan Raven & Co. 1-893183-11-4

Alberto Villoldo reference to "Dance of the Four Winds". Destiny Books 0-89281-514-0

Stephen Harrod Buhner from "The Secret Teachings of Plants", Bear & Co. 1-591-43035-6 .

Deepak Chopra "Quantum Healing". Bantam New Age 0-553-17332-4

Bruce Lipton from "The Biology of Belief", Mountain of Love Productions 0-9759914-7-7

Jack Cohen and Ian Stewart "The collapse of Chaos" Penguin 0-14-017874-0, while not quoted directly was very influential on our approach to genetics. Funny too.

Douglas Hofstaedter "Godel, Escher Bach". Deep intellectual entertainment. Penguin 0-14-005579-7

Mae-Wan Ho is quoted from her website www.i-sis.org.uk   Her book "The Rainbow and the Worm" is published by World Scientific 981-02-3427-9

Richard Dawkins is quoted from "The Blind Watchmaker" Penguin Books 0-14-008056-2

Stephen Jay Gould's "Life's Grandeur" is referred to. Even if we disagree, a great and interesting science writer. Jonathan Cape 0-224-04312-0

Robert M Pirsig "Zen and the Art of Motorcycle Maintenance" is one of the best, most profound books of the last 100 years. Vintage 0-099-32261-7

Elisabet Sahtouris from "EarthDance", iuniverse.com 0-595-13067-4

Rupert Sheldrake "A New Science of Life" Grafton Books Paladin 0-586-08583-1

Richard P. Feynmann "QED" Penguin 0-14-1-2505-1

Raymond Trevor Bradley. Overview of Heart Coherence material in World Futures of General Evolution 63:61-97, 2007 and on http://noosphere.princeton.edu/papers/pdf/bradley.intuition.2007.pdf

Danah Zohar. "The Quantum Self". Bloomsbury 0-7475-0271. See also "Spiritual Intelligence" and "Spiritual Capital".

David Bohm "Wholeness and the Implicate Order" Ark 0-7448-0000-5

Michael Talbot "The Holographic Universe" Harper Collins 0-586-09171-8 Not quoted, but a full – even holographic - account of that theory.

Lyall Watson "Lightning Bird" Hodder & Stoughton 0-340-27999-0 Out of print

Don Beck and Christopher Cowan. "Spiral Dynamics". Blackwell 1-55786-940-5

George A. Kelly "A Theory of Personality". The Norton Library 0-393-00152-0

Humberto Maturana and Francisco Varela "The Tree of Knowledge" Shambala 0-87773-642-1

Gregory Bateson. "Mind and Nature" Flamingo 0-00-654119-4 "Steps to an Ecology of Mind" University of Chicago 0-226-03905-6

Bede Griffiths reference to "The Marriage of East and West" Mediomedia 0-972-56271-0

William Bloom reference to "SOULution: The Holistic Manifesto" Hay House 1-4-19-0342-8.

Stephen W. Hawking reference to "A Brief History of Time" Bantam Press 0-593-01518-5

Recommended sources of further information

Don Beck's CD's (www.soundstrue.com) are probably the most accessible complete introduction to the detail of Spiral Dynamics integral. Jon's shorter but quite comprehensive free introduction is to be found at www.che-hub.com. Other overview material is available via www.spiraldynamics.net and www.wie.org/spiral . The Beck and Cowan text listed above is full and authoritative. The work in the Middle-East can be found on www.che-mideast.org

All of Caroline Myss' work has exceptional depth and authenticity. We would particularly recommend "Sacred Contracts". Much of her work is also available through audio on Soundstrue. www.myss.com

Jean Houston's work is expansive. Her psychological understanding and her ability to interweave the mythological and archetypal layer of human consciousness bring special richness to this picture, as does her interweaving of artistic perception. Her work on expansion of human capability is dynamic. Try "Jump Time" and "The Possible Human". Seeing her live is an entertainment as well as an inspiration. Her websites are www.jeanhouston.org and www.socialartistry.com

Brenda Dunne's work continues at the laboratory for International Consciousness Research. www.icrl.org .

Mae-Wan Ho's Institute for Science in Society website www.i-sis.org.uk is a huge source of understanding for those who want to keep track of scientific abuses in areas such as genetic engineering as well as giving access to her DVD's and other work.

William Bloom. Authentic, open-hearted and thoroughly grounded spirituality. To be found at www.williambloom.com . His regular thoughts are to be found in the Cygnus books newsletter. Cygnus are a great source of well-priced alternative books www.cygnus-books.co.uk . Check out "The Endorphin Effect".

Howard Martin and Rollin McCraty can be found through www.heartmath.com

Nassim Haramcin's work was essential. His DVD "Black Whole" and other material on www.resonanceproject.org unlocked doors.

We value our experiences with Juliet and Jiva Carter and their insights into sacred geometry and connection with the consciousness of the sun and planets. www.template.org

Our viewpoints have been influenced significantly, and our lives greatly benefited, by the insights and toolkits offered by Access Consciousness and by Gary Douglas and Dain Heer in particular. Expansive and practical at the same time. www.accessconsciousness.com

Alberto Villoldo can be found on www.thefourwinds.com Lots of excellent material.

Machaelle Small Wright is at www.perelandra.com. First-class plant-derived energy-based health products. Inspirational manuals on gardening with the plant spirits.

David Hawkins. Deep understanding about the nature of Truth and our ability to discern what it is and live by it. "Power vs Force" Hay House 1-56170-933-6. A man who giggles and chuckles at life. Greatly misunderstood and misrepresented. Read him or listen to him but don't read about him. www.veritaspub.com

Juliana's website is www.julianafreeman.co.uk

Jon's website is www.jonfreeman.co.uk